PETERSON'S STRESS CONCENTRATION FACTORS

Second Edition

PETERSON'S STRESS CONCENTRATION FACTORS

Second Edition

WALTER D. PILKEY

A Wiley-Interscience Publication
JOHN WILEY & SONS, INC.

New York / Chichester / Weinheim / Brisbane / Singapore / Toronto

This text is printed on acid-free paper.

Copyright © 1997 by John Wiley & Sons, Inc.

All rights reserved. Published simultaneously in Canada.

Reproduction or translation of any part of this work beyond that permitted by Section 107 or 108 of the 1976 United States Copyright Act without the permission of the copyright owner is unlawful. Requests for permission or further information should be addressed to the Permissions Department, John Wiley & Sons, Inc., 605 Third Avenue, New York, NY 10158-0012.

Library of Congress Cataloging in Publication Data:
Pilkey, Walter D.
    Peterson's stress concentration factors / Walter D. Pilkey.—2nd ed.
      p. cm.
    Includes index.
    ISBN 0-471-53849-3 (cloth : alk. paper)
    1. Stress concentration.   2. Structural analysis (Engineering)
I. Title.
TA417.6.P43   1997
624.1'76—dc20   96-27514

Printed in the United States of America

10 9 8 7 6 5 4 3 2 1

To Mother
A Pillar of Stability

CONTENTS

INDEX TO THE STRESS CONCENTRATION FACTORS   xv

PREFACE   xxxi

1 DEFINITIONS AND DESIGN RELATIONS   1

    1.1 Notation / 1
    1.2 Stress Concentration / 3
        1.2.1 Selection of Nominal Stresses / 6
        1.2.2 Accuracy of Stress Concentration Factors / 9
    1.3 Stress Concentration as a Two-Dimensional Problem / 10
    1.4 Stress Concentration as a Three-Dimensional Problem / 11
    1.5 Plane and Axisymmetric Problems / 13
    1.6 Local and Nonlocal Stress Concentration / 15
        1.6.1 Examples of "Reasonable" Approximations / 19
    1.7 Multiple Stress Concentration / 21
    1.8 Theories of Strength and Failure / 24
        1.8.1 Maximum Stress Criterion / 25
        1.8.2 Mohr's Theory / 26
        1.8.3 Maximum Shear Theory / 28
        1.8.4 von Mises Criterion / 28
        1.8.5 Observations on the Use of the Theories of Failure / 30

1.8.6 Stress Concentration Factors under Combined Loads, Principle of Superposition / 32
1.9 Notch Sensitivity / 36
1.10 Design Relations For Static Stress / 41
    1.10.1 Ductile Materials / 41
    1.10.2 Brittle Materials / 43
1.11 Design Relations for Alternating Stress / 44
    1.11.1 Ductile Materials / 44
    1.11.2 Brittle Materials / 45
1.12 Design Relations for Combined Alternating and Static Stresses / 45
    1.12.1 Ductile Materials / 46
    1.12.2 Brittle Materials / 49
1.13 Limited Number of Cycles of Alternating Stress / 49
1.14 Stress Concentration Factors and Stress Intensity Factors / 50
References / 55

2 NOTCHES AND GROOVES 59

2.1 Notation / 59
2.2 Stress Concentration Factors / 60
2.3 Notches in Tension / 62
    2.3.1 Opposite Deep Hyperbolic Notches in an Infinite Thin Element; Shallow Elliptical, Semicircular, U-Shaped or Keyhole-Shaped Notches in Semi-Infinite Thin Elements; Equivalent Elliptical Notch / 62
    2.3.2 Opposite Single Semicircular Notches in a Finite-Width Thin Element / 63
    2.3.3 Opposite Single U-Shaped Notches in a Finite-Width Thin Element / 63
    2.3.4 "Finite-Width Correction Factors" for Opposite Narrow Single Elliptical Notches in a Finite-Width Thin Element / 64
    2.3.5 Opposite Single V-Shaped Notches in a Finite-Width Thin Element / 65
    2.3.6 Single Notch on One Side of a Thin Element / 65
    2.3.7 Notches with Flat Bottoms / 65
    2.3.8 Multiple Notches in a Thin Element / 66
2.4 Depressions in Tension / 67
    2.4.1 Hemispherical Depression (Pit) in the Surface of a Semi-infinite Body / 67
    2.4.2 Hyperboloid Depression (Pit) in the Surface of a Finite-Thickness Element / 67
    2.4.3 Opposite Shallow Spherical Depressions (Dimples) in a Thin Element / 69
2.5 Grooves in Tension / 69

2.5.1 Deep Hyperbolic Groove in an Infinite Member (Circular Net Section) / 69
2.5.2 U-Shaped Circumferential Groove in a Bar of Circular Cross Section / 69
2.5.3 Flat-Bottom Grooves / 70
2.6 Bending of Thin Beams with Notches / 70
2.6.1 Opposite Deep Hyperbolic Notches in an Infinite Thin Element / 70
2.6.2 Opposite Semicircular Notches in a Flat Beam / 70
2.6.3 Opposite U-Shaped Notches in a Flat Beam / 70
2.6.4 V-Shaped Notches in a Flat Beam Element / 70
2.6.5 Notch on One Side of a Thin Beam / 70
2.6.6 Single or Multiple Notches with Semicircular or Semielliptical Notch Bottoms / 71
2.6.7 Notches with Flat Bottoms / 71
2.7 Bending of Plates with Notches / 72
2.7.1 Various Edge Notches in an Infinite Plate in Transverse Bending / 72
2.7.2 Notches in Finite-Width Plate in Transverse Bending / 72
2.8 Bending of Solids with Grooves / 72
2.8.1 Deep Hyperbolic Groove in an Infinite Member / 72
2.8.2 U-Shaped Circumferential Groove in a Bar of Circular Cross Section / 72
2.8.3 Flat-Bottom Grooves in Bars of Circular Cross Section / 73
2.9 Direct Shear and Torsion / 74
2.9.1 Deep Hyperbolic Notches in an Infinite Thin Element in Direct Shear / 74
2.9.2 Deep Hyperbolic Groove in an Infinite Member / 74
2.9.3 U-Shaped Circumferential Groove in a Bar of Circular Cross Section / 74
2.9.4 V-Shaped Circumferential Groove in a Bar of Circular Cross Section / 76
2.9.5 Shaft in Torsion with Grooves with Flat Bottoms / 76
2.10 Test Specimen Design for Maximum K_t for a Given r/D or r/H / 76
References / 78
Charts / 81

3 SHOULDER FILLETS 135

3.1 Notation / 135
3.2 Stress Concentration Factors / 137
3.3 Tension (Axial Loading) / 137
3.3.1 Opposite Shoulder Fillets in a Flat Bar / 137
3.3.2 Effect of Shoulder Geometry in a Flat Member / 138

3.3.3 Effect of a Trapezoidal Protuberance on the Edge of a Flat Bar / 139
3.3.4 Fillet of Noncircular Contour in a Flat Stepped Bar / 139
3.3.5 Stepped Bar of Circular Cross Section with a Circumferential Shoulder Fillet / 142
3.3.6 Tubes / 142
3.3.7 Stepped Pressure Vessel Wall with Shoulder Fillets / 143
3.4 Bending / 143
3.4.1 Opposite Shoulder Fillets in a Flat Bar / 143
3.4.2 Effect of Shoulder Geometry in a Flat Thin Member / 143
3.4.3 Elliptical Shoulder Fillet in a Flat Member / 143
3.4.4 Stepped Bar of Circular Cross Section with a Circumferential Shoulder Fillet / 143
3.5 Torsion / 144
3.5.1 Stepped Bar of Circular Cross Section with a Circumferential Shoulder Fillet / 144
3.5.2 Stepped Bar of Circular Cross Section with a Circumferential Shoulder Fillet and a Central Axial Hole / 144
3.5.3 Compound Fillet / 145
3.6 Methods of Reducing Stress Concentration at a Shoulder / 147
References / 148
Charts / 150

4 HOLES   175

4.1 Notation / 175
4.2 Stress Concentration Factors / 177
4.3 Circular Holes with In-Plane Stresses / 180
4.3.1 Single Circular Hole in an Infinite Thin Element in Uniaxial Tension / 180
4.3.2 Single Circular Hole in an Infinite Thin Element under Biaxial In-plane Stresses / 184
4.3.3 Single Circular Hole in a Cylindrical Shell with Tension or Internal Pressure / 186
4.3.4 Circular or Elliptical Hole in a Spherical Shell with Internal Pressure / 188
4.3.5 Reinforced Hole Near the Edge of a Semi-infinite Element in Uniaxial Tension / 188
4.3.6 Symmetrically Reinforced Hole in Finite-Width Element in Uniaxial Tension / 190
4.3.7 Nonsymmetrically Reinforced Hole in Finite-Width Element in Uniaxial Tension / 191
4.3.8 Symmetrically Reinforced Circular Hole in a Biaxially Stressed Wide, Thin Element / 192

4.3.9   Circular Hole with Internal Pressure / 199
   4.3.10 Two Circular Holes of Equal Diameter in a Thin Element in Uniaxial Tension or Biaxial In-plane Stresses / 200
   4.3.11 Two Circular Holes of Unequal Diameter in a Thin Element in Uniaxial Tension or Biaxial In-plane Stresses / 205
   4.3.12 Single Row of Equally Distributed Circular Holes in an Element in Tension / 207
   4.3.13 Double Row of Circular Holes in a Thin Element in Uniaxial Tension / 208
   4.3.14 Symmetrical Pattern of Circular Holes in a Thin Element in Uniaxial Tension or Biaxial In-plane Stresses / 208
   4.3.15 Radially Stressed Circular Element with a Ring of Circular Holes, with or without a Central Circular Hole / 209
   4.3.16 Thin Element with Circular Holes with Internal Pressure / 210
4.4 Elliptical Holes in Tension / 211
   4.4.1   Single Elliptical Hole in Infinite- and Finite-Width Thin Elements in Uniaxial Tension / 213
   4.4.2   Width Correction Factor for a Cracklike Central Slit in a Tension Panel / 215
   4.4.3   Single Elliptical Hole in an Infinite, Thin Element Biaxially Stressed / 215
   4.4.4   Infinite Row of Elliptical Holes in Infinite- and Finite-Width Thin Elements in Uniaxial Tension / 224
   4.4.5   Elliptical Hole with Internal Pressure / 224
   4.4.6   Elliptical Holes with Bead Reinforcement in an Infinite Thin Element under Uniaxial and Biaxial Stresses / 225
4.5 Various Configurations with In-Plane Stresses / 225
   4.5.1   Thin Element with an Ovaloid; Two Holes Connected by a Slit under Tension; Equivalent Ellipse / 225
   4.5.2   Circular Hole with Opposite Semicircular Lobes in a Thin Element in Tension / 226
   4.5.3   Infinite Thin Element with a Rectangular Hole with Rounded Corners Subject to Uniaxial or Biaxial Stress / 227
   4.5.4   Finite-Width Tension Thin Element with Round-Cornered Square Hole / 228
   4.5.5   Square Holes with Rounded Corners and Bead Reinforcement in an Infinite Panel under Uniaxial and Biaxial Stresses / 228
   4.5.6   Round-Cornered Equilateral Triangular Hole in an Infinite Thin Element under Various States of Tension / 228
   4.5.7   Uniaxially Stressed Tube or Bar of Circular Cross Section with a Transverse Circular Hole / 229
   4.5.8   Round Pin Joint in Tension / 229
   4.5.9   Inclined Round Hole in an Infinite Panel Subjected to Various States of Tension / 231

        4.5.10 Pressure Vessel Nozzle (Reinforced Cylindrical Opening) / 232
        4.5.11 Spherical or Ellipsoidal Cavities / 232
        4.5.12 Spherical or Ellipsoidal Inclusions / 234
        4.5.13 Cylindrical Tunnel / 236
        4.5.14 Intersecting Cylindrical Holes / 236
        4.5.15 Other Configurations / 238
4.6 Bending / 238
        4.6.1 Bending of a Beam with a Central Hole / 239
        4.6.2 Bending of a Beam with a Circular Hole Displaced from the Center Line / 240
        4.6.3 Bending of a Beam with an Elliptical Hole; Slot with Semicircular Ends (Ovaloid); or Round-Cornered Square Hole / 240
        4.6.4 Bending of an Infinite- and of a Finite-Width Plate with a Single Circular Hole / 240
        4.6.5 Bending of an Infinite Plate with a Row of Circular Holes / 241
        4.6.6 Bending of an Infinite Plate with a Single Elliptical Hole / 241
        4.6.7 Bending of an Infinite Plate with a Row of Elliptical Holes / 242
        4.6.8 Tube or Bar of Circular Cross Section with a Transverse Hole / 242
4.7 Shear, Torsion / 243
        4.7.1 Shear Stressing of Infinite Thin Element with Circular or Elliptical Hole, Unreinforced and Reinforced / 243
        4.7.2 Shear Stressing of an Infinite Thin Element with a Round-Cornered Rectangular Hole, Unreinforced and Reinforced / 244
        4.7.3 Two Circular Holes of Unequal Diameter in a Thin Element in Pure Shear / 244
        4.7.4 Shear Stressing of an Infinite Thin Element with Two Circular Holes or a Row of Circular Holes / 244
        4.7.5 Shear Stressing of an Infinite Thin Element with an Infinite Pattern of Circular Holes / 244
        4.7.6 Twisted Infinite Plate with a Circular Hole / 244
        4.7.7 Torsion of a Cylindrical Shell with a Circular Hole / 245
        4.7.8 Torsion of a Tube or Bar of Circular Cross Section with a Transverse Circular Hole / 245
References / 247
Charts / 256

5 MISCELLANEOUS DESIGN ELEMENTS   377

5.1 Notation / 377
5.2 Shaft with Keyseat / 378
        5.2.1 Bending / 379
        5.2.2 Torsion / 380

		5.2.3	Torque Transmitted through a Key / 380

- 5.2.3 Torque Transmitted through a Key / 380
- 5.2.4 Combined Bending and Torsion / 381
- 5.2.5 Effect of Proximitiy of Keyseat to Shaft Shoulder Fillet / 381
- 5.2.6 Fatigue Failures / 381
- 5.3 Splined Shaft in Torsion / 383
- 5.4 Gear Teeth / 383
- 5.5 Press-Fitted or Shrink-Fitted Members / 385
- 5.6 Bolt and Nut / 387
- 5.7 Bolt Head, Turbine-Blade, or Compressor-Blade Fastening (T-Head) / 389
- 5.8 Lug Joint / 391
  - 5.8.1 Lugs with $h/d < 0.5$ / 392
  - 5.8.2 Lugs with $h/d > 0.5$ / 393
- 5.9 Curved Bar / 394
- 5.10 Helical Spring / 395
  - 5.10.1 Round or Square Wire Compression or Tension Spring / 395
  - 5.10.2 Rectangular Wire Compression or Tension Spring / 397
  - 5.10.3 Helical Torsion Spring / 398
- 5.11 Crankshaft / 398
- 5.12 Crane Hook / 399
- 5.13 U-Shaped Member / 399
- 5.14 Angle and Box Sections / 400
- 5.15 Rotating Disk with Hole / 400
- 5.16 Ring or Hollow Roller / 402
- 5.17 Pressurized Cylinder / 402
- 5.18 Cylindrical Pressure Vessel with Torispherical Ends / 403
- 5.19 Pressurized Thick Cylinder with a Circular Hole in the Cylinder Wall / 403

References / 404
Charts / 408

6 STRESS CONCENTRATION ANALYSIS AND DESIGN   441

- 6.1 Computational Methods / 441
- 6.2 Finite Element Analysis / 445
  - 6.2.1 Principle of Virtual Work / 445
  - 6.2.2 Element Equations / 448
  - 6.2.3 Shape Functions / 450
  - 6.2.4 Mapping Functions / 454
  - 6.2.5 Numerical Integration / 455
  - 6.2.6 System Equations / 457
  - 6.2.7 Stress Computation / 460
- 6.3 Design Sensitivity Analysis / 467

       6.3.1   Finite Differences / 468
       6.3.2   Discrete Systems / 468
       6.3.3   Continuum Systems / 471
       6.3.4   Stresses / 474
       6.3.5   Structural Volume / 474
       6.3.6   Design Velocity Field / 475
6.4   Design Modification / 484
       6.4.1   Sequential Linear Programming / 487
       6.4.2   Sequential Quadratic Programming / 488
       6.4.3   Conservative Approximation / 489
       6.4.4   Equality Constraints / 490
       6.4.5   Minimum Weight Design / 491
       6.4.6   Minimum Stress Design / 492
References / 497

INDEX   **501**

INDEX TO THE STRESS CONCENTRATION FACTORS

CHAPTER 2

Form of Stress Raiser	Load Case	Shape of Stress Raiser	Section and Equation Number	Chart Number	Page Number of Chart
Single notch in semi-infinite thin element	Tension	U-Shaped	2.3.1	2.2	82
		Hyperbolic	2.3.6	2.8	88
		Elliptical	2.3.1	2.2	82
		Flat bottom	2.3.7	2.11	91
	Bending (in-plane)	Hyperbolic	2.6.5	2.29	109
	Bending (out-of-plane)	V-shaped	2.7.1	2.36	117
		Flat bottom	2.7.1	2.36	117
		Elliptical	2.7.1	2.37	118
Multiple notches in semi-infinite thin element	Bending (out-of-plane)	Semicircular	2.7.1	2.38	119
Opposite notches in infinite thin element	Tension	Hyperbolic	2.3.1	2.1	81
	Bending (in-plane)	Hyperbolic	2.6.1	2.23	103
	Bending (out-of-plane)	Hyperbolic	2.7.1	2.35	116
	Shear	Hyperbolic	2.9.1	2.45	126
Single notch in finite width thin element	Tension	U-Shaped	2.3.6	2.9	89
		Flat bottom	2.3.8	2.14	94
	Bending (in-plane)	U-Shaped	2.6.5	2.30	110
		V-Shaped	2.6.4	2.28	108
		Various shaped notches in impact test	2.6.5	2.31	112
		Semi-Elliptical	2.6.6	2.32	113

Form of Stress Raiser	Load Case	Shape of Stress Raiser	Section and Equation Number	Chart Number	Page Number of Chart
Multiple notches on one side of finite width thin element	Tension	Semicircular	2.3.8	2.14 2.15 2.16	94 95 96
	Bending (in-plane)	Semi-Elliptical	2.6.6	2.32	113
	Bending (out-of-plane)	Semicircular	2.7.1	2.38	119
Opposite single notches in finite width thin element	Tension	U-Shaped	2.3.3 Eq. (2.1)	2.4 2.5 2.6 2.53	84 85 86 134
		Semicircular	2.3.2	2.3	83
		V-Shaped	2.3.5	2.7	87
		Flat bottom	2.3.7	2.10	90
	Bending (in-plane)	Semicircular	2.6.2	2.24	104
		U-Shaped	2.6.3	2.25 2.26 2.27 2.53	105 106 107 134
		Flat bottom	2.6.7	2.33	114
	Bending (out-of-plane)	Arbitrary shaped	2.7.2	2.39	120
Opposite multiple notches in finite width thin element	Tension	Semicircular	2.3.8	2.12 2.13	92 93
Depressions in opposite sides of a thin element	Uniaxial tension	Spherical	2.4.3	2.17	97
		Cylindrical groove			

Form of Stress Raiser	Load Case	Shape of Stress Raiser	Section and Equation Number	Chart Number	Page Number of Chart
Depression in the surface of a semi-infinite body	Uniaxial tension	Hemispherical	2.4.1		
		Hyperboloid	2.4.2		
Groove in infinite medium	Tension	Hyperbolic	2.5.1	2.18	98
	Bending	Hyperbolic	2.8.1	2.40	121
	Torsion	Hyperbolic	2.9.2	2.46	127
Circumferential groove in shaft of circular cross section	Tension	U-Shaped	2.5.2	2.19 2.20 2.21 2.53	99 100 101 134
		Flat bottom	2.5.3	2.22 2.34	102 115
	Bending	U-Shaped	2.8.2	2.41 2.42 2.43 2.53	122 123 124 134
		Flat bottom	2.6.7 2.8.3	2.34 2.44	115 125
	Tension and bending	Flat bottom	2.6.7	2.34	115
	Torsion	U-Shaped	2.9.3	2.47 2.48 2.49 2.50 2.53	128 129 130 131 134
		V-Shaped	2.9.4	2.51	132
		Flat bottom	2.9.5	2.52	133

CHAPTER 3

Form of Stress Raiser	Load Case	Shape of Stress Raiser	Section and Equation Number	Chart Number	Page Number of Chart
Shoulder fillets in thin element	Tension	Single radius	3.3.1 Eq. (3.1)	3.1	150
		Tapered	3.3.4		
	Bending	Single radius	3.4.1	3.7	159
		Elliptical	3.4.3	3.9	163
		Tapered	3.3.4		
	Torsion	Tapered	3.3.4		
Shoulder fillets in thin element	Tension	Single radius	3.3.2	3.2	151
		Trapezoidal protuberance	3.3.3	3.3	154
	Bending	Single radius	3.4.2	3.8	160
Shoulder fillet in bar of circular cross section	Tension	Single radius	3.3.5	3.4	156
	Bending	Single radius	3.4.4	3.10 3.11	164 165
	Torsion	Single radius	3.5.1	3.12 3.13	166 167
		Compound radius	3.5.3	3.16 3.17	172 174
Shoulder fillet in bar of circular cross section with axial hole	Tension	Single radius	3.3.6	3.5	157
	Torsion	Single radius	3.5.2	3.14 3.15	168 169
Stepped pressure vessel	Internal pressure	Stepped ring	3.3.7	3.6	158

CHAPTER 4

Form of Stress Raiser	Load Case	Shape of Stress Raiser	Section and Equation Number	Chart Number	Page Number of Chart
Hole in infinite thin element	Uniaxial tension	Circular	4.3.1 Eqs. (4.6), (4.7), and (4.8)		
		Elliptical	4.4.1 Eqs. (4.57) and (4.58)	4.50	319
		Elliptical hole with Inclusion	4.5.12	4.50 4.75	319 351
		Circular hole with opposite semicircular lobes	4.5.2	4.60	331
		Rectangular	4.5.3 4.5.4	4.62a	333
		Equilateral triangular	4.5.6	4.65	340
		Inclined	4.5.9	4.70	346
	Internal pressure	Circular, elliptical, and other shapes	4.3.9, 4.3.16, 4.4.5 Eqs. (4.41) and (4.68)		
	Biaxial stress (in-plane)	Circular	4.3.2 Eqs. (4.17) and (4.18)		
		Rectangular	4.5.3	4.62	333
		Various shapes	4.5.1 4.5.3	4.63	337
		Equilateral triangular	4.5.6	4.65	340
		Elliptical	4.4.3 Eqs. (4.68), (4.69), (4.70), (4.71)	4.54 4.55	323 324
		Inclined	4.5.9	4.69	345
	Bending (out-of-plane)	Circular	4.6.4 Eqs. (4.103) and (4.104)	4.82	358
		Elliptical	4.6.6 Eqs. (4.106) and (4.107)	4.85	361
	Shear	Circular or Elliptical	4.7.1	4.88	364
		Rectangular	4.7.2	4.90	366
	Twist	Circular	4.7.6	4.97	374

INDEX TO THE STRESS CONCENTRATION FACTORS

Form of Stress Raiser	Load Case	Shape of Stress Raiser	Section and Equation Number	Chart Number	Page Number of Chart
Hole in finite width thin element	Uniaxial tension	Circular	4.3.1 Eq. (4.9)	4.1	256
		Eccentrically located circular	4.3.1 Eq. (4.14)	4.3	258
		Elliptical	4.4.1, 4.4.2	4.51	320
		Circular hole with opposite semi-circular lobes	4.5.2 Eqs. (4.78) and (4.79)	4.61	331
		Slot with semicircular or semielliptical end	4.5.1	4.59	330
	Internal pressure	Various shapes	4.3.16 4.4.5		
	Bending (in-plane)	Circular	4.6.1, 4.6.2 Eqs. (4.98), (4.99), (4.100), and (4.101)	4.79 4.80	355 356
		Elliptical	4.6.3	4.81	357
		Ovaloids, square	4.6.3 Eq. (4.102)		
	Bending (out-of-plane)	Circular	4.6.4 Eq. (4.103)	4.83	359
Hole in semi-infinite thin element	Uniaxial tension	Circular	4.3.1 Eq. (4.12)	4.2	257
		Elliptical	4.4.1	4.52	321
	Internal pressure	Various shapes	4.3.16 4.4.5		
Hole in cylindrical shell, pipe, or bar	Tension	Circular	4.3.3, 4.5.7 Eqs. (4.71), and (4.73)	4.4 4.66	259 342
	Internal pressure	Circular	4.3.3	4.5	260
	Bending	Circular	4.6.8 Eqs. (4.108), (4.109), and (4.110)	4.87	363
	Torsion	Circular	4.7.7, 4.7.8 Eqs. (4.113), (4.114), (4.116), (4.117), (4.118), and (4.119)	4.98 4.99	375 376

Form of Stress Raiser	Load Case	Shape of Stress Raiser	Section and Equation Number	Chart Number	Page Number of Chart
Row of holes in infinite thin element	Uniaxial tension	Circular	4.3.12	4.32, 4.33	299, 300
		Elliptical	4.4.4	4.56	325
		Elliptical holes with inclusions	4.5.12	4.76	352
	Biaxial stresses (in-plane)	Circular	4.3.12	4.34	301
	Bending (out-of-plane)	Circular	4.6.5	4.84	360
		Elliptical	4.6.7	4.86	362
	Shear	Circular	4.7.4	4.93	370
Row of holes in finite width thin element	Uniaxial tension	Elliptical	4.4.4	4.57	326
Double row of holes in infinite thin element	Uniaxial tension	Circular	4.3.13 Eqs. (4.46) and (4.47)	4.35 4.36	302 303
Triangular pattern of holes in infinite thin element	Uniaxial tension	Circular	4.3.14	4.37, 4.38, 4.39	304, 305, 306
	Biaxial stresses (in-plane)	Circular	4.3.14	4.37, 4.38, 4.39, 4.41	304, 305, 306, 310
	Shear	Circular	4.7.5	4.94	371
Square pattern of holes in infinite thin element	Uniaxial tension	Circular	4.3.14	4.40 4.43	309 312
	Biaxial stresses (in-plane)	Circular	4.3.14	4.40, 4.41, 4.42	309, 310, 311
	Shear	Circular	4.7.5	4.94, 4.95	371, 372
Diamond pattern of holes in infinite thin element	Uniaxial tension	Circular	4.3.14	4.44 4.45	313 314
	Shear	Circular	4.7.5	4.96	373

INDEX TO THE STRESS CONCENTRATION FACTORS

Form of Stress Raiser	Load Case	Shape of Stress Raiser	Section and Equation Number	Chart Number	Page Number of Chart
Hole in wall of thin spherical shell	Internal pressure	Circular or elliptical	4.3.4	4.6	261
Reinforced hole in infinite thin element	Biaxial stress (in-plane)	Circular	4.3.8	4.13, 4.14, 4.15, 4.16, 4.17, 4.18, 4.19	270, 275, 276, 277, 278, 279 280
		Elliptical	4.4.6	4.58	327
		Square	4.5.5	4.64	338
	Shear	Elliptical	4.7.1	4.89	365
		Square	4.7.2	4.91	367
Reinforced hole in semi-infinite thin element	Uniaxial tension	Circular	4.3.5	4.7	262
		Square	4.5.5	4.64a	338
Reinforced hole in finite width thin element	Uniaxial tension	Circular	4.3.6 4.3.7 Eq. (4.26)	4.8 4.9 4.10 4.11 4.12	263 266 267 268 269
Hole in panel	Internal pressure	Circular	4.3.9	4.20	281
Two holes in infinite thin element	Uniaxial tension	Circular	4.3.10 4.3.11 Eqs. (4.42), (4.44), and (4.45)	4.21, 4.22, 4.23, 4.25, 4.26, 4.27, 4.29, 4.30, 4.31	282, 283, 284, 286, 293, 294, 296, 297, 298
	Biaxial stresses (in-plane)	Circular	4.3.10 4.3.11 Eqs. (4.43), (4.44), and (4.45)	4.24 4.25 4.28	285 286 295
	Shear	Circular	4.7.3, 4.7.4	4.92	368

INDEX TO THE STRESS CONCENTRATION FACTORS   **XXV**

Form of Stress Raiser	Load Case	Shape of Stress Raiser	Section and Equation Number	Chart Number	Page Number of Chart
Ring of holes in circular thin element	Radial in-plane stresses	Circular	4.3.15 Table 4.1	4.46	315
	Internal pressure	Circular	4.3.16 Table 4.2	4.47	316
Hole in circular thin element	Internal pressure	Circular	4.3.16 Table 4.2	4.48	317
Circular pattern of holes in circular thin element	Internal pressure	Circular	4.3.16 Table 4.2	4.49	318
Pin joint with closely fitting pin	Tension	Circular	4.5.8 Eqs. (4.83) and (4.84)	4.67	343
Pinned or riveted joint with multiple holes	Tension	Circular	4.5.8	4.68	344
Cavity in infinite body	Tension	Circular cavity of elliptical cross section	4.5.11	4.71	347
		Elliptical cavity of circular cross section	4.5.11	4.72	348
Cavities in infinite panel and cylinder	Uniaxial tension or biaxial stresses	Spherical cavity	4.5.11 Eqs. (4.77), (4.78), and (4.79)	4.73	349
Row of cavities in infinite element	Tension	Elliptical cavity	4.5.11	4.74	350

Form of Stress Raiser	Load Case	Shape of Stress Raiser	Section and Equation Number	Chart Number	Page Number of Chart
Crack in thin tension element	Uniaxial tension	Narrow crack	4.4.2 Eqs. (4.62), (4.63), and (4.64)	4.53	322
Tunnel	Hydraulic pressure	Circular	4.5.13 Eqs. (4.97)	4.77 4.78	353 354

CHAPTER 5

Form of Stress Raiser	Load Case	Shape of Stress Raiser	Section and Equation Number	Chart Number	Page Number of Chart
Keyseat	Bending	Semicircular end	5.2.1	5.1	408
		Sled runner	5.2.1		
	Torsion	Semicircular end	5.2.2 5.2.3	5.2	409
	Combined bending and torsion	Semicircular end	5.2.4	5.3	410
Splined shaft	Torsion		5.3	5.4	411
Gear tooth	Bending		5.4 Eqs. (5.3) and (5.4)	5.5 5.6 5.7 5.8	412 413 414 415
Short beam	Bending	Shoulder fillets	5.4 Eq. (5.5)	5.9	416
Press-fitted member	Bending		5.5 Tables 5.1 and 5.2		
Bolt and nut	Tension		5.6		
T-head	Tension and bending	Shoulder fillets	5.7 Eqs. (5.7) and (5.8)	5.10	417

Form of Stress Raiser	Load Case	Shape of Stress Raiser	Section and Equation Number	Chart Number	Page Number of Chart
Lug joint	Tension	Square ended	5.8	5.11 5.13	422 424
		Round ended	5.8	5.12 5.13	423 424
Curced bar	Bending	Uniform bar	5.9 Eq. (5.11)	5.14	425
		Nonuniform: crane hook	5.12		
Helical spring	Tension or compression	Round or square wire	5.10.1 Eqs. (5.17) (5.18)	5.15	426
		Rectangular wire	5.10.2 Eq. (5.23)	5.16	427
	Torsional	Round or rectangular wire	5.10.3	5.17	428
Crankshaft	Bending		5.11 Eq. (5.26)	5.18 5.19	429 430
U-shaped member	Tension and bending		5.13 Eqs. (5.27) and (5.28)	5.20 5.21	431 432
Angle or box sections	Torsion		5.14	5.22	433

Form of Stress Raiser	Load Case	Shape of Stress Raiser	Section and Equation Number	Chart Number	Page Number of Chart
Disk	Rotating centrifugal inertial force	Central hole	5.15	5.23	434
		Noncentral hole	5.15	5.24	435
Ring	Diametrically opposite internal concentrated loads		5.16 Eq. (5.34)	5.25	436
	Diametrically opposite external concentrated loads		5.16 Eq. (5.35)	5.26	437
Thick cylinder	Internal pressure	No hole in cylinder wall	5.17 Eqs. (5.37) and (5.38)	5.27	438
		Hole in cylinder wall	5.19 Eq. (5.39)	5.29	440
Cylindrical pressure vessel	Internal pressure	Torispherical ends	5.18	5.28	439

PREFACE

Rudolph Earl Peterson (1901–1982) has been *Mr. Stress Concentration* for the past half century. Only after preparing this edition of this book has it become evident how much effort he put into his two previous books: *Stress Concentration Design Factors* (1953) and *Stress Concentration Factors* (1974). These were carefully crafted treatises. Much of the material from these books has been retained intact for the present edition. Stress concentration charts not retained are identified in the text so that interested readers can refer to the earlier editions.

The present book contains some recently developed stress concentration factors, as well as most of the charts from the previous editions. Moreover there is considerable material on how to perform computer analyses of stress concentrations and how to design to reduce stress concentration. Example calculations on the use of the stress concentration charts have been included in this edition.

One of the objectives of application of stress concentration factors is to achieve better balanced designs[1] of structures and machines. This can lead to conserving materials, cost reduction, and achieving lighter and more efficient apparatus. Chapter 6, with the computational formulation for design of systems with potential stress concentration problems, is intended to be used to assist in the design process.

---

[1] Balanced design is delightfully phrased in the poem, "The Deacon's Masterpiece, or the Wonderful One Hoss Shay" by Oliver Wendell Holmes (1858):

"Fur", said the Deacon, " 't 's mighty plain
That the weakes' place mus' stan' the strain,
"N the way t' fix it, uz I maintain,
Is only jest
T' make that place uz strong the rest"

After "one hundred years to the day" the shay failed "all at once, and nothing first."

The universal availability of general purpose structural analysis computer software has revolutionized the field of stress concentrations. No longer are there numerous photoelastic stress concentration studies being performed. The development of new experimental stress concentration curves has slowed to a trickle. Often structural analysis is performed computationally in which the use of stress concentration factors is avoided, since a high-stress region is simply part of the computer analysis.

Graphical keys to the stress concentration charts are provided to assist the reader in locating a desired chart.

Major contributions to this revised book were made by Huiyan Wu, Weize Kang, and Uwe Schramm. Drs. Wu and Kang were instrumental throughout in developing new material and, in particular, in securing new stress concentration charts from Japanese, Chinese, and Russian stress concentration books, all of which were in the Chinese language. Dr. Schramm was a principal contributor to Chapter 6 on computational analysis and design.

Special thanks are due to Brett Matthews and Wei Wei Ding, who skillfully prepared the text and figures, respectively. Many polynomial representations of the stress concentration factors were prepared by Debbie Pilkey. Several special figures were drawn by Charles Pilkey.

Even though the manuscript has been checked and rechecked, experience indicates that with so much material, it is almost impossible to completely eliminate errors. An errata list will be set up that will be linked to the author's home page. The address for the home page is:

http://watt.seas.virginiacomputer.edu/~wdp

If you find errors, please inform the author. A program to calculate stress concentration factors has been developed to accompany this book. Please see the author's home page for information on the availability of the computer program.

WALTER D. PILKEY

CHAPTER 1

DEFINITIONS AND DESIGN RELATIONS

1.1 NOTATION

A = area (or point)
a = radius of hole
a = material constant for evaluating notch sensitivity factor
a = major axis of ellipse
a = half crack length
b = minor axis of ellipse
d = diameter (or width)
D = larger diameter
H = larger width
h = thickness
K = stress concentration factor
K_e = effective stress concentration factor
K_f = fatigue notch factor for normal stress
K_{fs} = fatigue notch factor for shear stress
K_t = theoretical stress concentration factor for normal stress
K_{t2}, K_{t3} = stress concentration factors for two, three dimensional problems
K_{tf} = estimated fatigue notch factor for normal stress
K_{tsf} = estimated fatigue notch factor for shear stress
K_{tg} = stress concentration factor with the nominal stress based on gross area
K_{tn} = stress concentration factor with the nominal stress based on net area
$K_{tx}, K_{t\theta}$ = stress concentration factors in x, θ directions

K_{ts} = stress concentration factor for shear stress
K_t' = stress concentration factor using a theory of failure
K_I = mode I stress intensity factor
l, m = direction cosines of normal to boundary
L_b = limit design factor for bending
M = moment
n = factor of safety
p = pressure
p_x, p_y = surface forces per unit area in x, y directions
P = load, force
$\overline{p}_{Vx}, \overline{p}_{Vy}, \overline{p}_{Vr}$ = body forces per unit volume in x, y, r directions
q = notch sensitivity factor
r = radius of curvature of hole, arc, notch
R = radius of hole, circle, or radial distance
t = depth of groove, notch
T = torque
u, v, w = displacements in x, y, z directions (or in r, θ, x directions in cylindrical coordinates)
x, y, z = rectangular coordinates
r, θ = polar coordinates
r, θ, x = cylindrical coordinates
$\varepsilon_r, \varepsilon_\theta, \varepsilon_x, \gamma_{rx}$ = strain components
ν = Poisson's ratio
ρ = mass density
ρ' = material constant for evaluating notch sensitivity factor
σ = normal stress
σ_a = alternating normal stress amplitude
σ_n = normal stress based on net area
σ_{eq} = equivalent stress
σ_f = fatigue strength (endurance limit)
σ_{max} = maximum normal stress
σ_{nom} = nominal or reference normal stress
σ_0 = static normal stress
$\sigma_x, \sigma_y, \tau_{xy}$ = stress components
$\sigma_r, \sigma_\theta, \tau_{r\theta}, \tau_{\theta x}$ = stress components
σ_y = yield strength (tension)
σ_{ut} = ultimate tensile strength
σ_{uc} = ultimate compressive strength
$\sigma_1, \sigma_2, \sigma_3$ = principal stresses
τ = shear stress
τ_a = alternating shear stress
τ_f = fatigue limit in torsion

τ_y = yield strength in torsion
τ_{max} = maximum shear stress
τ_{nom} = nominal or reference shear stress
τ_0 = static shear stress
ω = angular rotating speed

1.2 STRESS CONCENTRATION

The elementary stress formulas used in the design of structural members are based on the members having a constant section or a section with gradual change of contour (Fig. 1.1). Such conditions, however, are hardly ever attained throughout the highly stressed region of actual machine parts or structural members. The presence of shoulders, grooves, holes, keyways, threads, and so on, results in modifications of the simple stress distributions of Fig. 1.1 so that localized high stresses occur as shown in Figs. 1.2 and 1.3. This localization of high stress is known as *stress concentration*, measured by the *stress concentration factor*. The stress concentration factor K can be defined as the ratio of the peak stress in the body (or stress in the perturbed region) to some other stress (or stresslike quantity) taken as a

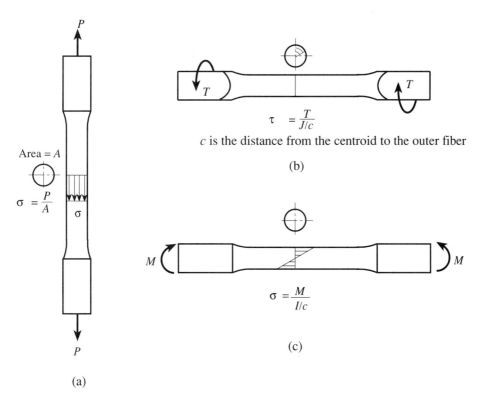

Figure 1.1 Elementary stress cases for specimens of constant cross section or with a gradual cross-sectional change: (*a*) Tension; (*b*) torsion; (*c*) bending.

4 DEFINITIONS AND DESIGN RELATIONS

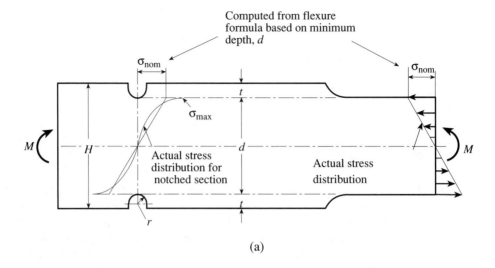

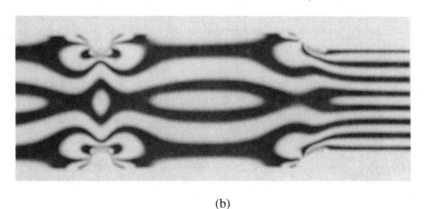

Figure 1.2 Stress concentrations introduced by a notch and a cross-sectional change which is not gradual: (*a*) Bending of specimen; (*b*) photoelastic fringe photograph (Peterson 1974).

reference stress:

$$K_t = \frac{\sigma_{max}}{\sigma_{nom}} \quad \text{for normal stress (tension or bending)} \tag{1.1}$$

$$K_{ts} = \frac{\tau_{max}}{\tau_{nom}} \quad \text{for shear stress (torsion)} \tag{1.2}$$

where the stresses σ_{max}, τ_{max} represent the maximum stresses to be expected in the member under the actual loads and the *nominal stresses* σ_{nom}, τ_{nom} are reference normal and shear stresses. The subscript t indicates that the stress concentration factor is a theoretical factor. That is to say, the peak stress in the body is based on the theory of elasticity, or it is derived from a laboratory stress analysis experiment. The subscript s of Eq. (1.2) is often ignored. In the case of the theory of elasticity, a two-dimensional stress distribution of a homogeneous elastic body under known loads is a function only of the body geometry and is not dependent on the material properties. This book deals primarily with elastic stress concentration

STRESS CONCENTRATION 5

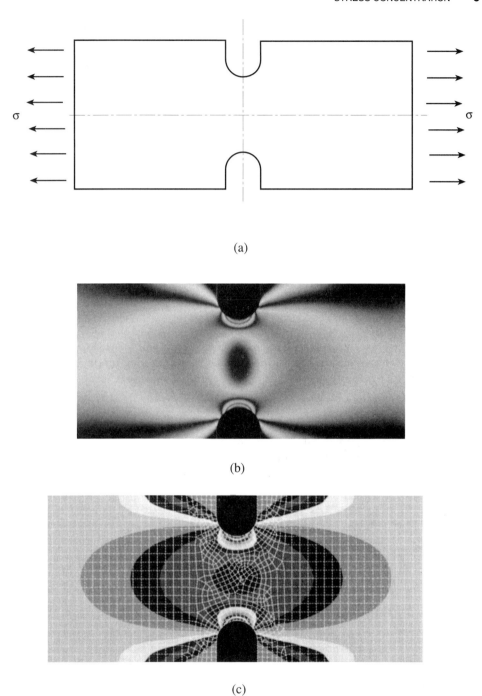

Figure 1.3 Tension bar with notches: (*a*) The specimen; (*b*) photoelastic fringe photograph (Doz. Dr.-Ing. habil. K. Fethke, Universität Rostock); (*c*) finite element solution (Guy Nerad, University of Virginia).

factors. In the plastic range one must consider separate stress and strain concentration factors that depend on the shape of the stress-strain curve and the stress or strain level. Sometimes K_t is also referred to as a *form factor*. The subscript t distinguishes factors derived from theoretical or computational calculations, or experimental stress analysis methods such as photoelasticity, or strain gage tests from factors obtained through mechanical damage tests such as impact tests. For example, the fatigue notch factor K_f is determined using a fatigue test. It will be described later.

The universal availability of powerful, effective computational capabilities, usually based on the finite element method, has altered the use of and the need for stress concentration factors. Often a computational stress analysis of a mechanical device, including highly stressed regions, is performed as shown in Fig. 1.3c, and the explicit use of stress concentration factors is avoided. Alternatively, a computational analysis can provide the stress concentration factor, which is then available for traditional design studies. The use of experimental techniques such as photoelasticity (Fig. 1.3b) to determine stress concentration factors has been virtually replaced by the more flexible and more efficient computational techniques.

1.2.1 Selection of Nominal Stresses

The definitions of the reference stresses σ_{nom}, τ_{nom} depend on the problem at hand. It is very important to properly identify the reference stress for the stress concentration factor of interest. In this book the reference stress is usually defined at the same time that a particular stress concentration factor is presented. Consider several examples to explain the selection of reference stresses.

Example 1.1   Tension Bar with Hole   Uniform tension is applied to a bar with a single circular hole, as shown in Fig. 1.4a. The maximum stress occurs at point A, and the stress distribution can be shown to be as in Fig. 1.4a. Suppose that the thickness of the plate is h, the width of the plate is H, and the diameter of the hole is d. The reference stress could be defined in two ways:

a. Use the stress in a cross section far from the circular hole as the reference stress. The area at this section is called the *gross cross-sectional area*. Thus define

$$\sigma_{\text{nom}} = \frac{P}{Hh} = \sigma \tag{1}$$

so that the stress concentration factor becomes

$$K_{tg} = \frac{\sigma_{\max}}{\sigma_{\text{nom}}} = \frac{\sigma_{\max}}{\sigma} = \frac{\sigma_{\max} Hh}{P} \tag{2}$$

b. Use the stress based on the cross section at the hole, which is formed by removing the circular hole from the gross cross section. The corresponding area is referred to as the *net cross-sectional area*. If the stresses at this cross section are uniformly distributed and equal to σ_n:

$$\sigma_n = \frac{P}{(H-d)h} \tag{3}$$

STRESS CONCENTRATION 7

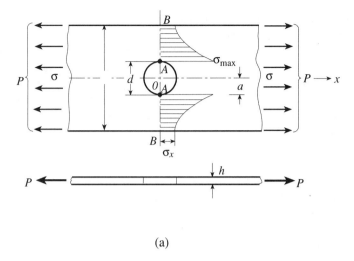

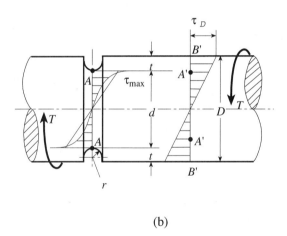

Figure 1.4 Examples: (*a*) Tension bar with hole; (*b*) torsion bar with groove.

The stress concentration factor based on the reference stress σ_n, namely, $\sigma_{\text{nom}} = \sigma_n$, is

$$K_{tn} = \frac{\sigma_{\max}}{\sigma_{\text{nom}}} = \frac{\sigma_{\max}}{\sigma_n} = \frac{\sigma_{\max}(H-d)h}{P} = K_{tg}\frac{H-d}{H} \qquad (4)$$

In general, K_{tg} and K_{tn} are different. Both are plotted in Chart 4.1. Observe that as d/H increases from 0 to 1, K_{tg} increases from 3 to ∞, whereas K_{tn} decreases from 3 to 2. Either K_{tn} or K_{tg} can be used in calculating the maximum stress. It would appear that K_{tg} is easier to determine as σ is immediately evident from the geometry

8 DEFINITIONS AND DESIGN RELATIONS

of the bar. But the value of K_{tg} is hard to read from a stress concentration plot for $d/H > 0.5$, since the curve becomes very steep. In contrast, the value of K_{tn} is easy to read, but it is necessary to calculate the net cross-sectional area to find the maximum stress. Since the stress of interest is usually on the net cross section, K_{tn} is the more generally used factor. In addition, in a fatigue analysis, only K_{tn} can be used to calculate the stress gradient correctly. In conclusion, normally it is more convenient to give stress concentration factors using reference stresses based on the net area rather than the gross area. However, if a fatigue analysis is not involved and $d/H < 0.5$, the user may choose to use K_{tg} to simplify calculations.

Example 1.2   Torsion Bar with Groove   A bar of circular cross section, with a U-shaped circumferential groove, is subject to an applied torque T. The diameter of the bar is D, the radius of the groove is r, and the depth of the groove is t. The stress distribution for the cross section at the groove is shown in Fig. 1.4*b*, with the maximum stress occurring at point A at the bottom of the groove. Among the alternatives to define the reference stress are:

   a. Use the stress at the outer surface of the bar cross section B'–B', which is far from the groove, as the reference stress. According to basic strength of materials (Pilkey 1994), the shear stress is linearly distributed along the radial direction and

$$\tau_{B'} = \tau_D = \frac{16T}{\pi D^3} = \tau_{\text{nom}} \tag{1}$$

   b. Consider point A' in the cross section B'–B'. The distance of A' from the central axis is same as that of point A, that is, $d = D - 2t$. If the stress at A' is taken as the reference stress, then

$$\tau_{A'} = \frac{16Td}{\pi D^4} = \tau_{\text{nom}} \tag{2}$$

   c. Use the surface stress of a grooveless bar of diameter $d = D - 2t$ as the reference stress. This corresponds to a bar of cross section measured at A–A of Fig. 1.4*b*. For this area $\pi d^2/4$, the maximum torsional stress taken as a reference stress would be

$$\tau_A = \frac{16T}{\pi d^3} = \tau_{\text{nom}} \tag{3}$$

In fact this stress based on the net area is an assumed value and never occurs at any point of interest in the bar with a U-shaped circumferential groove. However, since it is intuitively appealing and easy to calculate, it is more often used than the other two reference stresses.

Example 1.3   Cylinder with Eccentric Hole   A cylinder with an eccentric circular hole is subjected to internal pressure p as shown in Fig. 1.5. An elastic solution for stress is difficult to find. It is convenient to use the pressure p as the reference stress

$$\sigma_{\text{nom}} = p$$

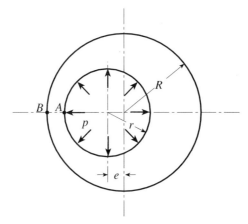

Figure 1.5 Circular cylinder with eccentric hole.

so that

$$K_t = \frac{\sigma_{\max}}{p}$$

These examples illustrate that there are many options for selecting a reference stress. In this book the stress concentration factors are given based on a variety of reference stresses, each of which is noted on the appropriate graph of the stress concentration factor. Sometimes more than one stress concentration factor is plotted on a single chart. The reader should select the type of factor that appears to be the most convenient.

1.2.2 Accuracy of Stress Concentration Factors

Stress concentration factors are obtained analytically from the elasticity theory, computationally from the finite element method, and experimentally using methods such as photoelasticity or strain gages. For torsion, the membrane analogy (Pilkey and Wunderlich 1993) can be employed. When the experimental work is conducted with sufficient precision, excellent agreement is often obtained with well-established analytical stress concentration factors.

Unfortunately, use of stress concentration factors in analysis and design is not on as firm a foundation as the theoretical basis for determining the factors. The theory of elasticity solutions are based on formulations that include such assumptions as that the material must be isotropic and homogeneous. However, in actuality materials may be neither uniform nor homogeneous, and may even have defects. More data are necessary because, for the required precision in material tests, statistical procedures are often necessary. Directional effects in materials must also be carefully taken into account. It is hardly necessary to point out that the designer cannot wait for exact answers to all of these questions. As always, existing information must be reviewed and judgment used in developing reasonable approximate procedures for design, tending toward the safe side in doubtful cases. In time, advances will take place and revisions in the use of stress concentration factors will need to be made accordingly. On the other hand, it can be said that our limited experience in using these methods has been satisfactory.

1.3 STRESS CONCENTRATION AS A TWO-DIMENSIONAL PROBLEM

Consider a thin element lying in the x, y plane, loaded by in-plane forces applied in the x, y plane at the boundary (Fig. 1.6a). For this case the stress components σ_z, τ_{xz}, τ_{yz} can be assumed to be equal to zero. This state of stress is called *plane stress*, and the stress components σ_x, σ_y, τ_{xy} are functions of x and y only. If the dimension in the z direction of a long cylindrical or prismatic body is very large relative to its dimensions in the x, y plane and the applied forces are perpendicular to the longitudinal direction (z direction) (Fig. 1.6b), it may be assumed that at the midsection the z direction strains ε_z, γ_{xz}, and γ_{yx} are equal to zero. This is called the *plane strain* state. These two-dimensional problems are referred to as *plane problems*.

The differential equations of equilibrium together with the compatibility equation for the stresses σ_x, σ_y, τ_{xy} in a plane elastic body are (Pilkey and Wunderlich 1993)

$$\frac{\partial \sigma_x}{\partial x} + \frac{\partial \tau_{xy}}{\partial y} + \overline{p}_{Vx} = 0$$
$$\frac{\partial \tau_{xy}}{\partial x} + \frac{\partial \sigma_y}{\partial y} + \overline{p}_{Vy} = 0 \tag{1.3}$$

$$\left(\frac{\partial^2}{\partial x^2} + \frac{\partial^2}{\partial y^2} \right)(\sigma_x + \sigma_y) = -f(\nu)\left(\frac{\partial \overline{p}_{Vx}}{\partial x} + \frac{\partial \overline{p}_{Vy}}{\partial y} \right) \tag{1.4}$$

where $\overline{p}_{Vx}$, $\overline{p}_{Vy}$ denote the components of the applied body force per unit volume in the x, y directions and $f(\nu)$ is a function of Poisson's ratio:

$$f(\nu) = \begin{cases} 1 + \nu & \text{for plane stress} \\ \dfrac{1}{1 - \nu} & \text{for plane strain} \end{cases}.$$

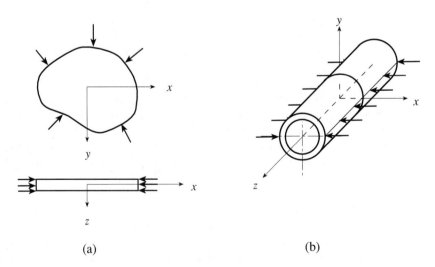

Figure 1.6 Plane stress and plane strain: (*a*) Plane stress; (*b*) plane strain.

The surface conditions are

$$p_x = l\sigma_x + m\tau_{xy}$$
$$p_y = l\tau_{xy} + m\sigma_y \qquad (1.5)$$

where p_x, p_y are the components of the surface force per unit area at the boundary in the x, y directions. Also, l, m are the direction cosines of the normal to the boundary. For constant body forces, $\partial \overline{p}_{Vx}/\partial x = \partial \overline{p}_{Vy}/\partial y = 0$, and Eq. (1.4) becomes

$$\left(\frac{\partial^2}{\partial x^2} + \frac{\partial^2}{\partial y^2}\right)(\sigma_x + \sigma_y) = 0 \qquad (1.6)$$

Equations (1.3), (1.5), and (1.6) are usually sufficient to determine the stress distribution for two-dimensional problems with constant body forces. These equations do not contain material constants. For plane problems, if the body forces are constant, the stress distribution is a function of the body shape and loadings acting on the boundary and not of the material. This implies for plane problems that stress concentration factors are functions of the geometry and loading and not of the type of material. Of practical importance is that stress concentration factors can be found using experimental techniques such as photoelasticty that utilize material different from the structure of interest.

1.4 STRESS CONCENTRATION AS A THREE-DIMENSIONAL PROBLEM

For three-dimensional problems, there are no simple relationships similar to Eqs. (1.3), (1.5), and (1.6) for plane problems that show the stress distribution to be a function of body shape and applied loading only. In general, the stress concentration factors will change with different materials. For example, Poisson's ratio ν is often involved in a three-dimensional stress concentration analysis. In this book most of the charts for three-dimensional stress concentration problems not only list the body shape and load but also the Poisson's ratio ν for the case. The influence of Poisson's ratio on the stress concentration factors varies with the configuration. For example, in the case of a circumferential groove in a round bar under torsional load (Fig. 1.7), the stress distribution and concentration factor do not depend on Poisson's ratio. This is because the shear deformation due to torsion does not change the volume of the element, namely the cross-sectional areas remain unchanged.

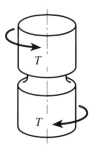

Figure 1.7 Round bar with circumferential groove and torsional loading.

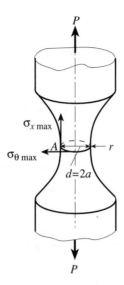

Figure 1.8 Hyperbolic circumferential groove in a round bar.

As another example, consider a hyperbolic circumferential groove in a round bar under tension load P (Fig. 1.8). The stress concentration factor in the axial direction is (Neuber 1958)

$$K_{tx} = \frac{\sigma_{x\,\max}}{\sigma_{\text{nom}}} = \frac{1}{(a/r) + 2\nu C + 2}\left[\frac{a}{r}(C + \nu + 0.5) + (1 + \nu)(C + 1)\right] \quad (1.7)$$

and in the circumferential direction is

$$K_{t\theta} = \frac{\sigma_{\theta\,\max}}{\sigma_{\text{nom}}} = \frac{a/r}{(a/r) + 2\nu C + 2}(\nu C + 0.5) \quad (1.8)$$

where r is the radius of curvature at the base of the groove, C is $\sqrt{(a/r) + 1}$, and the reference stress σ_{nom} is $P/(\pi a^2)$. Obviously K_{tx} and $K_{t\theta}$ are functions of ν. Table 1.1 lists the stress concentration factors for different Poisson's ratios for the hyperbolic circumferential groove when $a/r = 7.0$. From this table it can be seen that as the value of ν increases, K_{tx} decreases slowly whereas $K_{t\theta}$ increases relatively rapidly. When $\nu = 0$, $K_{tx} = 3.01$ and $K_{t\theta} = 0.39$. It is interesting that when Poisson's ratio is equal to zero (there is no transverse contraction in the round bar), the maximum circumferential stress $\sigma_{\theta\,\max}$ is not equal to zero.

TABLE 1.1 Stress Concentration Factor as a Function of Poisson's Ratio for a Shaft in Tension with Groove

ν	0.0	0.1	0.2	0.3	0.4	≈ 0.5
K_{tx}	3.01	2.95	2.89	2.84	2.79	2.75
$K_{t\theta}$	0.39	0.57	0.74	0.88	1.01	1.13

Definition: Shaft has a hyperbolic circumferential groove with $a/r = 7.0$.

1.5 PLANE AND AXISYMMETRIC PROBLEMS

For a solid of revolution deformed symmetrically with respect to the axis of revolution, it is convenient to use cylindrical coordinates (r, θ, x). The stress components are independent of the angle θ and $\tau_{r\theta}$, $\tau_{\theta x}$ are equal to zero. The equilibrium and compatibility equations for the axisymmetrical case are (Timoshenko and Goodier 1970)

$$\frac{\partial \sigma_r}{\partial r} + \frac{\partial \tau_{rx}}{\partial x} + \frac{\sigma_r - \sigma_\theta}{r} + \bar{p}_{Vr} = 0$$
$$\frac{\partial \tau_{rx}}{\partial r} + \frac{\partial \sigma_x}{\partial x} + \frac{\tau_{rx}}{r} + \bar{p}_{Vx} = 0$$
(1.9)

$$\frac{\partial^2 \varepsilon_r}{\partial x^2} + \frac{\partial^2 \varepsilon_x}{\partial r^2} = \frac{\partial^2 \gamma_{rx}}{\partial r \partial x}$$
(1.10)

The strain components are

$$\varepsilon_r = \frac{\partial u}{\partial r}, \quad \varepsilon_\theta = \frac{u}{r}, \quad \varepsilon_x = \frac{\partial w}{\partial x}, \quad \gamma_{rx} = \frac{\partial u}{\partial x} + \frac{\partial w}{\partial r}$$
(1.11)

where u and w are the displacements in the r (radial) and x (axial) directions, respectively. The axisymmetric stress distribution in a solid of revolution is quite similar to the stress distribution for a two-dimensional plane element, the shape of which is the same as a longitudinal section of the solid of revolution (see Fig. 1.9). Strictly speaking, their stress distributions and stress concentration factors should not be equal. But under certain cir-

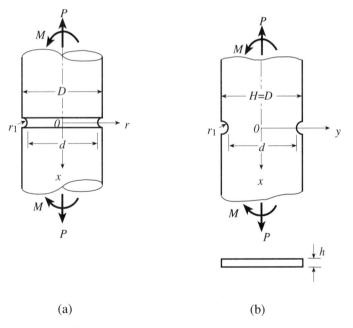

Figure 1.9 Shaft with a circumferential groove and a plane element with the same longitudinal sectional shape: (*a*) Shaft, K_{t3}; (*b*) plane, K_{t2}.

14 DEFINITIONS AND DESIGN RELATIONS

cumstances, their stress concentration factors are very close. To understand the relationship between plane and axisymmetric problems, consider the following cases.

CASE 1. A Shaft with a Circumferential Groove and with the Stress Raisers Far from the Central Axis of Symmetry Consider a shaft with a circumferential groove under tension (or bending) load, and suppose the groove is far from the central axis, $d/2 \gg r_1$, as shown in Fig. 1.9a. A plane element with the same longitudinal section under the same loading is shown in Fig 1.9b. Let K_{t3} and K_{t2} denote the stress concentration factors for the axisymmetric solid body and the corresponding plane problem, respectively. Since the groove will not affect the stress distribution in the area near the central axis, the distributions of stress components σ_x, σ_r, τ_{xr} near the groove in the axisymmetric shaft are almost the same as those of the stress components σ_x, σ_y, τ_{xy} near the notch in the plane element, so that $K_{t3} \approx K_{t2}$.

For the case where a small groove is a considerable distance from the central axis of the shaft, the same conclusion can be explained as follows. Set the terms with $1/r$ equal to 0 (since the groove is far from the central axis, r is very large), and note that differential Eqs. (1.9) reduce to

$$\frac{\partial \sigma_r}{\partial r} + \frac{\partial \tau_{rx}}{\partial x} + \overline{p}_{Vr} = 0$$

$$\frac{\partial \tau_{rx}}{\partial r} + \frac{\partial \sigma_x}{\partial x} + \overline{p}_{Vx} = 0 \qquad (1.12)$$

and Eq. (1.11) becomes

$$\varepsilon_r = \frac{\partial u}{\partial r}, \quad \varepsilon_\theta = 0, \quad \varepsilon_x = \frac{\partial w}{\partial x}, \quad \gamma_{rx} = \frac{\partial u}{\partial x} + \frac{\partial w}{\partial r} \qquad (1.13)$$

Introduce the material law

$$\varepsilon_r = \frac{1}{E}[\sigma_r - \nu(\sigma_\theta + \sigma_x)], \quad \varepsilon_\theta = 0 = \frac{1}{E}[\sigma_\theta - \nu(\sigma_x + \sigma_r)]$$

$$\varepsilon_x = \frac{1}{E}[\sigma_x - \nu(\sigma_r + \sigma_\theta)], \quad \gamma_{rx} = \frac{1}{G}\tau_{rx}$$

into Eq. (1.10) and use Eq. (1.12). For constant body forces this leads to an equation identical (with y replaced by r) to that of Eq. (1.6). This means that the governing equations are the same. However, the stress σ_θ is not included in the governing equations and it can be derived from

$$\sigma_\theta = \nu(\sigma_r + \sigma_x) \qquad (1.14)$$

When $\nu = 0$, the stress distribution of a shaft is identical to that of the plane element with the same longitudinal section.

CASE 2. General Case of an Axisymmetrical Solid with Shallow Grooves and Shoulders In general, for a solid of revolution with shallow grooves or shoulders under tension or bending as shown in Fig. 1.10, the stress concentration factor K_{t3} can be obtained in terms

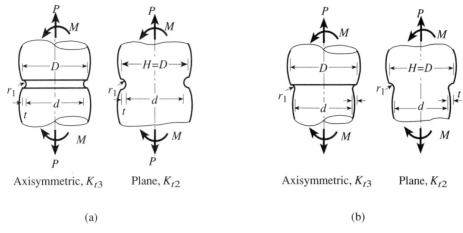

Figure 1.10 Shallow grooves and shoulders: (*a*) Groove; (*b*) shoulder.

of the plane case factor K_{t2} using (Nishida 1976)

$$\left(1 + \frac{2t}{d}\right) K_{t3} - K_{t2} = \frac{t}{d}\left(1 + \sqrt{\frac{2t}{r_1}}\right) \quad (1.15)$$

where r_1 is the radius of the groove and $t = (D - d)/2$ is the depth of the groove (or shoulder). The effective range for Eq. (1.15) is $0 \leq t/d \leq 7.5$. If the groove is far from the central axis, $t/d \to 0$ and $K_{t3} = K_{t2}$, which is consistent with the results discussed in Case 1.

CASE 3. Deep Hyperbolic Groove   As mentioned in Section 1.4, Neuber (1958) provided formulas for bars with deep hyperbolic grooves. For the case of an axisymmetric shaft under tensile load, for which the minimum diameter of the shaft d (Fig. 1.8) is smaller than the depth of the groove, the following empirical formula is available (Nishida 1976):

$$K_{t3} = 0.75 K_{t2} + 0.25 \quad (1.16)$$

Equation (1.16) is close to the theoretical value over a wide range and is useful in engineering analysis. This equation not only applies to tension loading but also to bending and shearing load. However, the error tends to be relatively high in the latter cases.

1.6 LOCAL AND NONLOCAL STRESS CONCENTRATION

If the dimensions of a stress raiser are much smaller than those of the structural member, its influence is usually limited to a localized area (or volume for a 3-dimensional case). That is, the global stress distribution of the member except for the localized area is the same as that for the member without the stress raiser. This kind of problem is referred to as *localized stress concentration*. Usually stress concentration theory deals with the *localized stress concentration* problems. The simplest way to solve these problems is to separate this

16   DEFINITIONS AND DESIGN RELATIONS

localized part from the member, then to determine K_t by using the formulas and curves of a simple case with a similar raiser shape and loading. If a wide stress field is affected, the problem is called *nonlocal stress concentration* and can be quite complicated. Then a full-fledged stress analysis of the problem may be essential, probably with general purpose structural analysis computer software.

Example 1.4 Rotating Disk A disk rotating at speed ω has a central hole and two additional symmetrically located holes as shown in Fig. 1.11. Suppose that $R_1 = 0.24R_2$, $a = 0.06R_2$, $R = 0.5R_2$, $\nu = 0.3$. Determine the stress concentration factor near the small circle O_1.

Since $R_2 - R_1$ is more than 10 times greater than a, it can be reasoned that the existence of the small O_1 hole will not affect the general stress distribution. That is to say, the disruption in stress distribution due to circle O_1 is limited to a local area. This qualifies then as localized stress concentration.

For a rotating disk with a central hole, the theory of elasticity gives the stress components (Pilkey 1994)

$$\sigma_r = \frac{3+\nu}{8}\rho\omega^2\left(R_2^2 + R_1^2 - \frac{R_1^2 R_2^2}{R_x^2} - R_x^2\right)$$

$$\sigma_\theta = \frac{3+\nu}{8}\rho\omega^2\left(R_2^2 + R_1^2 + \frac{R_1^2 R_2^2}{R_x^2} - \frac{1+3\nu}{3+\nu}R_x^2\right) \quad (1)$$

where ω is the speed of rotation (rad/s), ρ is the mass density, and R_x is the radius at which σ_r, σ_θ are to be calculated.

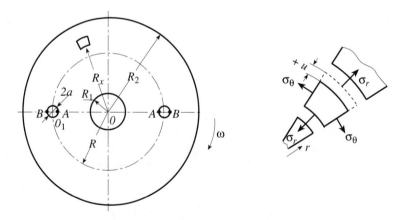

R_1 Radius of central hole R_2 Outer radius of the disk

R Distance between the center of the disk and the center of O_1

R_x Radius at which σ_r, σ_θ are to be calculated.

a Radius of hole O_1 r, θ Polar coordinates

Figure 1.11 Rotating disk with a central hole and two symmetrically located holes.

The O_1 hole may be treated as if it were in an infinite region and subjected to biaxial stresses σ_r, σ_θ as shown in Fig. 1.12a. For point $A, R_A = R - a = 0.5R_2 - 0.06R_2 = 0.44R_2$, and the elasticity solution of (1) gives

$$\sigma_{rA} = \frac{3+\nu}{8}\rho\omega^2 R_2^2 \left(1 + 0.24^2 - \frac{0.24^2}{0.44^2} - 0.44^2\right) = 0.566 \left(\frac{3+\nu}{8}\rho\omega^2 R_2^2\right)$$

$$\sigma_{\theta A} = \frac{3+\nu}{8}\rho\omega^2 R_2^2 \left(1 + 0.24^2 + \frac{0.24^2}{0.44^2} - \frac{1+3\nu}{3+\nu}0.44^2\right) \tag{2}$$

$$= 1.244 \left(\frac{3+\nu}{8}\rho\omega^2 R_2^2\right)$$

Substitute $\alpha = \sigma_{rA}/\sigma_{\theta A} = 0.566/1.244 = 0.455$ into the stress concentration factor formula for the case of an element with a circular hole under biaxial tensile load (Eq. 4.18) giving

$$K_{tA} = \frac{\sigma_{A\max}}{\sigma_{\theta A}} = 3 - 0.455 = 2.545$$

and the maximum stress at point A is

$$\sigma_{A\max} = K_{tA} \cdot \sigma_{\theta A} = 3.1660 \left(\frac{3+\nu}{8}\rho\omega^2 R_2^2\right) \tag{3}$$

Similarly, at point B, $R_B = R + a = 0.5R_2 + 0.06R_2 = 0.56R_2$,

$$\sigma_{rB} = 0.56 \left(\frac{3+\nu}{8}\rho\omega^2 R_2^2\right) \tag{4}$$

$$\sigma_{\theta B} = 1.061 \left(\frac{3+\nu}{8}\rho\omega^2 R_2^2\right)$$

Substitute $\alpha = \sigma_{rB}/\sigma_{\theta B} = 0.56/1.061 = 0.528$ into the stress concentration factor formula of Eq. (4.18):

$$K_{tB} = \frac{\sigma_{B\max}}{\sigma_{\theta B}} 3 - 0.528 = 2.472$$

and the maximum stress at point B becomes

$$\sigma_{B\max} = K_{tB} \cdot \sigma_{\theta B} = 2.6228 \left(\frac{3+\nu}{8}\rho\omega^2 R_2^2\right) \tag{5}$$

To calculate the stress at the edge of the central hole, substitute $R_x = R_1 = 0.24R_2$ into σ_θ of (1):

$$\sigma_{\theta 1} = 2.204 \left(\frac{3+\nu}{8}\rho\omega^2 R_2^2\right) \tag{6}$$

18 DEFINITIONS AND DESIGN RELATIONS

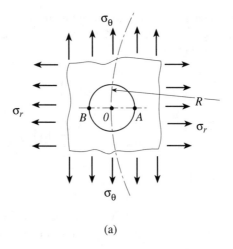

(a)

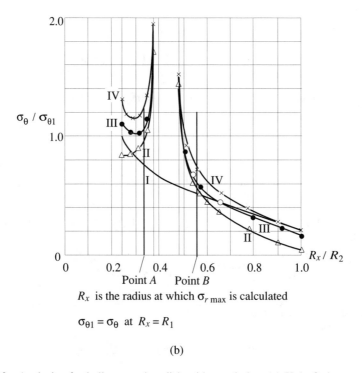

R_x is the radius at which $\sigma_{r\,max}$ is calculated

$\sigma_{\theta 1} = \sigma_\theta$ at $R_x = R_1$

(b)

Figure 1.12 Analysis of a hollow rotating disk with two holes: (*a*) Hole 0_1 is treated as being subjected to biaxial stresses σ_r, σ_θ; (*b*) results from Ku (1960). (I) No central hole; (II) approximate solution; (III) exact solution; (IV) photoelastic results (Newton 1940).

Equations (3) and (5) give the maximum stresses at points A and B of an infinite region as shown in Fig. 1.12a. If $\sigma_{\theta 1}$ is taken as the reference stress, the corresponding stress concentration factors are

$$K_{t1A} = \frac{\sigma_{A\max}}{\sigma_{\theta 1}} = \frac{3.1660}{2.204} = 1.44$$

$$K_{t1B} = \frac{\sigma_{B\max}}{\sigma_{\theta 1}} = \frac{2.6228}{2.204} = 1.19$$
(7)

This approximation of treating the hole as if it were in an infinite region and subjected to biaxial stresses is based on the assumption that the influence of circle O_1 is limited to a local area. The results are very close to the theoretical solution. Ku (1960) analyzed the case with $R_1 = 0.24R_2$, $R = 0.435R_2$, $a = 0.11R_2$. Although the circle O_1 is larger than that of this example, he still obtained reasonable approximations by treating the hole as if it were in an infinite region and subjected to biaxial stresses. The results are given in Fig. 1.12b, in which $\sigma_{\theta 1}$ on the central circle ($R_1 = 0.24R_2$) was taken as the reference stress. Curve II was obtained by the approximation of this example and curve III is from the theoretical solution (Howland 1930). For point A, $r/R_2 = 0.335$ and for point B, $r/R_2 = 0.545$. From Fig. 1.12b, it can be seen that at points A and B of the edge of hole O_1, the results from curves II and III are very close.

The method used in the above example can be summarized as follows: First, find the stress field in the member without the stress raiser at the position where the stress raiser occurs. This analysis provides the loading condition at this local point. Second, find a formula or curve from the charts in this book that applies to the loading condition and the stress raiser shape. Finally, use the formula or curve to evaluate the maximum stress. It should be remembered that this method is only applicable for localized stress concentration.

1.6.1 Examples of "Reasonable" Approximations

Consider now the concept of localized stress concentration for the study of the stress caused by notches and grooves. Begin with a thin flat element with a shallow notch under uniaxial tension load as shown in Fig. 1.13a. Since the notch is shallow, the bottom edge

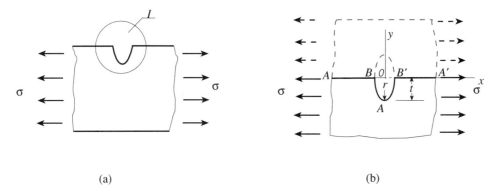

Figure 1.13 Shallow groove: (a) Shallow groove I; (b) model of I.

of the element is considered to be a substantial distance from the notch. It is a local stress concentration problem in the vicinity of the notch. Consider another element with an elliptic hole loaded by uniaxial stress σ as indicated in Fig. 1.13b. (The solution for this problem can be derived from Eq. 4.58.) Cut the second element with the symmetrical axis A–A'. The normal stresses on section A–A' are small and can be neglected. Then the solution for an element with an elliptical hole (Eq. 4.58 with a replaced by t)

$$K_t = 1 + 2\sqrt{\frac{t}{r}} \qquad (1.17)$$

can be taken as an approximate solution for an element with a shallow notch. According to this approximation, the stress concentration factor for a shallow notch is a function only of the depth t and radius of curvature r of the notch.

For a deep notch in a plane element under uniaxial tension load (Fig. 1.14a), the situation is quite different. For the enlarged model of Fig. 1.14b, the edge A–A' is considered to be a substantial distance from bottom edge B–B', and the stresses near the A–A' edge are almost zero. Such a low stress area probably can be safely neglected. The local areas that should be considered are the bottom of the groove and the straight line edge B–B' close to the groove bottom. Thus the deep notch problem, which might appear to be a nonlocal stress concentration problem, can also be considered as localized stress concentration. Furthermore the bottom part of the groove can be approximated by a hyperbola, since it is a small segment. Because of symmetry (Fig. 1.14) it is reasoned that the solution to this problem is same as that of a plane element with two opposing hyperbola notches. The equation for the stress concentration factor is (Durelli 1982)

$$K_t = \frac{2\left(\dfrac{d}{r} + 1\right)\sqrt{\dfrac{d}{r}}}{\left(\dfrac{d}{r} + 1\right)\arctan\sqrt{\dfrac{d}{r}} + \sqrt{\dfrac{d}{r}}} \qquad (1.18)$$

where d is the distance between the notch and edge B–B' (Fig. 1.14b). It is evident that the stress concentration factor of the deep notch is a function of the radius of curvature r of the bottom of the notch and the minimum width d of the element (Fig. 1.14). For notches of intermediate depth, refer to the Neuber method (see Eq. 2.1).

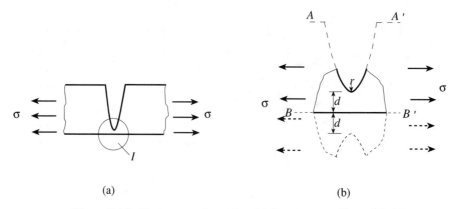

Figure 1.14 Deep groove in tension: (a) Deep groove; (b) model of I.

1.7 MULTIPLE STRESS CONCENTRATION

Two or more stress concentrations occurring at the same location in a structural member are said to be in a state of *multiple stress concentration*. Multiple stress concentration problems occur often in engineering design. An example would be a uniaxially tension-loaded plane element with a circular hole, supplemented by a notch at the edge of the hole as shown in Fig. 1.15. The notch will lead to a higher stress than would occur with the hole alone. Use K_{t1} to represent the stress concentration factor of the element with a circular hole and K_{t2} to represent the stress concentration factor of a thin, flat tension element with a notch on an edge. In general, the multiple stress concentration factor of the element $K_{t1,2}$ cannot be deduced directly from K_{t1} and K_{t2}. The two different factors will interact with each other and produce a new stress distribution. Because of it's importance in engineering design, considerable effort has been devoted to finding solutions to the multiple stress concentration problems. Some special cases of these problems follow.

CASE 1. The Geometrical Dimension of One Stress Raiser Much Smaller Than That of the Other   Assume that $d/2 \gg r$ in Fig. 1.15, where r is the radius of curvature of the notch. Notch r will not significantly influence the global stress distribution in the element with the circular hole. However, the notch can produce a local disruption in the stress field of the element with the hole. For an infinite element with a circular hole, the stress concentration factor K_{t1} is 3.0, and for the element with a semicircular notch K_{t2} is 3.06 (Chapter 2). Since the notch does not affect significantly the global stress distribution near the circular hole, the stress around the notch region is approximately $K_{t1}\sigma$. Thus the notch can be considered to be located in a tensile specimen subjected to a tensile load $K_{t1}\sigma$ (Fig. 1.15b). Therefore the peak stress at the tip of the notch is $K_{t2} \cdot K_{t1}\sigma$. It can be concluded that the multiple stress concentration factor at point A is equal to the product of K_{t1} and K_{t2},

$$K_{t1,2} = K_{t1} \cdot K_{t2} = 9.18 \tag{1.19}$$

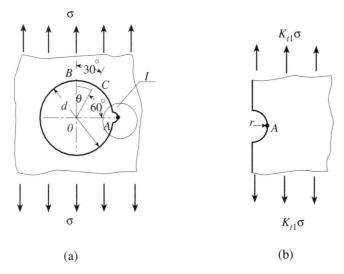

Figure 1.15 Multiple stress concentration: (*a*) Small notch at the edge of a circular hole; (*b*) enlargement of *I*.

22 DEFINITIONS AND DESIGN RELATIONS

which is close to the value displayed in Chart 4.60 for $r/d \to 0$. If the notch is relocated to point B instead of A, the multiple stress concentration factor will be different. Since at point B the stress concentration factor due to the hole is -1.0 (refer to Fig. 4.5), $K_{t1,2} = -1.0 \cdot 3.06 = -3.06$. Using the same argument, when the notch is situated at point C ($\theta = \pi/6$), $K_{t2} = 0$ (refer to Section 4.3.1 and Fig. 4.5) and $K_{t1,2} = 0 \cdot 3.06 = 0$. It is evident that the stress concentration factor can be effectively reduced by placing the notch at point C.

Consider a shaft with a circumferential groove subject to a torque T, and suppose that there is a small radial cylindrical hole at the bottom of the groove as shown in Fig. 1.16. (If there were no hole, the state of stress at the bottom of the groove would be one of pure shear, and K_{s1} for this location could be found from Chart 2.47.) The stress concentration near the small radial hole can be modeled using an infinite element with a circular hole under shearing stress. Designate the corresponding stress concentration factor as K_{s2}. (Then K_{s2} can be found from Chart 4.88, with $a = b$.) The multiple stress concentration factor at the edge of the hole is

$$K_{t1,2} = K_{s1} \cdot K_{s2} \tag{1.20}$$

CASE 2. *The Size of One Stress Raiser Not Much Different from the Size of the Other Stress Raiser* Under such circumstances the multiple stress concentration factor cannot be calculated as the product of the separate stress concentration factors as in Eqs. (1.19) or (1.20). In the case of Fig. 1.17, for example, the maximum stress location A_1 for stress concentration factor 1 does not coincide with the maximum stress location A_2 for stress concentration factor 2. In general, the multiple stress concentration factor adheres to the relationship (Nishida 1976)

$$\max(K_{t1}, K_{t2}) < K_{t1,2} \leq K_{t1} \cdot K_{t2} \tag{1.21}$$

Some approximate formulas are available for special cases. For the three cases of Fig. 1.18—that is, a shaft with double circumferential grooves under torsion load (Fig. 1.18*a*), a semi-infinite element with double notches under tension (Fig. 1.18*b*),

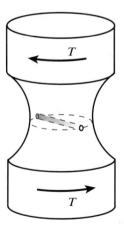

Figure 1.16 Small radial hole through a groove.

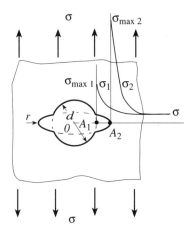

Figure 1.17 Two stress raisers of almost equal magnitude in an infinite two-dimensional element.

and an infinite element with circular and elliptical holes under tension (Fig. 1.18c)—an empirical formula (Nishida 1976)

$$K_{t1,2} \approx K_{t1c} + (K_{t2e} - K_{t1c})\sqrt{1 - \frac{1}{4}\left(\frac{d}{b}\right)^2} \qquad (1.22)$$

was developed. Under the loading conditions corresponding to Figs. 1.18a, b, and c, as appropriate, K_{t1c} is the stress concentration factor for an infinite element with a circular hole and K_{t2e} is the stress concentration factor for an element with the elliptical notch. This approximation is quite close to the theoretical solution of the cases of Figs. 1.18a and b. For the case of Fig. 1.18c, the error is somewhat larger, but the approximation is still adequate.

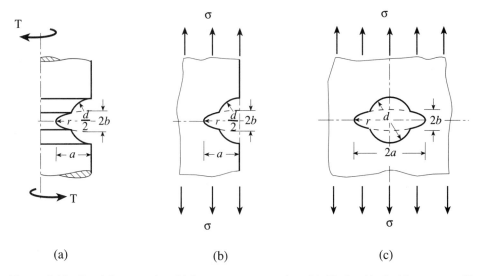

Figure 1.18 Special cases of multiple stress concentration: (a) Shaft with double grooves; (b) semi-infinite element with double notches; (c) circular hole with elliptical notches.

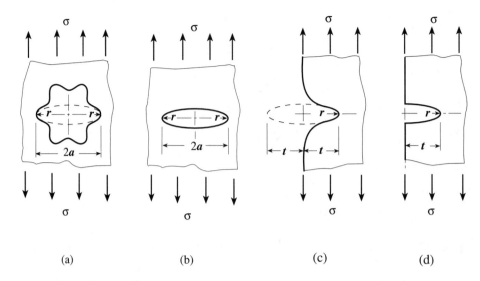

Figure 1.19 Equivalent ellipses: (*a*) Element with a hexagonal hole; (*b*) element with an equivalent ellipse; (*c*) semi-infinite element with a groove; (*d*) semi-infinite element with the equivalent elliptic groove.

Another effective method is to use the *equivalent ellipse* concept. To illustrate the method, consider a flat element with a hexagonal hole (Fig. 1.19*a*). An ellipse of major semiaxes a and minimum radius of curvature r is the enveloping curve of two ends of the hexagonal hole. This ellipse is called the "equivalent ellipse" of the hexagonal hole. The stress concentration factor of a flat element with the equivalent elliptical hole (Fig. 1.19*b*) is (Eq. 4.58)

$$K_t = 2\sqrt{\frac{a}{r}} + 1 \quad (1.23)$$

which is very close to the K_t for the flat element in Fig. 1.19*a*. Although this is an approximate method, the calculation is simple and the results are within an error of 10%. Similarly the stress concentration factor for a semi-infinite element with a groove under uniaxial tensile loading (Fig. 1.19*c*) can be estimated by finding K_t of the same element with the equivalent elliptical groove of Fig. 1.19*d* for which (Nishida 1976) (Eq. (4.58))

$$K_t = 2\sqrt{\frac{t}{r}} + 1 \quad (1.24)$$

1.8 THEORIES OF STRENGTH AND FAILURE

If our design problems involved only uniaxial stress problems, we would need to give only limited consideration to the problem of strength and failure of complex states of stress. However, even very simple load conditions may result in biaxial stress systems. An example is a thin spherical vessel subjected to internal pressure, resulting in biaxial tension acting on an element of the vessel. Another example is a bar of circular cross section subjected to tension, resulting in biaxial tension and compression acting at 45°.

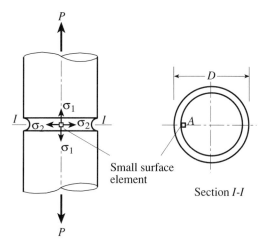

Figure 1.20 Biaxial stress in a notched tensile member.

From the standpoint of stress concentration cases, it should be noted that such simple loading as an axial tension produces biaxial surface stresses in a grooved bar (Fig. 1.20). Axial load P results in axial tension σ_1 and circumferential tension σ_2 acting on a surface element of the groove.

A considerable number of theories have been proposed relating uniaxial to biaxial or triaxial stress systems (Pilkey 1994); only the theories ordinarily utilized for design purposes are considered here. These are, for *brittle materials*,[1] maximum-stress criterion and Mohr's theory and, for *ductile materials*, maximum-shear theory and the von Mises criterion.

For the following theories it is assumed that the tension or compressive critical stresses (strength level, yield stress, or ultimate stress) are available. Also it is necessary to understand that any state of stress can be reduced through a rotation of coordinates to a state of stress involving only the principal stresses σ_1, σ_2, and σ_3.

1.8.1 Maximum Stress Criterion

The *maximum stress criterion* (or *normal stress* or *Rankine criterion*) can be stated as follows: failure occurs in a multiaxial state of stress when either a principal tensile stress reaches the uniaxial tensile strength σ_{ut} or a principal compressive stress reaches the uniaxial compressive strength σ_{uc}. For a brittle material σ_{uc} is usually considerably greater than σ_{ut}. In Fig. 1.21, which represents biaxial conditions (σ_1 and σ_2 principal stresses, $\sigma_3 = 0$), the maximum stress criterion is represented by the square $CFHJ$.

The strength of a bar under uniaxial tension σ_{ut} is OB in Fig. 1.21. Note that according to the maximum stress criterion, the presence of an additional stress σ_2 at right angles does not affect the strength.

For torsion of a bar, only shear stresses appear on the cross section (i.e., $\sigma_x = \sigma_y = 0$, $\tau_{xy} = \tau$) and the principal stress (Pilkey 1994) $\sigma_2 = -\sigma_1 = \tau$ (line AOE). Since these

---

[1] The distinction between brittle and ductile materials is arbitrary, sometimes an elongation of 5% is considered to be the division between the two (Soderberg 1930).

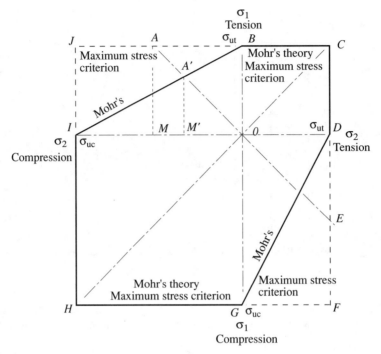

Figure 1.21 Biaxial conditions for strength theories for brittle materials.

principal stresses are equal in magnitude to the applied shear stress τ, the maximum stress condition of failure for torsion (MA, Fig. 1.21) is

$$\tau_u = \sigma_{ut} \tag{1.25}$$

In other words, according to the maximum stress criterion, the torsion and tension strength values should be equal.

1.8.2 Mohr's Theory

The condition of failure of brittle materials according to *Mohr's theory* (or the *Coulomb-Mohr theory* or *internal friction theory*) is illustrated in Fig. 1.22. Circles of diameters σ_{ut} and σ_{uc} are drawn as shown. A stress state, for which the Mohr's circle just contacts the line of tangency[2] of the σ_{ut} and σ_{uc} circles, represents a condition of failure (Pilkey 1994). See the Mohr's circle (dotted) of diameter $\sigma_1 - \sigma_2$ of Fig. 1.22. The resultant plot for biaxial conditions is shown in Fig. 1.21. The conditions of failure are as follows:

For $\sigma_1 \geq 0$ and $\sigma_2 \geq 0$ (first quadrant), with $\sigma_1 \geq \sigma_2$

$$\sigma_1 = \sigma_{ut} \tag{1.26}$$

---

[2]The straight line is a special case of the more general Mohr's theory, which is based on a curved envelope.

THEORIES OF STRENGTH AND FAILURE 27

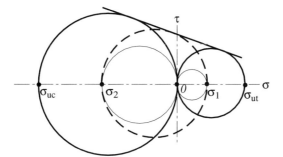

Figure 1.22 Mohr's theory of failure of brittle materials.

For $\sigma_1 \geq 0$ and $\sigma_2 \leq 0$ (second quadrant)

$$\frac{\sigma_1}{\sigma_{ut}} - \frac{\sigma_2}{\sigma_{uc}} = 1 \tag{1.27}$$

For $\sigma_1 \leq 0$ and $\sigma_2 \leq 0$ (third quadrant)

$$\sigma_2 = -\sigma_{uc} \tag{1.28}$$

For $\sigma_1 \leq 0$ and $\sigma_2 \geq 0$ (fourth quadrant)

$$-\frac{\sigma_1}{\sigma_{uc}} + \frac{\sigma_2}{\sigma_{ut}} = 1 \tag{1.29}$$

As will be seen later (Fig. 1.23) this is similar to the representation for the maximum shear theory, except for nonsymmetry.

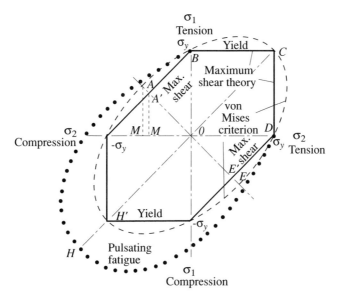

Figure 1.23 Biaxial conditions for strength theories for ductile materials.

Certain tests of brittle materials seem to substantiate the maximum stress criterion (Draffin and Collins 1938), whereas other tests and reasoning lead to a preference for Mohr's theory (Marin 1952). The maximum stress criterion gives the same results in the first and third quadrants. For the torsion case ($\sigma_2 = -\sigma_1$), use of Mohr's theory is on the "safe side," since the limiting strength value used is $M'A'$ instead of MA (Fig. 1.21). The following can be shown for $M'A'$ of Fig. 1.21:

$$\tau_u = \frac{\sigma_{ut}}{1 + (\sigma_{ut}/\sigma_{uc})} \tag{1.30}$$

1.8.3 Maximum Shear Theory

The *maximum shear theory* (or *Tresca's* or *Guest's theory*) was developed as a criterion for yield or failure, but it has also been applied to fatigue failure, which in ductile materials is thought to be initiated by the maximum shear stress (Gough 1933). According to the maximum shear theory, failure occurs when the maximum shear stress in a multiaxial system reaches the value of the shear stress in a uniaxial bar at failure. In Fig. 1.23, the maximum shear theory is represented by the six-sided figure. For principal stresses σ_1, σ_2, and σ_3, the maximum shear stresses are (Pilkey 1994)

$$\frac{\sigma_1 - \sigma_2}{2}, \quad \frac{\sigma_1 - \sigma_3}{2}, \quad \frac{\sigma_2 - \sigma_3}{2} \tag{1.31}$$

The actual maximum shear stress is the peak value of the expressions of Eq. (1.31). The value of the shear failure stress in a simple tensile test is $\sigma/2$, where σ is the tensile failure stress (yield σ_y or fatigue σ_f) in the tensile test. Suppose that fatigue failure is of interest and that σ_f is the uniaxial fatigue limit in alternating tension and compression. For the biaxial case set $\sigma_3 = 0$, and suppose that σ_1 is greater than σ_2 for both in tension. Then failure occurs when $(\sigma_1 - 0)/2 = \sigma_f/2$ or $\sigma_1 = \sigma_f$. This is the condition represented in the first quadrant of Fig. 1.23 where σ_y rather than σ_f is displayed. However, in the second and fourth quadrants, where the biaxial stresses are of opposite sign, the situation is different. For $\sigma_2 = -\sigma_1$, represented by line AE of Fig. 1.23, failure occurs in accordance with the maximum shear theory when $[\sigma_1 - (-\sigma_1)]/2 = \sigma_f/2$ or $\sigma_1 = \sigma_f/2$, namely $M'A' = OB/2$ in Fig. 1.23.

In the torsion test $\sigma_2 = -\sigma_1 = \tau$,

$$\tau_f = \frac{\sigma_f}{2} \tag{1.32}$$

This is half the value corresponding to the maximum stress criterion.

1.8.4 von Mises Criterion

The following expression was proposed by R. von Mises (1913), as representing a criterion of failure by yielding:

$$\sigma_y = \sqrt{\frac{(\sigma_1 - \sigma_2)^2 + (\sigma_2 - \sigma_3)^2 + (\sigma_1 - \sigma_3)^2}{2}} \tag{1.33}$$

where σ_y is the yield strength in a uniaxially loaded bar. For another failure mode, such as fatigue failure, replace σ_y by the appropriate stress level, such as σ_f. The quantity on the right-hand side of Eq. (1.33), which is sometimes available as output of structural analysis software, is often referred to as the *equivalent stress* σ_{eq}:

$$\sigma_{eq} = \sqrt{\frac{(\sigma_1 - \sigma_2)^2 + (\sigma_2 - \sigma_3)^2 + (\sigma_1 - \sigma_3)^2}{2}} \qquad (1.34)$$

This theory, which is also called the *Maxwell-Huber-Hencky-von Mises theory*, *octahedral shear stress theory* (Eichinger 1926; Nadai 1937), and *maximum distortion energy theory* (Hencky 1924), states that failure occurs when the energy of distortion reaches the same energy for failure in tension[3]. If $\sigma_3 = 0$, Eq. (1.34) reduces to

$$\sigma_{eq} = \sqrt{\sigma_1^2 - \sigma_1\sigma_2 + \sigma_2^2} \qquad (1.35)$$

This relationship is shown by the dashed ellipse of Fig. 1.23 with $OB = \sigma_y$. Unlike the six-sided figure, it does not have the discontinuities in slope, which seem unrealistic in a physical sense. Sachs (1928) and Cox and Sopwith (1937) maintain that close agreement with the results predicted by Eq. (1.33) is obtained if one considers the statistical behavior of a randomly oriented aggregate of crystals.

For the torsion case with $\sigma_2 = -\sigma_1 = \tau_y$, the von Mises criterion becomes

$$\tau_y = \frac{\sigma_y}{\sqrt{3}} = 0.577\sigma_y \qquad (1.36)$$

or $MA = (0.577)OB$ in Fig. 1.23, where τ_y is the yield strength of a bar in torsion. Note from Figs. 1.21 and 1.23 that all the foregoing theories are in agreement at C, representing equal tensions, but they differ along AE, representing tension and compression of equal magnitudes (torsion).

Yield tests of ductile materials have shown that the von Mises criterion interprets well the results of a variety of biaxial conditions. It has been pointed out (Prager and Hodge 1951) that although the agreement must be regarded as fortuitous, the von Mises criterion would still be of practical interest because of it's mathematical simplicity even if the agreement with test results had been less satisfactory.

There is evidence (Peterson 1974; Nisihara and Kojima 1939) that for ductile materials the von Mises criterion also gives a reasonably good interpretation of fatigue results in the upper half (*ABCDE*) of the ellipse of Fig. 1.23 for completely alternating or pulsating tension cycling. As shown in Fig. 1.24, results from alternating tests are in better agreement with the von Mises criterion (upper line) than with the maximum shear theory (lower line). If yielding is considered the criterion of failure, the ellipse of Fig. 1.23 is symmetrical about AE. With regard to the region below AE (compression side), there is evidence that for pulsating compression (e.g., 0 to maximum compression) this area is considerably enlarged

---

[3]The proposals of both von Mises and Hencky were to a considerable extent anticipated by Huber in 1904. Although limited to mean compression and without specifying mode of failure; his paper in the Polish language did not attract international attention until 20 years later.

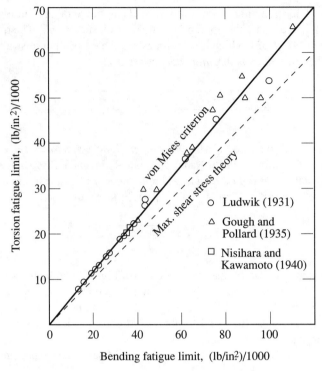

Figure 1.24 Comparison of torsion and bending fatigue limits for ductile materials.

(Newmark et al. 1951; Nishihara and Kojima 1939; Rôs and Eichinger 1950). For the cases treated here we deal primarily with the upper area.[4]

1.8.5 Observations on the Use of the Theories of Failure

If a member is in a uniaxial stress state (i.e., $\sigma_{max} = \sigma_1$, $\sigma_2 = \sigma_3 = 0$), the maximum stress can be used directly in $\sigma_{max} = K_t \sigma_{nom}$ for a failure analysis. However, when the location of the maximum stress is in a biaxial or triaxial stress state, it is important to consider not only the effects of σ_1 but also of σ_2 and σ_3, according to one of the theories of strength (failure). For example, for a shaft with a circumferential groove under tensile loading, a point at the bottom of the groove is in a biaxial stress state; that is, the point is subjected to axial stress σ_1 and circumferential stress σ_2 as shown in Fig. 1.20. If the von Mises theory is used in a failure analysis, then (Eq. 1.35)

$$\sigma_{eq} = \sqrt{\sigma_1^2 - \sigma_1 \sigma_2 + \sigma_2^2} \tag{1.37}$$

---

[4]It will be noted that all representations in Figs. 1.21 and 1.23 are symmetrical about line HC. In some cases, such as forgings and bars, strong directional effects can exist (i.e., transverse strength can be considerably less than longitudinal strength). Findley (1951) gives methods for taking anisotropy into account in applying strength theories.

To combine the stress concentration and the von Mises strength theory, introduce a factor K'_t:

$$K'_t = \frac{\sigma_{eq}}{\sigma} \qquad (1.38)$$

where $\sigma = 4P/(\pi D^2)$ is the reference stress. Substitute Eq. (1.37) into Eq. (1.38),

$$K'_t = \frac{\sigma_1}{\sigma}\sqrt{1 - \frac{\sigma_2}{\sigma_1} + \left(\frac{\sigma_2}{\sigma_1}\right)^2} = K_t\sqrt{1 - \frac{\sigma_2}{\sigma_1} + \left(\frac{\sigma_2}{\sigma_1}\right)^2} \qquad (1.39)$$

where $K_t = \sigma_1/\sigma$ is defined as the stress concentration factor at point A that can be read from a chart of this book. Usually $0 < \sigma_2/\sigma_1 < 1$, so that $K'_t < K_t$. In general, K'_t is about 90% to 95% of the value of K_t and not less than 85%.

Consider the case of a three-dimensional block with a spherical cavity under uniaxial tension σ. The two principal stresses at point A on the surface of the cavity (Fig. 1.25) are (Nishida 1976)

$$\sigma_1 = \frac{3(9 - 5\nu)}{2(7 - 5\nu)}\sigma, \qquad \sigma_2 = \frac{3(5\nu - 1)}{2(7 - 5\nu)}\sigma \qquad (1.40)$$

From these relationships

$$\frac{\sigma_2}{\sigma_1} = \frac{5\nu - 1}{9 - 5\nu} \qquad (1.41)$$

Substitute Eq. (1.41) into Eq. (1.39):

$$K'_t = K_t\sqrt{1 - \frac{5\nu - 1}{9 - 5\nu} + \left(\frac{5\nu - 1}{9 - 5\nu}\right)^2} \qquad (1.42)$$

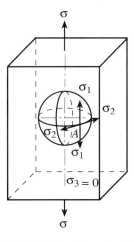

Figure 1.25 Block with a spherical cavity.

For $\nu = 0.4$,

$$K'_t = 0.94 K_t$$

and when $\nu = 0.3$,

$$K'_t = 0.97 K_t$$

It is apparent that K'_t is lower than and quite close to K_t. It can be concluded that the usual design using K_t is on the safe side and will not be accompanied by significant errors. Therefore charts for K'_t are not included in this book.

1.8.6 Stress Concentration Factors under Combined Loads, Principle of Superposition

In practice, a structural member is often under the action of several types of loads, instead of being subjected to a single type of loading as represented in the graphs of this book. In such a case, evaluate the stress for each type of load separately. and superimpose the individual stresses. Since superposition presupposes a linear relationship between the applied loading and resulting response, it is necessary that the maximum stress be less than the elastic limit of the material. The following examples illustrate this procedure.

Example 1.5 Tension and Bending of a Two-dimensional Element A notched thin element is under combined loads of tension and in-plane bending as shown in Fig. 1.26. Find the maximum stress.

For tension load P, the stress concentration factor K_{tn1} can be found from Chart 2.3 and the maximum stress is

$$\sigma_{\max 1} = K_{tn1} \sigma_{\text{nom}1} \qquad (1)$$

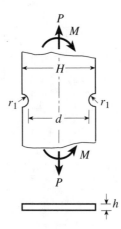

Figure 1.26 Element under tension and bending loading.

in which $\sigma_{\text{nom1}} = P/(dh)$. For the in-plane bending moment M, the maximum bending stress is (the stress concentration factor can be found from Chart 2.25)

$$\sigma_{\max 2} = K_{tn2}\sigma_{\text{nom2}} \tag{2}$$

where $\sigma_{\text{nom2}} = 6M/(d^2h)$ is the stress at the base of the groove. Stresses $\sigma_{\max 1}$ and $\sigma_{\max 2}$ are both normal stresses that occur at the same point, namely at the base of the groove. Hence, when the element is under these combined loads, the maximum stress at the notch is

$$\sigma_{\max} = \sigma_{\max 1} + \sigma_{\max 2} = K_{tn1}\sigma_{\text{nom1}} + K_{tn2}\sigma_{\text{nom2}} \tag{3}$$

Example 1.6 Tension, Bending, and Torsion of a Grooved Shaft A shaft of circular cross section with a circumferential groove is under the combined loads of axial force P, bending moment M, and torque T, as shown in Fig. 1.27. Calculate the maximum stresses corresponding to the various failure theories.

The maximum stress is (the stress concentration factor of this shaft due to axial force P can be found from Chart 2.19)

$$\sigma_{\max 1} = K_{tn1}\frac{4P}{\pi d^2} \tag{1}$$

The maximum stress corresponding to the bending moment (from Chart 2.41) is

$$\sigma_{\max 2} = K_{tn2}\frac{32M}{\pi d^3} \tag{2}$$

The maximum torsion stress due to torque T is obtained from Chart 2.47 as

$$\tau_{\max 3} = K_{ts}\frac{16T}{\pi d^3} \tag{3}$$

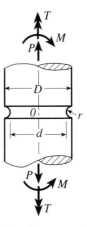

Figure 1.27 Grooved shaft subject to tension, bending, and torsion.

The maximum stresses of Eqs. (1)–(3) occur at the same location, namely at the base of the groove, and the principal stresses are calculated using the familiar formulas (Pilkey 1994, sect. 3.3)

$$\sigma_1 = \frac{1}{2}(\sigma_{\max 1} + \sigma_{\max 2}) + \frac{1}{2}\sqrt{(\sigma_{\max 1} + \sigma_{\max 2})^2 + 4\tau_{\max 3}^2} \qquad (4)$$

$$\sigma_2 = \frac{1}{2}(\sigma_{\max 1} + \sigma_{\max 2}) - \frac{1}{2}\sqrt{(\sigma_{\max 1} + \sigma_{\max 2})^2 + 4\tau_{\max 3}^2} \qquad (5)$$

The various failure criteria for the base of the groove can now be formulated.

Maximum Stress Criterion

$$\sigma_{\max} = \sigma_1 \qquad (6)$$

Mohr's Theory From Eqs. (4) and (5), it is easy to prove that $\sigma_1 > 0$ and $\sigma_2 < 0$. The condition of failure is (Eq. 1.27)

$$\frac{\sigma_1}{\sigma_{ut}} - \frac{\sigma_2}{\sigma_{uc}} = 1 \qquad (7)$$

where σ_{ut} is the uniaxial tensile strength and σ_{uc} is the uniaxial compressive strength.

Maximum Shear Theory Since $\sigma_1 > 0$, $\sigma_2 < 0$, $\sigma_3 = 0$, the maximum shear stress is

$$\tau_{\max} = \frac{\sigma_1 - \sigma_2}{2} = \frac{1}{2}\sqrt{(\sigma_{\max 1} + \sigma_{\max 2})^2 + 4\tau_{\max 3}^2} \qquad (8)$$

von Mises Criterion From Eq. (1.34),

$$\sigma_{eq} = \sqrt{\sigma_1^2 - \sigma_1\sigma_2 + \sigma_2^2} = \sqrt{(\sigma_{\max 1} + \sigma_{\max 2})^2 + 3\tau_{\max 3}^2} \qquad (9)$$

Example 1.7 An Infinite Element with a Circular Hole with Internal Pressure   Find the stress concentration factor for an infinite element subjected to internal pressure p on it's circular hole edge as shown in Fig. 1.28a.

This example can be solved by superimposing two configurations. The loads on the element can be assumed to consist of two cases: (1) biaxial tension $\sigma = p$ (Fig. 1.28b); (2) biaxial compression $\sigma = -p$, with pressure on the circular hole edge (Fig. 1.28c).

For case 1, $\sigma = p$, the stresses at the edge of the hole are (Eq. 4.16)

$$\sigma_{r1} = 0$$
$$\sigma_{\theta 1} = 2p \qquad (1)$$
$$\tau_{r\theta 1} = 0$$

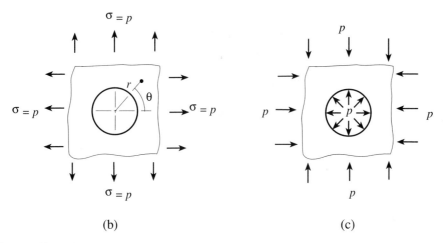

Figure 1.28 Infinite element subjected to internal pressure p on a circular hole edge: (*a*) Element subjected to pressure p; (*b*) element under biaxial tension at area remote from the hole; (*c*) element under biaxial compression.

For case 2 the stresses at the edge of the hole (hydrostatic pressure) are

$$\sigma_{r2} = -p$$
$$\sigma_{\theta 2} = -p \qquad (2)$$
$$\tau_{r\theta 2} = 0$$

The stresses for both cases can be derived from the formulas of Little (1973). The total stresses at the edge of the hole can be obtained by superposition

$$\sigma_r = \sigma_{r1} + \sigma_{r2} = -p$$
$$\sigma_\theta = \sigma_{\theta 1} + \sigma_{\theta 2} = p \qquad (3)$$
$$\tau_{r\theta} = \tau_{r\theta 1} + \tau_{r\theta 2} = 0$$

The maximum stress is $\sigma_{\max} = p$. If p is taken as the nominal stress (Example 1.3), the corresponding stress concentration factor can be defined as

$$K_t = \frac{\sigma_{\max}}{\sigma_{\text{nom}}} = \frac{\sigma_{\max}}{p} = 1 \qquad (4)$$

36 DEFINITIONS AND DESIGN RELATIONS

1.9 NOTCH SENSITIVITY

As noted at the beginning of this chapter, the theoretical stress concentration factors apply mainly to ideal elastic materials and depend on the geometry of the body and the loading. Sometimes a more realistic model is preferable. When the applied loads reach a certain level, plastic deformations may be involved. The actual strength of structural members may be quite different from that derived using theoretical stress concentration factors, especially for the cases of impact and alternating loads.

It is reasonable to introduce the concept of the *effective stress concentration factor K_e*. This is also referred to as the factor of stress concentration at rupture or the notch rupture strength ratio (ASTM 1994). The magnitude of K_e is obtained experimentally. For instance, K_e for a round bar with a circumferential groove subjected to a tensile load P' (Fig. 1.29a) is obtained as follows: (1) Prepare two sets of specimens of the actual material, the round bars of the first set having circumferential grooves, with d as the diameter at the root of the groove (Fig. 1.29a). The round bars of the second set are of diameter d without grooves (Fig. 1.29b). (2) Perform a tensile test for the two sets of specimens, the rupture load for the first set is P', while the rupture load for second set is P. (3) The effective stress concentration factor is defined as

$$K_e = \frac{P}{P'} \qquad (1.43)$$

In general, $P' < P$ so that $K_e > 1$. The effective stress concentration factor is a function not only of geometry but also of material properties. Some characteristics of K_e for static loading of different materials are discussed briefly below.

1. *Ductile material.* Consider a tensile loaded plane element with a V-shaped notch. The material law for the material is sketched in Fig. 1.30. If the maximum stress at the root of the notch is less than the yield strength $\sigma_{max} < \sigma_y$, the stress distributions near the notch would appear as in curves 1 and 2 in Fig. 1.30. The maximum stress

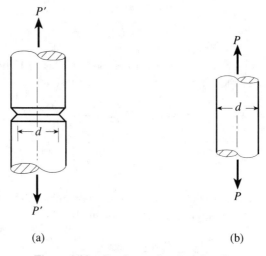

Figure 1.29 Specimens for obtaining K_e.

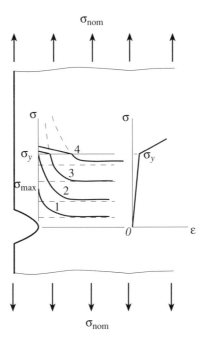

Figure 1.30 Stress distribution near a notch for a ductile material.

value is

$$\sigma_{\max} = K_t \sigma_{\text{nom}} \tag{1.44}$$

As the $\sigma_{\max}$ exceeds σ_y, the strain at the root of the notch continues to increase but the maximum stress increases only slightly. The stress distributions on the cross section will be of the form of curves 3 and 4 in Fig. 1.30. Equation (1.44) no longer applies to this case. As σ_{nom} continues to increase, the stress distribution at the notch becomes more uniform and the effective stress concentration factor K_e is close to unity.

2. *Brittle material.* Most brittle materials can be treated as elastic bodies. When the applied load increases, the stress and strain retain their linear relationship until damage occurs. The effective stress concentration factor K_e is the same as K_t.

3. *Gray cast iron.* Although gray cast irons belong to brittle materials, they contain flake graphite dispersed in the steel matrix and a number of small cavities, which produce much higher stress concentrations than would be expected from the geometry of the discontinuity. In such a case the use of the stress concentration factor K_t may result in significant error and K_e can be expected to approach unity, since the stress raiser has a smaller influence on the strength of the member than that of the small cavities and flake graphite.

It can be reasoned from these three cases that the effective stress concentration factor depends on the characteristics of the material and the nature of the load, as well as the geometry of the stress raiser. Also $1 \leq K_e \leq K_t$. The maximum stress at rupture can be defined to be

$$\sigma_{\max} = K_e \sigma_{\text{nom}} \tag{1.45}$$

To express the relationship between K_e and K_t, introduce the concept of *notch sensitivity* q (Boresi et al. 1993):

$$q = \frac{K_e - 1}{K_t - 1} \tag{1.46}$$

or

$$K_e = q(K_t - 1) + 1 \tag{1.47}$$

Substitute Eq. (1.47) into Eq. (1.45):

$$\sigma_{\max} = [q(K_t - 1) + 1]\sigma_{\text{nom}} \tag{1.48}$$

If $q = 0$, then $K_e = 1$, meaning that the stress concentration does not influence the strength of the structural member. If $q = 1$, then $K_e = K_t$, implying that the theoretical stress concentration factor should be fully invoked. The notch sensitivity is a measure of the agreement between K_e and K_t.

The concepts of the effective stress concentration factor and notch sensitivity are used primarily for fatigue strength design. For fatigue loading, replace K_e in Eq. (1.43) by K_f or K_{fs}, defined as

$$K_f = \frac{\text{Fatigue limit of unnotched specimen (axial or bending)}}{\text{Fatigue limit of notched specimen (axial or bending)}} = \frac{\sigma_f}{\sigma_{nf}} \tag{1.49}$$

$$K_{fs} = \frac{\text{Fatigue limit of unnotched specimen (shear stress)}}{\text{Fatigue limit of notched specimen (shear stress)}} = \frac{\tau_f}{\tau_{nf}} \tag{1.50}$$

where K_f is the *fatigue notch factor* for normal stress and K_{fs} is the fatigue notch factor for shear stress, such as torsion. The notch sensitivities for fatigue become

$$q = \frac{K_f - 1}{K_t - 1} \tag{1.51}$$

or

$$q = \frac{K_{fs} - 1}{K_{ts} - 1} \tag{1.52}$$

where K_{ts} is defined in Eq. (1.2). The values of q vary from $q = 0$ for no notch effect ($K_f = 1$) to $q = 1$ for the full theoretical effect ($K_f = K_t$).

Equations (1.51) and (1.52) can be rewritten in the following form for design use:

$$K_{tf} = q(K_t - 1) + 1 \tag{1.53}$$

$$K_{tsf} = q(K_{ts} - 1) + 1 \tag{1.54}$$

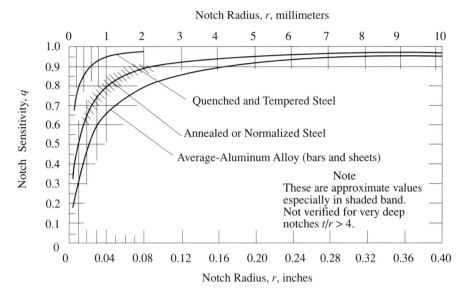

Figure 1.31 Average fatigue notch sensitivity.

where K_{tf} is the *estimated* fatigue notch factor for normal stress, a calculated factor using an average q value obtained from Fig. 1.31 or a similar curve, and K_{tsf} is the *estimated* fatigue notch factor for shear stress.

If no information on q is available, as would be the case for newly developed materials, it is suggested that the full theoretical factor, K_t or K_{ts}, be used. It should be noted in this connection that if notch sensitivity is not taken into consideration at all in design ($q = 1$), the error will be on the safe side ($K_{tf} = K_t$ in Eq. (1.53)).

In plotting K_f for geometrically similar specimens, it was found that typically K_f decreased as the specimen size decreased (Peterson 1933a, 1933b, 1936, 1943). For this reason it is not possible to obtain reliable comparative q values for different materials by making tests of a standardized specimen of fixed dimension (Peterson 1945). Since the local stress distribution (stress gradient,[5] volume at peak stress) is more dependent on the notch radius r than on other geometrical variables (Peterson 1938; Neuber 1958; von Phillipp 1942), it was apparent that it would be more logical to plot q versus r rather than q versus d (for geometrically similar specimens the curve shapes are of course the same). Plotted q versus r curves (Peterson 1950, 1959) based on available data (Gunn 1952; Lazan and Blatherwick 1953; Templin 1954; Fralich 1959) were found to be within reasonable scatter bands.

A q versus r chart for design purposes is given in Fig. 1.31; it averages the previously mentioned plots. Note that the chart is not verified for notches having a depth greater than four times the notch radius because data are not available. Also note that the curves are to be considered as approximate (see shaded band).

Notch sensitivity values for radii approaching zero still must be studied. It is, however, well known that tiny holes and scratches do not result in a strength reduction corresponding

---

[5] The stress is approximately linear in the peak stress region (Peterson 1938; Leven 1955).

to theoretical stress concentration factors. In fact, in steels of low tensile strength, the effect of very small holes or scratches is often quite small. However, in higher-strength steels the effect of tiny holes or scratches is more pronounced. Much more data are needed, preferably obtained from statistically planned investigations. Until better information is available, Fig. 1.31 provides reasonable values for design use.

Several expressions have been proposed for the q versus r curve. Such a formula could be useful in setting up a computer design program. Since it would be unrealistic to expect failure at a volume corresponding to the point of peak stress becuase of the plastic deformation (Peterson 1938), formulations for K_f are based on failure over a distance below the surface (Neuber 1958; Peterson 1974). From the K_f formulations, q versus r relations are obtained. These and other variations are found in the literature (Peterson 1945). All of the formulas yield acceptable results for design purposes. One must, however, always remember the approximate nature of the relations. In Fig. 1.31 the following simple formula (Peterson 1959) is used:[6]

$$q = \frac{1}{1 + \alpha/r} \tag{1.55}$$

where α is a material constant and r is the notch radius.

In Fig. 1.31, $\alpha = 0.0025$ for quenched and tempered steel, $\alpha = 0.01$ for annealed or normalized steel, $\alpha = 0.02$ for aluminum alloy sheets and bars (avg.). In Peterson (1959) more detailed values are given, including the following approximate design values for steels as a function of tensile strength:

$\sigma_{ut}/1000$	α
50	0.015
75	0.010
100	0.007
125	0.005
150	0.0035
200	0.0020
250	0.0013

where σ_{ut} = tensile strength in pounds per square inch. In using the foregoing α values, one must keep in mind that the curves represent averages (see shaded band in Fig. 1.31).

A method has been proposed by Neuber (1968) wherein an equivalent larger radius is used to provide a lower K factor. The increment to the radius is dependent on the stress state, the kind of material, and its tensile strength. Application of this method gives results that are in reasonably good agreement with the calculations of other methods (Peterson 1953).

---

[6]The corresponding Kuhn-Hardrath formula (Kuhn and Hardrath 1952) based on Neuber relations is

$$q = \frac{1}{1 + \sqrt{\rho'/r}}$$

Either formula may be used for design purposes (Peterson 1959). The quantities α or ρ', a material constant, are determined by test data.

1.10 DESIGN RELATIONS FOR STATIC STRESS

1.10.1 Ductile Materials

As discussed in Section 1.8, under ordinary conditions a ductile member loaded with a steadily increasing uniaxial stress does not suffer loss of strength due to the presence of a notch, since the notch sensitivity q usually lies in the range 0 to 0.1. However, if the function of the member is such that the amount of inelastic strain required for the strength to be insensitive to the notch is restricted, the value of q may approach 1.0 ($K_e = K_t$). If the member is loaded statically and is also subjected to shock loading, or if the part is to be subjected to high (Davis and Manjoine 1952) or low temperature, or if the part contains sharp discontinuities, a ductile material may behave in the manner of a brittle material, which should be studied with fracture mechanics methods. These are special cases. If there is doubt, K_t should be applied ($q = 1$). Ordinarily, for static loading of a ductile material, set $q = 0$ in Eq. (1.48), namely $\sigma_{max} = \sigma_{nom}$.[7]

Traditionally design safety is measured by the *factor of safety n*. It is defined as the ratio of the load that would cause failure of the member to the working stress on the member. For ductile material the failure is assumed to be caused by yielding and the equivalent stress σ_{eq} can be used as the working stress (von Mises criterion of failure, Section 1.8). For axial loading (normal, or direct, stress $\sigma_1 = \sigma_{0d}$, $\sigma_2 = \sigma_3 = 0$):

$$n = \frac{\sigma_y}{\sigma_{0d}} \quad (1.56)$$

where σ_y is the yield strength and σ_{0d} is the static normal stress $= \sigma_{eq} = \sigma_1$. For bending ($\sigma_1 = \sigma_{0b}$, $\sigma_2 = \sigma_3 = 0$),

$$n = \frac{L_b \sigma_y}{\sigma_{0b}} \quad (1.57)$$

where L_b is the limit design factor for bending and σ_{0b} is the static bending stress.

In general, the limit design factor L is the ratio of the load (force or moment) needed to cause complete yielding throughout the section of a bar to the load needed to cause initial yielding at the "extreme fiber" (Van den Broek 1942), assuming no stress concentration. For tension, $L = 1$; for bending of a rectangular bar, $L_b = 3/2$; for bending of a round bar, $L_b = 16/(3\pi) = 1.70$; for torsion of a round bar, $L_s = 4/3$; for a tube, it can be shown that for bending and torsion, respectively,

$$\begin{aligned} L_b &= \frac{16}{3\pi} \left[\frac{1 - (d_i/d_0)^3}{1 - (d_i/d_0)^4} \right] \\ L_s &= \frac{4}{3} \left[\frac{1 - (d_i/d_0)^3}{1 - (d_i/d_0)^4} \right] \end{aligned} \quad (1.58)$$

---

[7] This consideration is on the basis of strength only. Stress concentration does not ordinarily reduce the strength of a notched member in a static test, but usually it does reduce total deformation to rupture. This means lower "ductility," or, expressed in a different way, less area under the stress-strain diagram (less energy expended in producing complete failure). It is often of major importance to have as much energy-absorption capacity as possible (cf. metal versus plastic for an automobile body). However, this is a consideration depending on consequence of failure, and so on, and is not within the scope of this book, which deals only with strength factors. Plastic behavior is involved in a limited way in the use of the factor L, as is discussed in this section.

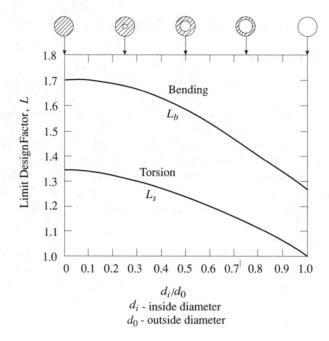

Figure 1.32 Limit design factors for tubular members.

where d_i and d_0 are the inside and outside diameters, respectively, of the tube. These relations are plotted in Fig. 1.32.

Criteria other than complete yielding can be used. For a rectangular bar in bending, L_b values have been calculated (Steele et al. 1952), yielding to 1/4 depth $L_b = 1.22$, and yielding to 1/2 depth $L_b = 1.375$; for 0.1% inelastic strain in steel with yield point of 30,000 psi, $L_b = 1.375$. For a circular bar in bending, yielding to 1/4 depth, $L_b = 1.25$, and yielding to 1/2 depth, $L_b = 1.5$. For a tube $d_i/d_0 = 3/4$: yielding 1/4 depth, $L_b = 1.23$, and yielding 1/2 depth, $L_b = 1.34$.

All the foregoing L values are based on the assumption that the stress-strain diagram becomes horizontal after the yield point is reached, that is, the material is *elastic, perfectly plastic*. This is a reasonable assumption for low- or medium-carbon steel. For other stress-strain diagrams which can be represented by a sloping line or curve beyond the elastic range, a value of L closer to 1.0 should be used (Van den Broek 1942). For design $L\sigma_y$ should not exceed the tensile strength σ_{ut}.

For torsion of a round bar (shear stress), using Eq. (1.36) obtains

$$n = \frac{L_s \tau_y}{\tau_0} = \frac{L_s \sigma_y}{\sqrt{3}\tau_0} \tag{1.59}$$

where τ_y is the yield strength in torsion and τ_0 is the static shear stress.

For combined normal (axial and bending) and shear stress the principal stresses are

$$\sigma_1 = \frac{1}{2}\left(\sigma_{0d} + \frac{\sigma_{0b}}{L_b}\right) + \frac{1}{2}\sqrt{\left[\sigma_{0d} + (\sigma_{0b}/L_b)\right]^2 + 4(\tau_0/L_s)^2}$$

$$\sigma_2 = \frac{1}{2}\left[\sigma_{0d} + (\sigma_{0b}/L_b)\right] - \frac{1}{2}\sqrt{\left[\sigma_{0d} + (\sigma_{0b}/L_b)\right]^2 + 4(\tau_0/L_s)^2}$$

where σ_{0d} is the static axial stress and σ_{0b} is the static bending stress. Since $\sigma_3 = 0$, the formula for the von Mises theory is given by (Eq. 1.35)

$$\sigma_{eq} = \sqrt{\sigma_1^2 - \sigma_1\sigma_2 + \sigma_2^2}$$

so that

$$n = \frac{\sigma_y}{\sigma_{eq}} = \frac{\sigma_y}{\sqrt{[\sigma_{0d} + (\sigma_{0b}/L_b)]^2 + 3\left(\tau_0/L_s\right)^2}} \quad (1.60)$$

1.10.2 Brittle Materials

It is customary to apply the full K_t factor in the design of members of brittle materials. The use of the full K_t factor for cast iron may be considered, in a sense, as penalizing this material unduly, since experiments show that the full effect is usually not obtained (Roark et al. 1938). The use of the full K_t factor may be partly justified as compensating, in a way, for the poor shock resistance of brittle materials. Since it is difficult to design rationally for shock or mishandling in transportation and installation, the larger sections obtained by the preceding rule may be a means of preventing some failures that might otherwise occur. However, notable designs of cast-iron members have been made (large paper-mill rolls, etc.) involving rather high stresses where full application of stress concentration factors would rule out this material. Such designs should be carefully made and may be viewed as exceptions to the rule. For ordinary design it seems wise to proceed cautiously in the treatment of notches in brittle materials, especially in critical load-carrying members.

The following factors of safety are based on the *maximum stress criterion* of failure of Section 1.8. For axial tension or bending (normal stress),

$$n = \frac{\sigma_{ut}}{K_t\sigma_0} \quad (1.61)$$

where σ_{ut} is the tensile ultimate strength, K_t is the stress concentration factor for normal stress, and σ_0 is the normal stress. For torsion of a round bar (shear stress),

$$n = \frac{\sigma_{ut}}{K_{ts}\tau_0} \quad (1.62)$$

where K_{ts} is the stress concentration factor for shear stress and τ_0 is the static shear stress.

The following factors of safety are based on *Mohr's theory* of failure of Section 1.8. Since the factors based on Mohr's theory are on the "safe side" compared to those based on the maximum stress criterion, they are suggested for design use. For axial tension or bending, Eq. (1.61) applies. For torsion of a round bar (shear stress), by Eq. (1.30),

$$n = \frac{\sigma_{ut}}{K_{ts}\tau_0}\left[\frac{1}{1 + (\sigma_{ut}/\sigma_{uc})}\right] \quad (1.63)$$

where σ_{ut} is the tensile ultimate strength and σ_{uc} is the compressive ultimate strength. For combined normal and shear stress,

$$n = \frac{2\sigma_{ut}}{K_t\sigma_0(1 - \sigma_{ut}/\sigma_{uc}) + (1 + \sigma_{ut}/\sigma_{uc})\sqrt{(K_t\sigma_0)^2 + 4(K_{ts}\tau_0)^2}} \quad (1.64)$$

1.11 DESIGN RELATIONS FOR ALTERNATING STRESS

1.11.1 Ductile Materials

For alternating (completely reversed cyclic) stress, the stress concentration effects must be considered. As explained in Section 1.9, the fatigue notch factor K_f is usually less than the stress concentration factor K_t. The factor K_{tf} represents a calculated estimate of the actual fatigue notch factor K_f. Naturally, if K_f is available from tests, one uses this, but a designer is very seldom in such a fortunate position. The expression for K_{tf} and K_{tsf}, Eqs. (1.53) and (1.54), respectively, are repeated here:

$$K_{tf} = q(K_t - 1) + 1 \tag{1.65}$$

$$K_{tsf} = q(K_{ts} - 1) + 1$$

The following expressions for factors of safety, are based on the von Mises criterion of failure as discussed in Section 1.8:

For axial or bending loading (normal stress),

$$n = \frac{\sigma_f}{K_{tf}\sigma_a} = \frac{\sigma_f}{[q(K_t - 1) + 1]\sigma_a} \tag{1.66}$$

where σ_f is the fatigue limit (endurance limit) in axial or bending test (normal stress) and σ_a is the alternating normal stress amplitude.

For torsion of a round bar (shear stress),

$$n = \frac{\tau_f}{K_{tsf}\tau_a} = \frac{\sigma_f}{\sqrt{3}K_{ts}\tau_a} = \frac{\sigma_f}{\sqrt{3}[q(K_{ts} - 1) + 1]\tau_a} \tag{1.67}$$

where τ_f is the fatigue limit in torsion and τ_a is the alternating shear stress amplitude.

For combined normal stress and shear stress,

$$n = \frac{\sigma_f}{\sqrt{(K_{tf}\sigma_a)^2 + 3(K_{tsf}\tau_a)^2}} \tag{1.68}$$

By rearranging Eq. (1.68), the equation for an ellipse is obtained,

$$\frac{\sigma_a^2}{(\sigma_f/nK_{tf})^2} + \frac{\tau_a^2}{(\sigma_f/n\sqrt{3}K_{tsf})^2} = 1 \tag{1.69}$$

where $\sigma_f/(nK_{tf})$ and $\sigma_f/(n\sqrt{3}K_{tsf})$ are the major and minor semiaxes. Fatigue tests of unnotched specimens by Gough and Pollard (1935) and by Nisihara and Kawamoto (1940) are in excellent agreement with the elliptical relation. Fatigue tests of notched specimens (Gough and Clenshaw 1951) are not in as good agreement with the elliptical relation as are the unnotched, but for design purposes the elliptical relation seems reasonable for ductile materials.

1.11.2 Brittle Materials

Since our knowledge in this area is very limited, it is suggested that unmodified K_t factors be used. Mohr's theory of Section 1.8, with $\sigma_{ut}/\sigma_{uc} = 1$, is suggested for design purposes for brittle materials subjected to alternating stress.

For axial or bending loading (normal stress),

$$n = \frac{\sigma_f}{K_t \sigma_a} \tag{1.70}$$

For torsion of a round bar (shear stress),

$$n = \frac{\tau_f}{K_{ts} \tau_a} = \frac{\sigma_f}{2K_{ts} \tau_a} \tag{1.71}$$

For combined normal stress and shear stress,

$$n = \frac{\sigma_f}{\sqrt{(K_t \sigma_a)^2 + 4(K_{ts} \tau_a)^2}} \tag{1.72}$$

1.12 DESIGN RELATIONS FOR COMBINED ALTERNATING AND STATIC STRESSES

The majority of important strength problems comprises neither simple static nor alternating cases, but involves fluctuating stress, which is a combination of both. A cyclic fluctuating stress (Fig. 1.33) having a maximum value σ_{max} and minimum value σ_{min} can be considered as having an *alternating component* of amplitude

$$\sigma_a = \frac{\sigma_{max} - \sigma_{min}}{2} \tag{1.73}$$

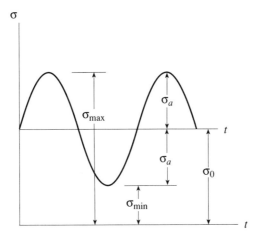

Figure 1.33 Combined alternating and steady stresses.

and a *steady* or *static component*

$$\sigma_0 = \frac{\sigma_{\max} + \sigma_{\min}}{2} \tag{1.74}$$

1.12.1 Ductile Materials

In designing parts to be made of ductile materials for normal temperature use, it is the usual practice to apply the stress concentration factor to the alternating component but not to the static component. This appears to be a reasonable procedure and is in conformity with test data (Houdremont and Bennek 1932) such as that shown in Fig. 1.34*a*. The limitations discussed in Section 1.10 still apply.

By plotting minimum and maximum limiting stresses in (Fig. 1.34*a*), the relative positions of the static properties, such as yield strength and tensile strength, are clearly shown. However, one can also use a simpler representation such as that of Fig. 1.34*b*, with the alternating component as the ordinate.

If, in Fig. 1.34*a*, the curved lines are replaced by straight lines connecting the end points σ_f and σ_u, σ_f/K_{tf} and σ_u, we have a simple approximation which is on the safe side for steel members.[8] From Fig. 1.34*b* we can obtain the following simple rule for factor of safety:

$$n = \frac{1}{(\sigma_0/\sigma_u) + (K_{tf}\sigma_a/\sigma_f)} \tag{1.75}$$

This is the same as the following Soderberg rule (Pilkey 1994), except that σ_u is used instead of σ_y. Soderberg's rule is based on the yield strength (see lines in Fig. 1.34 connecting σ_f and σ_y, σ_f/K_{tf} and σ_y):

$$n = \frac{1}{(\sigma_0/\sigma_y) + (K_{tf}\sigma_a/\sigma_f)} \tag{1.76}$$

By referring to Fig. 1.34*b*, it can be shown that $n = OB/OA$. Note that in Fig. 1.34*a*, the pulsating (0 to max) condition corresponds to $\tan^{-1} 2$, or 63.4°, which in Fig. 1.34*b* is 45°.

Equation (1.76) may be further modified to be in conformity with Eqs. (1.56) and (1.57), which means applying limit design for yielding, with the factors and considerations as stated in Section 1.10.1:

$$n = \frac{1}{(\sigma_{0d}/\sigma_y) + (\sigma_{0b}/L_b\sigma_y) + (K_{tf}\sigma_a/\sigma_f)} \tag{1.77}$$

As mentioned previously $L_b\sigma_y$ must not exceed σ_u. That is, the factor of safety n from Eq. (1.77) must not exceed n from Eq. (1.75).

---

[8]For steel members, a cubic relation (Peterson 1952; Nichols 1969) fits available data fairly well, $\sigma_a = [\sigma_f/(7K_{tf})]\{8 - [(\sigma_0/\sigma_u) + 1]^3\}$. This is the equation for the lower full curve of Fig. 1.34*b*. For certain aluminum alloys, the σ_a, σ_0 curve has a shape (Lazan and Blatherwick 1952) that is concave slightly below the σ_f/K_f, σ_u line at the upper end and is above the line at the lower end.

DESIGN RELATIONS FOR COMBINED ALTERNATING AND STATIC STRESSES 47

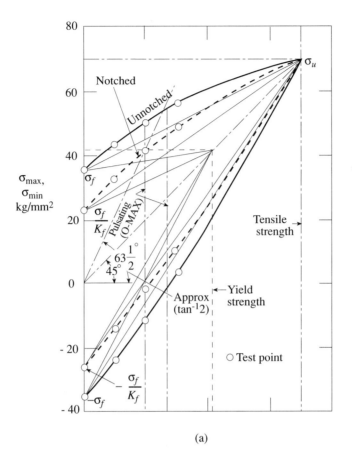

(a)

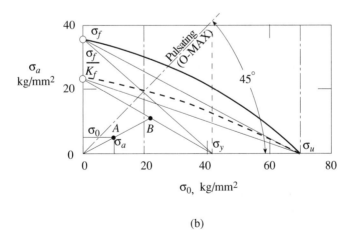

(b)

Figure 1.34 Limiting values of combined alternating and steady stresses for plain and notched specimens (data of Schenck, 0.7% C steel, Houdremont and Bennek 1932): (*a*) Limiting minimum and maximum values; (*b*) limiting alternating and steady components.

For torsion, the same assumptions and use of the von Mises criterion result in:

$$n = \frac{1}{\sqrt{3}\left[(\tau_0/L_s\sigma_y) + (K_{tsf}\tau_a/\sigma_f)\right]} \quad (1.78)$$

For notched specimens Eq. (1.78) represents a design relation, being on the safe edge of test data (Smith 1942). It is interesting to note that, for unnotched torsion specimens, static torsion (up to a maximum stress equal to the yield strength in torsion) does not lower the limiting alternating torsional range. It is apparent that further research is needed in the torsion region; however, since notch effects are involved in design (almost without exception), the use of Eq. (1.78) is indicated. Even in the absence of stress concentration, Eq. (1.78) would be on the "safe side," though by a large margin for relatively large values of statically applied torque.

For a combination of *static (steady) and alternating normal stresses plus static and alternating shear stresses* (alternating components in phase) the following relation, derived by Soderberg (1930), is based on expressing the shear stress on an arbitrary plane in terms of static and alternating components, assuming failure is governed by the maximum shear theory and a "straight-line" relation similar to Eq. (1.76) and finding the plane that gives a minimum factor of safety n (Peterson 1953):

$$n = \frac{1}{\sqrt{\left[(\sigma_0/\sigma_y) + (K_t\sigma_a/\sigma_f)\right]^2 + 4\left[(\tau_0/\sigma_y) + (K_{ts}\tau_a/\sigma_f)\right]^2}} \quad (1.79)$$

The following modifications are made to correspond to the end conditions represented by Eqs. (1.56), (1.57), (1.59), (1.66), and (1.67). Then Eq. (1.79) becomes

$$n = \frac{1}{\sqrt{\left[(\sigma_{0d}/\sigma_y) + (\sigma_{0b}/L_b\sigma_y) + (K_{tf}\sigma_a/\sigma_f)\right]^2 + 3\left[(\tau_0/L_s\sigma_y) + (K_{tsf}\tau_a/\sigma_f)\right]^2}} \quad (1.80)$$

For steady stress only, Eq. (1.80) reduces to Eq. (1.60).
For alternating stress only, Eq. (1.80) reduces to Eq. (1.68).
For normal stress only, Eq. (1.80) reduces to Eq. (1.77).
For torsion only, Eq. (1.80) reduces to Eq. (1.78).

In tests by Ono (1921, 1929) and by Lea and Budgen (1926) the alternating bending fatigue strength was found not to be affected by the addition of a static (steady) torque (less than the yield torque). Other tests reported in a discussion by Davies (1935) indicate a lowering of the bending fatigue strength by the addition of static torque. Hohenemser and Prager (1933) found that a static tension lowered the alternating torsional fatigue strength; Gough and Clenshaw (1951) found that steady bending lowered the torsional fatigue strength of plain specimens but that the effect was smaller for specimens involving stress concentration. Further experimental work is needed in this area of special combined stress combinations, especially in the region involving the additional effect of stress concentration. In the meantime, while it appears that use of Eq. (1.80) may be overly 'safe' in certain cases of alternating bending plus steady torque, it is believed that Eq. (1.80) provides a reasonable general design rule.

1.12.2 Brittle Materials

A "straight-line" simplification similar to that of Fig. 1.34 and Eq. (1.75) can be made for brittle material, except that the stress concentration effect is considered to apply also to the static (steady) component.

$$n = \frac{1}{K_t\left[(\sigma_0/\sigma_{ut}) + (\sigma_a/\sigma_f)\right]} \tag{1.81}$$

As previously mentioned, unmodified K_t factors are used for the brittle material cases.

For *combined shear and normal stresses*, data are very limited. For combined alternating bending and static torsion, Ono (1921) reported a decrease of the bending fatigue strength of cast iron as steady torsion was added. By use of the Soderberg method (Soderberg 1930) and basing failure on the normal stress criterion (Peterson 1953), we obtain

$$n = \frac{2}{K_t\left(\dfrac{\sigma_0}{\sigma_{ut}} + \dfrac{\sigma_a}{\sigma_f}\right) + \sqrt{K_t^2\left(\dfrac{\sigma_0}{\sigma_{ut}} + \dfrac{\sigma_a}{\sigma_f}\right)^2 + 4K_{ts}^2\left(\dfrac{\tau_0}{\sigma_{ut}} + \dfrac{\tau_a}{\sigma_f}\right)^2}} \tag{1.82}$$

A rigorous formula for combining Mohr's theory components of Eqs. (1.64) and (1.72) does not seem to be available. The following approximation which satisfies Eqs. (1.61), (1.63), (1.70), and (1.71) may be of use in design, in the absence of a more exact formula.

$$n = \frac{2}{K_t\left(\dfrac{\sigma_0}{\sigma_{ut}} + \dfrac{\sigma_a}{\sigma_f}\right)\left(1 - \dfrac{\sigma_{ut}}{\sigma_{uc}}\right) + \left(1 + \dfrac{\sigma_{ut}}{\sigma_{uc}}\right)\sqrt{K_t^2\left(\dfrac{\sigma_0}{\sigma_{ut}} + \dfrac{\sigma_a}{\sigma_f}\right)^2 + 4K_{ts}^2\left(\dfrac{\tau_0}{\sigma_{ut}} + \dfrac{\tau_a}{\sigma_f}\right)^2}} \tag{1.83}$$

For steady stress only, Eq. (1.83) reduces to Eq. (1.64).
For alternating stress only, with $\sigma_{ut}/\sigma_{uc} = 1$, Eq. (1.83) reduces to Eq. (1.72).
For normal stress only, Eq. (1.83) reduces to Eq. (1.81).
For torsion only, Eq. (1.83) reduces to

$$n = \frac{1}{K_{ts}\left(\dfrac{\tau_0}{\sigma_{ut}} + \dfrac{\tau_a}{\sigma_f}\right)\left(1 + \dfrac{\sigma_{ut}}{\sigma_{uc}}\right)} \tag{1.84}$$

This in turn can be reduced to the component cases of Eqs. (1.63) and (1.71).

1.13 LIMITED NUMBER OF CYCLES OF ALTERNATING STRESS

In *Stress Concentration Design Factors* (1953) Peterson presented formulas for a limited number of cycles (upper branch of the S-N diagram). These relations were based on an average of available test data and therefore apply to polished test specimens 0.2 to 0.3 in. diameter. If the member being designed is not too far from this size range, the formulas

50 DEFINITIONS AND DESIGN RELATIONS

may be useful as a rough guide, but otherwise they are questionable, since the number of cycles required for a crack to propagate to rupture of a member depends on the size of the member.

Fatigue failure consists of three stages: crack initiation, crack propagation, and rupture. Crack initiation is thought not to be strongly dependent on size, although from statistical considerations of the number of "weak spots," one would expect some effect. So much progress has been made in the understanding of crack propagation under cyclic stress, that it is believed that reasonable estimates can be made for a number of problems.

1.14 STRESS CONCENTRATION FACTORS AND STRESS INTENSITY FACTORS

Consider an elliptical hole of major axis $2a$ and minor axis $2b$ in a plane element (Fig. 1.35a). If $b \to 0$ (or $a \gg b$), the elliptical hole becomes a crack of length $2a$ (Fig. 1.35b). The *stress intensity factor K* represents the strength of the elastic stress fields surrounding the crack tip (Pilkey 1994). It would appear that there might be a relationship between the stress concentration factor and the stress intensity factor. Creager and Paris (1967) analyzed the stress distribution around the tip of a crack of length $2a$ using the coordinates shown in Fig. 1.36. The origin O of the coordinates is set a distance of $r/2$ from the tip, in which r is the radius of curvature of the tip. The stress σ_y in the y direction near the tip can be expanded as a power series in terms of the radial distance. Discarding all terms higher than second order, the approximation for mode I fracture (Pilkey 1994; sec. 7.2) becomes

$$\sigma_y = \sigma + \frac{K_I}{\sqrt{2\pi\rho}} \frac{r}{2\rho} \cos \frac{3\theta}{2} + \frac{K_I}{\sqrt{2\pi\rho}} \cos \frac{\theta}{2} \left(1 + \sin \frac{\theta}{2} \sin \frac{3\theta}{2}\right) \quad (1.85)$$

where σ is the tensile stress remote from the crack, (ρ, θ) are the polar coordinates of the crack tip with origin O (Fig. 1.36), K_I is the mode I stress intensity factor of the case in Fig. 1.35b. The maximum longitudinal stress occurs at the tip of the crack, that is, at

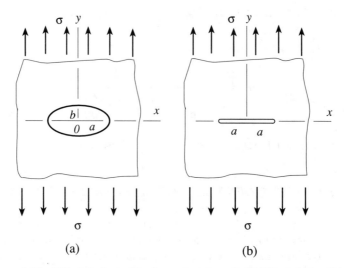

Figure 1.35 Elliptic hole model of a crack as $b \to 0$: (a) Elliptic hole; (b) crack.

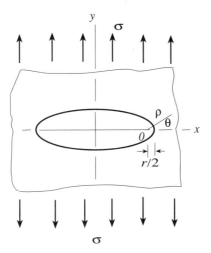

Figure 1.36 Coordinate system for stress at the tip of an ellipse.

$\rho = r/2$, $\theta = 0$. Substituting this condition into Eq. (1.85) gives

$$\sigma_{max} = \sigma + 2\frac{K_I}{\sqrt{\pi r}} \qquad (1.86)$$

However, the stress intensity factor can be written as (Pilkey 1994)

$$K_I = C\sigma\sqrt{\pi a} \qquad (1.87)$$

where C is a constant that depends on the shape and the size of the crack and the specimen. Substituting Eq. (1.87) into Eq. (1.86), the maximum stress is

$$\sigma_{max} = \sigma + 2C\sigma\sqrt{\frac{a}{r}} \qquad (1.88)$$

With σ as the reference stress, the stress concentration factor at the tip of the crack for a two-dimensional element subjected to uniaxial tension is

$$K_t = \sigma_{max}/\sigma_{nom} = 1 + 2C\sqrt{a/r} \qquad (1.89)$$

Equation (1.89) gives an approximate relationship between the stress concentration factor and the stress intensity factor. Due to the rapid development of fracture mechanics, a large number of crack configurations have been analyzed, and the corresponding results can be found in various handbooks. These results may be used to estimate the stress concentration factor for many cases. For instance, for a crack of length $2a$ in an infinite element under uniaxial tension, the factor C is equal to 1, so the corresponding stress concentration factor is

$$K_t = \frac{\sigma_{max}}{\sigma_{nom}} = 1 + 2\sqrt{\frac{a}{r}} \qquad (1.90)$$

Eq. (1.90) is the same as found in Chapter 4 (Eq. 4.58) for the case of a single elliptical hole in an infinite element in uniaxial tension. It is not difficult to apply Eq. (1.89) to other cases.

Example 1.8   An Element with a Circular Hole with Opposing Semicircular Lobes
Find the stress concentration factor of an element with a hole of diameter d and opposing semicircular lobes of radius r as shown in Fig. 1.37, which is under uniaxial tensile stress σ. Use known stress intensity factors. Suppose that $a/H = 0.1, r/d = 0.1$.

For this problem, choose the stress intensity factor for the case of radial cracks emanating from a circular hole in a rectangular panel as shown in Fig. 1.38. From Sih (1973) it is found that $C = 1.0249$ when $a/H = 0.1$. The crack length is $a = d/2 + r$ and $r/d = 0.1$, so

$$\frac{a}{r} = \frac{d/2 + r}{r} = 1 + \frac{d}{2r} = 1 + \frac{1}{2 \times 0.1} = 6 \tag{1}$$

Substitute $C = 1.0249$ and $a/r = 6$ into Eq. (1.89),

$$K_t = 1 + 2 \cdot 1.0249 \cdot \sqrt{6} = 6.02 \tag{2}$$

The stress concentration factor for this case also can be found from Chart 4.61. Corresponding to $a/H = 0.1, r/d = 0.1$, the stress concentration factor based on the net area is

$$K_{tn} = 4.80 \tag{3}$$

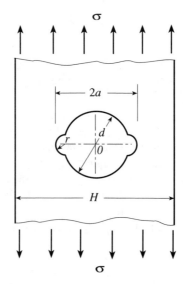

Figure 1.37   Element with a circular hole with two opposing semicircular lobes.

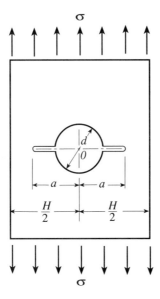

Figure 1.38 Element with a circular hole and a pair of equal length cracks.

The stress concentration factor based on the gross area is (Example 1.1)

$$K_{tg} = \frac{K_{tn}}{1 - (d/H)} = \frac{4.80}{1 - 0.2} = 6.00 \tag{4}$$

The results of (2) and (4) are very close.

Further results are listed below. It would appear that this kind of approximation is reasonable.

H	r/d	K_t from Eq. (1.89)	K_{tg} from Chart 4.61	% Difference
0.2	0.05	7.67	7.12	7.6
0.2	0.25	4.49	4.6	−2.4
0.4	0.1	6.02	6.00	0.33
0.6	0.1	6.2	6.00	.3
0.6	0.25	4.67	4.7	−0.6

Shin et. al. (1994) compared the use of Eq. (1.89) with the stress concentration factors obtained from handbooks and the finite element method. The conclusion is that in the range of practical engineering geometries where the notch tip is not too close to the boundary line

TABLE 1.2 Stress Concentration Factors for the Configurations of Fig. 1.39

a/l	a/r	e/f	C	K_t	K_t from Eq (1.89)	Discrepancy (90%)
0.34	87.1	0.556	0.9	17.84	17.80	−0.2
0.34	49	0.556	0.9	13.38	13.60	1.6
0.34	25	0.556	0.9	9.67	10.00	3.4
0.34	8.87	0.556	0.9	6.24	6.36	1.9
0.114	0.113	1.8	1.01	1.78	1.68	−6.0

Sources: Values for C from Shin et al. (1994); values for K_t from Murakami (1987.)

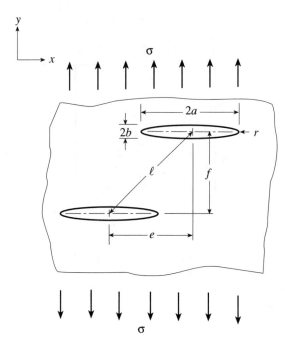

Figure 1.39 Infinite element with two identical ellipses that are not aligned in the y direction.

of the element, the discrepancy is normally within 10%. Table 1.2 provides a comparison for a case in which two identical parallel ellipses in an infinite element are not aligned in the axial loading direction (Fig. 1.39).

REFERENCES

ASTM, 1994, *Annual Book of ASTM Standards*, Vol. 03.01, ASTM, Philadelphia, PA.

Boresi, A. P., Schmidt, R. J., and Sidebottom, O. M., 1993, *Advanced Mechanics of Materials*, 5th ed., Wiley, New York.

Cox, H. L., and Sopwith, D. G., 1937, "The Effect of Orientation on Stresses in Single Crystals and of Random Orientation on the Strength of Polycrystalline Aggregates," *Proc. Phys. Soc.*, London, Vol. 49, p. 134.

Creager, M., and Paris, P. C., 1967, "Elastic Field Equations for Blunt Cracks with Reference to Stress Corrosion Cracking," *Int. J. Fract. Mech.*, Vol. 3, pp. 247–252.

Davis, E. A., and Manjoine, M. J., 1952, "Effect of Notch Geometry on Rupture Strength at Elevated Temperature," *Proc. ASTM*, Vol. 52.

Davies, V. C., 1935, discussion based on theses of S. K. Nimhanmimie and W. J. Huitt (Battersea Polytechnic), *Proc. Inst. Mech. Engrs.*, London, Vol. 131, p. 66.

Draffin, J. O., and Collins, W. L., 1938, "Effect of Size and Type of specimens on the Torsional Properties of Cast Iron," Proc. ASTM, Vol. 38, p. 235.

Durelli, A. J., 1982, "Stress Concentrations" Office of Naval Research, Washington, D.C., U.M. Project No. SF-CARS, School of Engineering, University of Maryland.

Eichinger, A., 1926, "Versuche zur Klärung der Frage der Bruchgefahr," *Proc. 2nd Intern. Congr. Appl. Mech.*, Zurich, p. 325.

Findley, W. N., 1951, discussion of "Engineering Steels under Combined Cyclic and Static Stresses" by Gough, H. J., 1949, *Trans. ASME, Applied Mechanics Section,* Vol. 73, p.211.

Fralich, R. W., 1959, "Experimental Investigation of Effects of Random Loading on the Fatigue Life of Notched Cantilever Beam Specimens of 7075-T6 Aluminum Alloy," NASA Memo 4-12-59L.

Gough, H. J., 1933, "Crystalline Structure in Relation to Failure of Metals," *Proc. ASTM*, Vol.33, Part 2, p. 3.

Gough, H. J., and Pollard, H. V., 1935, "Strength of Materials under Combined Alternating Stress," *Proc. Inst. Mech.* Engrs. London, Vol. 131, p. 1, Vol. 132, p. 549.

Gough, H. J., and Clenshaw, W. J., 1951, "Some Experiments on the Resistance of Metals to Fatigue under Combined Stresses," *Aeronaut. Research Counc. Repts. Memoranda 2522*, London, H. M. Stationery Office.

Gunn, N. J. F., 1952, "Fatigue Properties at Low Temperature on Transverse and Longitudinal Notched Specimens of DTD363A Aluminum Alloy," *Tech. Note Met.* 163, Royal Aircraft Establishment, Farnborough, England.

Hencky, H., 1924, "Zur Theorie Plastischer Deformationen und der hierdurch im Material hervorgerufenen Nebenspannungen," *Proc. 1st Intern. Congr. Appl. Mech.*, Delft, p. 312.

Hohenemser, K., and Prager, W., 1933, "Zur Frage der Ermüdungsfestigkeit bei mehrachsigen Spannungsuständen," *Metall*, Vol. 12, p. 342.

Houdremont, R., and Bennek, H., 1932, "Federstähle," *Stahl u. Eisen*, Vol. 52, p. 660.

Howland, R. C. J., 1930, "On the Stresses in the Neighborhood of a Circular hole in a Strip Under Tension," *Transactions, Royal Society of London, Series A*, Vol. 229, p. 67.

Ku, Ta-Cheng, 1960, "Stress Concentration in a Rotating Disk with a Central Hole and Two Additional Symmetrically Located Holes," *J. Appl. Mech.* Vol. 27, Ser. E, No.2, pp. 345–360.

Kuhn, P., and Hardrath, H. F., 1952, "An Engineering Method for Estimating Notch-Size Effect in Fatigue Tests of Steel," *NACA Tech. Note* 2805.

Lazan, B. J., and Blatherwick, A. A., 1952, "Fatigue Properties of Aluminum Alloys at Various Direct Stress Ratios," *WADC TR 52-306 Part I*, Wright-Patterson Air Force Base, Dayton, Ohio.

Lazan, B. J., and Blatherwick, A. A., 1953, "Strength Properties of Rolled Aluminum Alloys under Various Combinations of Alternating and Mean Axial Stresses," *Proc. ASTM*, Vol. 53, p. 856.

Lea, F. C., and Budgen, H. P., 1926, "Combined Torsional and Repeated Bending Stresses," *Engineering*, London, Vol. 122, p. 242.

Leven, M. M., 1955, "Quantitative Three-Dimensional Photoelasticity," *Proc. SESA*, Vol. 12, No. 2, p. 167.

Little, R. W., 1973, *Elasticity*, Prentice-Hall, New Jersey, p. 160.

Ludwik, P., 1931, "Kerb-und Korrosionsdauerfestigkeit," *Metall*, Vol. 10, p. 705.

Marin, J., 1952, *Engineering Materials*, Prentice-Hall, New York.

Murakami, Y., 1987, *Stress Intensity Factor Handbook*, Pergamon Press, New York.

Nadai, A., 1937, "Plastic Behavior of Metals in the Strain Hardening Range," *J. Appl. Phys.*, Vol. 8, p. 203.

Neuber, H., 1958, *Kerbspannungslehre*, 2nd ed. in German, Springer, Berlin (*Theory of Notch Stresses*, English Translation by Office of Technical Services, Dept. of Commerce, Washington, D.C., 1961, p.207).

Neuber, H., 1968, "Theoretical Determination of Fatigue Strength at Stress Concentration," *Report AFML-TR-68-20* Air Force Materials Lab, Wright-Patterson Air Force Base Dayton, Ohio.

Newmark, N. M., Mosborg, R. J., Munse, W. H., and Elling, R. E., 1951, "Fatigue Tests in Axial Compression," *Proc. ASTM*, Vol. 51, p. 792.

Newton, R. E., 1940, "A Photoelastic study of stresses in Rotating Disks," *J. Applied Mechanics*, Vol. 7, p. 57.

Nichols, R. W., Ed., 1969, Chapter 3, *A Manual of Pressure Vessel Technology*, Elsevier, London.

Nishida, M, 1976, *Stress Concentration*, Mori Kita Press, Tokyo, in Japanese.

Nisihara, T., and Kojima, K., 1939, "Diagram of Endurance Limit of Duralumin For Repeated Tension and Compression," *Trans. Soc. Mech. Engrs. Japan*, Vol. 5, No. 20, p. I-1.

Nisihara, T., and Kawamoto, A., 1940, "The Strength of Metals under Combined Alternating Stresses," *Trans. Soc. Mech. Engrs. Japan*, Vol. 6, No. 24, p. s-2.

Ono, A., 1921, "Fatigue of Steel under Combined Bending and Torsion," *Mem. Coll. Eng. Kyushu Imp. Univ.*, Vol. 2 No. 2.

Ono, A., 1929, "Some results of Fatigue Tests of Metals," *J. Soc. Mech. Engrs. Japan*, Vol. 32, p. 331.

Peterson, R. E., 1933a, "Stress Concentration Phenomena in Fatigue of Metals," *Trans. ASME, Applied Mechanics Section*, Vol. 55, p. 157.

Peterson, R. E., 1933b, "Model Testing as Applied to Strength of Materials," *Trans. ASME, Applied Mechanics Section*, Vol. 55, p. 79.

Peterson, R. E. and Wahl, A. M., 1936, "Two and Three Dimensional Cases of Stress Concentration, and Comparison with Fatigue Tests," *Trans. ASME, Applied Mechanics Section*, Vol. 57, p. A-15.

Peterson, R. E., 1938, "Methods of Correlating Data from Fatigue Tests of Stress Concentration Specimens," *Stephen Timoshenko Anniversary Volume*, Macmillan, New York, p. 179.

Peterson, R. E., 1943, "Application of Stress Concentration Factors in Design," *Proc. Soc. Exp. Stress Analysis*, Vol. 1, No. 1, p. 118.

Peterson, R. E., 1945, "Relation between Life Testing and Conventional Tests of Materials," *ASTM Bull*, p. 13.

Peterson, R. E., 1950, "Relation between Stress Analysis and Fatigue of Metals," *Proc. Soc. Exp. Stress Analysis*, Vol. 11, No. 2, p. 199.

Peterson, R. E., 1952, "Brittle Fracture and Fatigue in Machinery," *Fatigue and Fracture of Metals*, Wiley, New York, p. 74

Peterson, R. E., 1953, *Stress Concentration Design Factors*, Wiley, New York.

Peterson, R. E., 1959, "Analytical Approach to Stress Concentration Effect in Aircraft Materials," *U.S. Air Force-WADC Symposium on Fatigue of Metals, Technical Report 59-507*, Dayton, Ohio, p. 273.

Peterson, R. E., 1974, *Stress Concentration Factors*, Wiley, New York.

Pilkey, W. D., 1994, *Formulas For Stress, Strain, and Structural Matrices*, Wiley, New York

Pilkey, W. D., and Wunderlich, W., 1993, *Mechanics of Structures, Variational and Computational Methods*, CRC Press, Boca Raton, Florida.

Prager, W., and Hodge, P. G., 1951, *Theory of Perfectly Plastic Solids*, Wiley, New York.

Roark, R. J., Hartenberg, R. S., and Williams, R. Z., 1938, "The Influence of Form and Scale on Strength," *Univ. Wisconsin Expt. Sta. Bull.* 84.

Rôs, M., and Eichinger, A., 1950, "Die Bruchgefahr fester Körper," *Eidgenöss. Materialprüf. Ber.*, Vol. 173, Zurich.

Sachs, G., 1928, "Zur Ableitung einer Fliessbedingung," *Z. VDI*, Vol. 72, p. 734

Shin, C. S., Man, K. C., and Wang, C. M., 1994, "A Practical Method to Estimate the Stress Concentration of Notches," *Intern. J. of Fatigue*, Vol. 16, No.4, pp. 242–256.

Sih, G. C., 1973, *Handbook of Stress Intensity Factors*, Lehigh University, Bethlehem, Pennsylvania.

Smith, J. O., 1942, "The Effect of Range of Stress on the Fatigue Strength of Metals," *Univ. Illinois Expt. Sta. Bull.* 334

Soderberg, C. R., 1930, "Working Stress," *Trans. ASME*, Vol. 52, part 1, p. APM 52-2.

Steele, M. C., Liu, C.K., and Smith, J. O., 1952, "Critical Review and Interpretation of the Literature on Plastic (Inelastic) Behavior of Engineering Metallic Materials," *Research Report of Dept. of Theoretical and Applied Mechanics*, Univ. of Illinois, Urbana, Ill.

Templin, R. L., 1954, "Fatigue of Aluminum," *Proc. ASTM*, Vol. 54, p. 641.

Timoshenko, S., and Goodier, J. N., 1970, *Theory of Elasticity*, McGraw-Hill, New York.

Van den Broek, J. A., 1942, *Theory of Limit Design*, Wiley, New York.

Von Mises, R., 1913, "Mechanik der festen Körper im plastisch deformablen Zustand," *Nachr. Ges. Wiss. Göttingen Jahresber. Geschäftsjahr. Math-phys. Kl.*, p. 582

Von Phillipp, H. A., 1942, "Einfluss von Querschittsgrösse und Querschittsform auf die Dauerfestigkeit bei ungleichmässig verteilten Spannungen," *Forschung*, Vol. 13, p. 99.

Sources of Stress Concentration Factors

Neuber, H., 1958, *Theory of Notch Stresses*, 2nd ed., Springer, Berlin (English Translation by Office of Technical Services, Dept. of Commerce, Washington, D.C., 1961)

Nishida, M, 1976, *Stress Concentration*, Mori Kita Press, Tokyo, in Japanese.

Pilkey, W. D., 1994, *Formulas For Stress, Strain, and Structural Matrices*, Wiley, New York

Savin, G. N., 1961, *Stress Concentration Around Holes*, Pergamon, London (English Translation Editor, Johnson, W.)

Savin, G. N., and Tulchii, V. I., 1976, *Handbook on Stress Concentration*, Higher Education Publishing House. (Chinese translation, Heilongjiang Science and Technology Press, Harbin.)

Young, W. C., 1989, *ROARK'S Formulas for Stress & Strain*, 6th ed., McGraw-Hill, New York.

CHAPTER 2

NOTCHES AND GROOVES

2.1 NOTATION

Definition:
  Panel. A thin flat element with in-plane loading. This is a sheet that is sometimes referred to as a membrane.
Symbols:

- a = width of a notch or semimajor axis of an ellipse
- b = distance between notch centers or semiminor axis of an ellipse
- c = distance between notch centers
- d = minimum diameter (for three-dimensional) or minimum width (for two-dimensional) of member
- D = diameter of member
- h_o = minimum thickness of a thin element
- h = thickness of a thin element
- H = height or width of member
- K_t = stress concentration factor
- K_{tg} = stress concentration factor with the nominal stress based on gross area
- K_{tn} = stress concentration factor with the nominal stress based on net area
- L = length of member
- M = bending moment
- m = bending moment per unit length
- P = total applied in-plane force
- r = radius of a notch, notch bottom radius

T = torque

t = depth of a notch

α = open angle of notch

ν = Poisson's ratio

σ = normal stress

τ = shear stress

2.2 STRESS CONCENTRATION FACTORS

The U-shaped notch or circumferential groove is a geometrical shape of considerable interest in engineering. It occurs in machine elements such as in turbine rotors between blade rows and at seals. Other examples include in a variety of shafts (Fig. 2.1) as a shoulder relief groove or as a retainer for a spring washer.

The round-bottomed V-shaped notch or circumferential groove, and to a lesser extent the U-shaped notch, is a conventional contour shape for stress concentration test pieces in the areas of fatigue, creep-rupture, and brittle fracture. A threaded part may be considered an example of a multigrooved member.

As mentioned in Chapter 1, two basic K_t factors may be defined: K_{tg}, based on the larger (gross) section of width H and K_{tn}, based on the smaller (net) section of width d, see sketch and definitions on Chart 2.3. For tension (Chart 2.3), $K_{tg} = \sigma_{max}/\sigma$, where $\sigma = P/hH$ and $K_{tn} = \sigma_{max}/\sigma_{nom}$, where $\sigma_{nom} = P/hd$. Since design calculations are usually based on the critical section (width d) where σ_{max} is located, K_{tn} is the generally used factor. Unless otherwise specified, K_t refers to K_{tn} in this chapter.

Neuber (1958) found the theoretical stress concentration factors for the deep hyperbolic notch (Fig. 2.2a) and the shallow elliptical notch (Fig. 2.2b) in infinitely wide members under tension, bending and shear. These results will be given in this chapter. For finite width members, Neuber introduced the following ingenious simple relation for notches

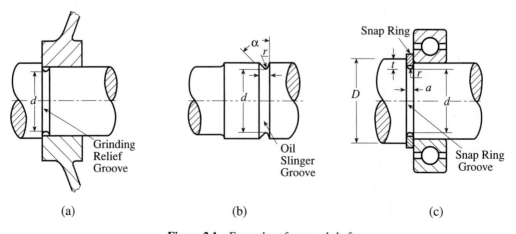

Figure 2.1 Examples of grooved shafts.

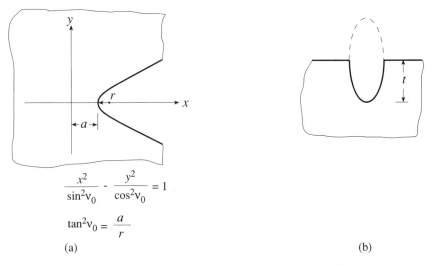

Figure 2.2 Notches: (*a*) Deep hyperbolic; (*b*) shallow elliptical.

with arbitrary shapes:

$$K_{tn} = 1 + \sqrt{\frac{(K_{te} - 1)^2 (K_{th} - 1)^2}{(K_{te} - 1)^2 + (K_{th} - 1)^2}} \qquad (2.1)$$

where K_{te} is the stress concentration factor for a shallow elliptical notch (with the same t/r as for the notch in the finite width member) in a semi-infinitely wide member, K_{th} is the stress concentration factor for a deep hyperbolic notch (with the same r/d as for the notch in the finite width member) in an infinitely wide member, t is the notch depth, r is the notch radius (minimum contour radius), d is a minimum diameter (three-dimensional) or minimum width (two-dimensional) of the member.

Consider, for example, a flat bar with opposite notches of arbitrary shape, with notch depth t, notch radius r, and a minimum bar width d. If K_{te} is the stress concentration factor of a shallow elliptical notch of Chart 2.2 with the same t/r and K_{th} is the stress concentration factor of a deep hyperbolic notch of Chart 2.1 with the same r/d, then K_{tn} of Eq. (2.1) is an estimate of the stress concentration factor of the flat bar with opposite notches.

Equation (2.1) is not exact. Recent investigations have provided more accurate values for the parameter ranges covered by the investigations, as will be presented in the following sections. If the actual member being designed has a notch or groove that is either very deep or shallow, the Neuber approximation will be close. However, for values of d/H in the region of $1/2$, the Neuber K_t can be as much as 12% too low, which is on the unsafe side from a design standpoint. More accurate values have been obtained over the most used ranges of parameters. These form the basis of some of the charts presented here. However, when a value for an extreme condition such as a very small or large r/d is sought, the Neuber method is the only means of obtaining a useful factor. Some charts covering the extreme ranges are also included in this book.

Another use of the charts of Neuber factors is in designing a test piece for maximum K_t, as detailed in Section 2.10.

The K_t factors for the flat members covered in this chapter are for two-dimensional states of stress (plane stresses) and apply strictly to very thin sheets, or more strictly to where $h/r \to 0$, where h = element thickness and r = notch radius. As h/r increases, a state of plane strain is approached, in which case the stress at the notch surface at the middle of the element thickness increases and the stress at the element surface decreases. Some guidance may be obtained by referring to the introductory remarks at the beginning of Chapter 4.

The K_t factor for a notch can be lowered by use of a reinforcing bead (Suzuki 1967).

2.3 NOTCHES IN TENSION

2.3.1 Opposite Deep Hyperbolic Notches in an Infinite Thin Element; Shallow Elliptical, Semicircular, U-Shaped or Keyhole-Shaped Notches in Semi-Infinite Thin Elements; Equivalent Elliptical Notch

In Chart 2.1, K_{tn} values are given for the deep $(d/H \to 0)$ hyperbolic notch in an infinite thin element (Neuber 1958; Peterson 1953). Chart 2.2 provides K_{tg} values for an elliptical or U-shaped notch in a semi-infinite thin element (Seika 1960; Bowie 1966; Barrata and Neal 1970). For the higher t/r values, K_{tg} for the U-notch is up to 1% higher than for the elliptical notch. For practical purposes the solid curve of Chart 2.2 covers both cases.

The semicircular notch $(t/r = 1)$ in a semi-infinite thin element has been studied by a number of investigators. Ling (1967) has provided the following summary of K_t factors:

1936	Maunsell	3.05
1940	Isibasi	3.06
1941	Weinel	3.063
1948	Ling	3.065
1965	Yeung	3.06
1965	Mitchell	3.08

Similar to the "equivalent elliptical hole" in an infinite panel (see Section 4.5.1), an "equivalent elliptical notch" in a semi-infinite thin element may be defined as an elliptical notch that has the same t/r and envelops the notch geometry under consideration. All such notches, U-shaped, keyhole (circular hole connected to edge by saw cut), and so on have very nearly the same K_t as the equivalent elliptical notch. The "equivalent elliptical notch" applies for tension. It is not applicable for shear.

Stress concentration factors have been approximated by splitting a thin element with a central hole axially through the middle of the hole (Fig. 2.3) and using the K_t for the hole to represent the resulting notches. From Eq. (1.90) the stress concentration factor for an elliptical hole in a thin element is

$$K_t = 1 + 2\sqrt{\frac{t}{r}} \tag{2.2}$$

where t is the semiaxis which is perpendicular to the tensile loading. Chart 2.2 shows this K_t also. The factors for the U-shaped slot (Isida 1955) are practically the same. A comparison of the curves for notches and holes in Chart 2.2 shows that the preceding approximation can be in error by as much as 10% for the larger values of t/r.

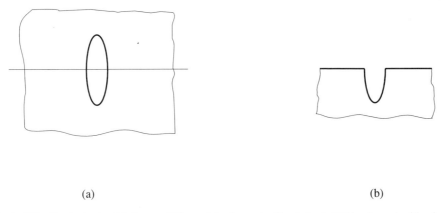

Figure 2.3 Equivalent notch from splitting a thin element with a hole: (*a*) Thin element with a hole; (*b*) Half of thin element with a notch.

2.3.2 Opposite Single Semicircular Notches in a Finite-Width Thin Element

For the tension case of opposite semicircular notches in a finite-width thin element, K_{tg} and K_{tn} factors (Isida 1953; Ling 1968; Appl and Koerner 1968; Hooke 1968) are given in Chart 2.3. The difference between K_{tg} and K_{tn} is illustrated in Chart 2.3.

Consider a bar of constant width H and a constant force P. As notches are cut deeper (increasing $2r/H$), K_{tn} decreases, reflecting a decreasing stress concentration (peak stress divided by average stress across d). This continues until $2r/H \rightarrow 1$, in effect a uniform stress tension specimen. Factor K_{tg} increases as $2r/H$ increases, reflecting the increase in σ_{max} owing to the loss of section. Slot (1972) found that with $2r/H = 1/2$ in a strip of length $1.5H$, good agreement was obtained with the stress distribution for σ applied at infinity.

2.3.3 Opposite Single U-Shaped Notches in a Finite-Width Thin Element

Strain gage tests (Kikukawa 1962), photoelastic tests (Flynn and Roll 1966), and mathematical analysis (Appl and Koerner 1969) provide consistent data for the opposite U-shaped notches in a flat bar (two-dimensional case) of Chart 2.4. An important check is provided by including in Chart 2.4 the curve representing mathematical results (Isida 1953; Ling 1968) for the semicircular notch (Chart 2.3), a special case of the U-notch. The agreement is excellent for values of $H/d \leq 2$. Photoelastic results (Wilson and White 1973) for $H/d = 1.05$ are also in good agreement.

Barrata (1972) has compared empirical formulas for K_{tn} with experimentally determined values and concludes that the following two formulas are satisfactory for predictive use.

Barrata and Neal (1970):

$$K_{tn} = \left(0.780 + 2.243\sqrt{\frac{t}{r}}\right)\left[0.993 + 0.180\left(\frac{2t}{H}\right)\right. \tag{2.3}$$
$$\left. -1.060\left(\frac{2t}{H}\right)^2 + 1.710\left(\frac{2t}{H}\right)^3\right]\left(1 - \frac{2t}{H}\right)$$

64 NOTCHES AND GROOVES

Heywood (1952):

$$K_{tn} = 1 + \left[\frac{t/r}{1.55\,(H/d) - 1.3} \right]^n \qquad (2.4)$$

$$n = \frac{H/d - 1 + 0.5\sqrt{t/r}}{H/d - 1 + \sqrt{t/r}}$$

with t the depth of a notch, $t = (H - d)/2$.

Referring to Chart 2.4, Eq. (2.3) gives values in good agreement with the solid curves for $r/d < 0.25$. Equation (2.4) is in better agreement for $r/d > 0.25$. For the dashed curves (not the dot-dash curve for semicircular notches), Eq. (2.3) gives lower values as r/d increases. The tests on which the formulas are based do not include parameter values corresponding to the dashed curves, which are uncertain owing to their determination by interpolation of r/d curves having H/d as abscissae. In the absence of better basic data, the dashed curves, representing higher values, should be used for design.

In Chart 2.4 the values of r/d are from 0 to 0.3, and the values of H/d are from 1 to 2, covering the most widely used parameter ranges. There is considerable evidence (Kikukawa 1962; Flynn and Roll 1966; Appl and Koerner 1969) that for greater values of r/d and H/d, the K_t versus H/d curve for a given r/d does not flatten out but reaches a peak value and then decreases slowly toward a somewhat lower K_t value as $H/d \to \infty$. The effect is small and is not shown on Chart 2.4.

In Chart 2.4 the range of parameters corresponds to the investigations of Kikukawa (1962), Flynn and Roll (1966), and Appl and Koerner (1969). For smaller and larger r/d values, the Neuber values (Eq. 2.1, Charts 2.5 and 2.6), although approximate, are the only wide-range values available and are useful for certain problems. The largest errors are at the midregion of d/H. For shallow or deep notches the error becomes progressively smaller. Some specific photoelastic tests (Liebowitz et al. 1967) with $d/H \approx 0.85$ and r/H varying from ≈ 0.001 to 0.02 gave higher K_{tn} values than does Chart 2.5.

2.3.4 "Finite-Width Correction Factors" for Opposite Narrow Single Elliptical Notches in a Finite-Width Thin Element

For the very narrow elliptical notch, approaching a crack, "finite-width correction" formulas have been proposed by Dixon (1962), Westergaard (1939), Irwin (1958), Bowie (1963), Brown and Strawley (1966), and Koiter (1965). Plots for opposite narrow edge notches are given in Peterson (1974).

The formula (Barrata and Neal 1970; Brown and Strawley 1966), based on Bowie's results for opposite narrow elliptical notches in a finite width thin element (Fig. 2.4), is satisfactory for values of $2t/H < 0.5$:

$$\frac{K_{tg}}{K_{t\infty}} = 0.993 + 0.180 \left(\frac{2t}{H} \right) - 0.160 \left(\frac{2t}{H} \right)^2 + 1.710 \left(\frac{2t}{H} \right)^3 \qquad (2.5)$$

where t is the crack length and $K_{t\infty}$ is K_t for $H = \infty$.

The following Koiter (1965) formula covers the entire $2t/H$ range. For the lower $2t/H$ range agreement with Eq. (2.5) is good. For the mid-$2t/H$ range, somewhat higher values

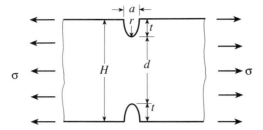

Figure 2.4 Opposite narrow edge notch.

than the Eq. (2.5) values are obtained.

$$\frac{K_{tg}}{K_{t\infty}} = \left[1 - 0.50\left(\frac{2t}{H}\right) - 0.0134\left(\frac{2t}{H}\right)^2 + 0.081\left(\frac{2t}{H}\right)^3\right]\left[1 - \frac{2t}{H}\right]^{-1/2} \quad (2.6)$$

These gross area factors K_{tg} are related to the net area factors K_{tn} by

$$\frac{K_{tn}}{K_{t\infty}} = \frac{K_{tg}}{K_{t\infty}}\left(1 - \frac{2t}{H}\right) \quad (2.7)$$

2.3.5 Opposite Single V-Shaped Notches in a Finite-Width Thin Element

Stress concentration factors $K_{t\alpha}$ have been obtained (Appl and Koerner 1969) for the flat tension bar with opposite V notches as a function of the V angle, α (Chart 2.7). The Leven-Frocht (1953) method of relating $K_{t\alpha}$ to the K_{tu} of a corresponding U notch as used in Chart 2.7, shows that for $H/d = 1.66$ the angle has little effect up to 90°. For $H/d = 3$ it has little effect up to 60°. In comparing these results with Chart 2.28, where the highest $H/d = 1.82$, the agreement is good, even though the two cases are different (symmetrical notches, in tension [Chart 2.7]; notch on one side, in bending [Chart 2.28]).

2.3.6 Single Notch on One Side of a Thin Element

Neuber (1958) has obtained approximate K_{tn} values for a semi-infinite thin element with a deep hyperbolic notch, wherein tension loading is applied along a midline through the minimum section (Chart 2.8). Chart 2.9 presents K_{tn} curves based on photoelastic tests (Cole and Braun 1958). Corresponding Neuber K_{tn} factors obtained by use of Chart 2.8 and of Eq. (2.1) are on the average 18% lower than the K_{tn} factors of Chart 2.9.

The curve for the semicircular notch is obtained by noting that for this case $r = H - d$ or $H/d = 1 + r/d$ and that $K_{tn} = 3.065$ at $r/d \to 0$.

2.3.7 Notches with Flat Bottoms

Chart 2.10 gives stress concentration factors K_{tn} for opposing notches in finite width thin elements, with flat bottoms (Neuber 1958; Hetényi and Liu 1956; Sobey 1965).

Chart 2.11 provides factors for a rectangular notch on the edge of a wide (semi-infinite) flat panel in tension (Rubenchik 1975; ESDU 1981). The maximum stress $\sigma_{\max}$ occurs at points A of the figure in Chart 2.11. When $a = 2r$, the notch base is semicircular, and two points A coincide at the base of the notch.

2.3.8 Multiple Notches in a Thin Element

It has long been recognized that a single notch represents a higher degree of stress concentration than a series of closely spaced notches of the same kind as the single notch. Considered from the standpoint of flow analogy, a smoother flow is obtained in Figs. 2.5*b* and 2.5*c*, than in Fig. 2.5*a*.

For the infinite row of semicircular edge notches, factors have been obtained mathematically (Atsumi 1958) as a function of notch spacing and the relative width of a bar, with results summarized in Charts 2.12 and 2.13. For infinite notch spacing, the K_{tn} factors are in agreement with the single-notch factors of Isida (1953) and Ling (1968), Chart 2.3.

For a specific case (Slot 1972) with $r/H = 1/4$ and $b/a = 3$, good agreement was obtained with the corresponding Atsumi (1958) value.

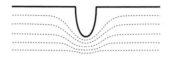

(a)

(b)

(c)

Figure 2.5 Multiple notches.

An analysis (Weber 1942) of a semi-infinite panel with an edge of wave form of depth t and minimum radius r gives $K_{tn} = 2.13$ for $t/r = 1$ and $b/a = 2$, which is in agreement with Chart 2.12.

Stress concentration factors K_{tn} are available for the case of an infinite row of circular holes in a panel stressed in tension in the direction of the row (Weber 1942; Schulz 1942, 1943–45). If we consider the panel as split along the axis of the holes, the K_{tn} values should be nearly the same (for the single hole, $K_{tn} = 3$; for the single notch, $K_{tn} = 3.065$). The K_{tn} curve for the holes as a function of b/a fits well (slightly below) the top curve of Chart 2.12.

For a finite number of multiple notches (Fig. 2.5b), the stress concentration of the intermediate notches is considerably reduced. The maximum stress concentration occurs at the end notches (Charts 2.14 and 2.15, Durelli et al. 1952), and this is also reduced (as compared to a single notch) but to a lesser degree than for the intermediate notches. Sometimes a member can be designed as in Fig. 2.5c, resulting in a reduction of stress concentration as compared to Fig. 2.5b.

Factors for two pairs of notches (Atsumi 1967) in a square pattern ($b/H = 1$) are included in Chart 2.13.

Photoelastic tests have been made for various numbers (up to six) of semicircular notches (Durelli et al. 1952; Hetényi 1943). The results (Charts 2.14, 2.15, and 2.16) are consistent with the recent mathematical factors for the infinite row (Atsumi 1958). Chart 2.16 provides, for comparison, K_{tg} for a groove that corresponds to a lower K_{tg} limit for any number of semicircular notches with overall edge dimension c.

For the Aero thread shape (semicircular notches with $b/a = 1.33$), two-dimensional photoelastic tests (Hetényi 1943) of six notches gave K_t values of 1.94 for the intermediate notch and 2.36 for the end notch. For the Whitworth thread shape (V notch with rounded bottom), the corresponding photoelastic tests (Hetényi 1943) gave K_t values of 3.35 and 4.43, respectively.

Fatigue tests (Moore 1926) of threaded specimens and specimens having a single groove of the same dimensions showed considerably higher strength for the threaded specimens.

2.4 DEPRESSIONS IN TENSION

2.4.1 Hemispherical Depression (Pit) in the Surface of a Semi-infinite Body

For a semi-infinite body with a hemispherical depression under equal biaxial stress (Fig. 2.6), $K_t = 2.23$ was found (Eubanks 1954) for Poisson's ratio $\nu = 1/4$. This is about 7% higher than for the corresponding case of a spherical cavity ($K_t = 2.09$ for $\nu = 1/4$), (Chart 4.73). Moreover the semicircular edge notch in tension ($K_t = 3.065$) (Chart 2.2) is about 2% higher than the circular hole in tension ($K_t = 3$) (Chart 4.1 and Eq. 4.18).

2.4.2 Hyperboloid Depression (Pit) in the Surface of a Finite-Thickness Element

The hyperboloid depression simulates the type of pit caused by meteoroid impact of an aluminum panel (Denardo 1968). For equal biaxial stress (Fig. 2.7) values of K_t in the range of 3.4 to 3.8 were obtained for individual specific geometries as reported by Reed and Wilcox (1970). The authors point out that the K_t values are higher than for complete penetration in the form of a circular hole ($K_t = 2$; see Eq. 4.17).

68 NOTCHES AND GROOVES

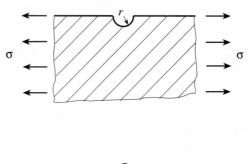

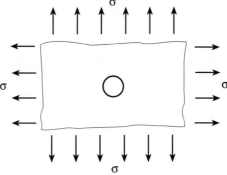

Figure 2.6 A semi-infinite body with a hemispherical depression under equal biaxial stress.

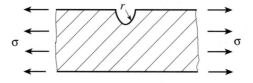

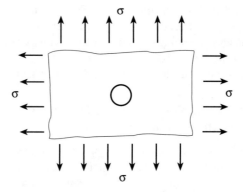

Figure 2.7 Hyperbolical depression in the surface of a finite thickness panel under equal biaxial stress.

2.4.3 Opposite Shallow Spherical Depressions (Dimples) in a Thin Element

The geometry of opposite dimples in a thin element has been suggested for a test piece in which a crack forming at the thinnest location can progress only into a region of lower stress (Cowper 1962). Dimpling is often used to remove a small surface defect. If the depth is small relative to the thickness (h_o/h approaching 1.0 in Chart 2.17), the stress increase is small. The $K_{tg} = \sigma_{\max}/\sigma$ values are shown for uniaxial stressing in Chart 2.17. These values also apply for equal biaxial stressing.

The calculated values of Chart 2.17 are for a shallow spherical depression having a diameter greater than four or five times the thickness of the element. In terms of the variables given in Chart 2.17, the spherical radius is

$$r = \frac{1}{4}\left[\frac{d^2}{(h - h_o)} + (h - h_o)\right] \qquad (2.8)$$

for $d \geq 5h$, $r/h_o > 25$. For such a relatively large radius, the stress increase for a thin section ($h_o/h \to 0$) is due to the thinness of section rather than stress concentration per se (i.e., stress gradient is not steep).

For comparison, for a groove having the same sectional contour as the dimple, K_{tg} is shown by the dashed line on Chart 2.17, the K_{tg} values being calculated from the K_{tn} values of Chart 2.6. Note that in Chart 2.6 the K_{tn} values for $r/d = 25$ represent a stress concentration of about 1%. The K_{tg} factors therefore essentially represent the loss of section.

Removal of a surface defect in a thick section by means of creating a relatively shallow spherical depression results in negligible stress concentration, on the order of 1%.

2.5 GROOVES IN TENSION

2.5.1 Deep Hyperbolic Groove in an Infinite Member (Circular Net Section)

Exact K_t values for Neuber's solution (Neuber 1958; Peterson 1953) are given in Chart 2.18 for a deep hyperbolic groove in an infinite member. Note that Poisson's ratio has only a relatively small effect.

2.5.2 U-Shaped Circumferential Groove in a Bar of Circular Cross Section

The K_{tn} values for a bar of circular cross section with a U groove (Chart 2.19) are obtained by multiplying the K_{tn} of Chart 2.4 by the ratio of the corresponding Neuber three-dimensional K_{tn} (Chart 2.18) over two-dimensional K_{tn} values (Chart 2.1). This is an approximation. However, after comparison with the bending and torsion cases, the results seem reasonable.

The maximum stress occurs at the bottom of the groove. Cheng (1970) has by a photoelastic test obtained $K_{tn} = 1.85$ for $r/d = 0.209$ and $D/d = 1.505$. The corresponding K_{tn} from Chart 2.19 is 1.92, which agrees fairly well with Cheng's value, which he believes to be somewhat low.

Approximate K_{tn} factors, based on the Neuber (1958) formula, are given in Chart 2.20 for smaller r/d values and in Chart 2.21 for larger r/d values (e.g., test specimens).

2.5.3 Flat-Bottom Grooves

Chart 2.22 gives K_{tn} for flat-bottom grooves under tension based on the Neuber formula (Peterson 1953; ESDU 1981).

2.6 BENDING OF THIN BEAMS WITH NOTCHES

2.6.1 Opposite Deep Hyperbolic Notches in an Infinite Thin Element

Exact K_{tn} values of Neuber's solution (Neuber 1958; Peterson 1953) are presented in Chart 2.23 for infinite thin elements subject to in-plane moments with opposite deep hyperbolic notches.

2.6.2 Opposite Semicircular Notches in a Flat Beam

Chart 2.24 provides K_{tn} values determined mathematically (Ling 1967; Isida 1953) for a thin beam element with opposite semicircular notches. Slot (1972) found that with $r/H = 1/4$ a strip of length $1.5H$, good agreement was obtained with the stress distribution for M applied at infinity.

2.6.3 Opposite U-Shaped Notches in a Flat Beam

The stress concentration factor K_{tn} for opposite U-shaped notches in a finite-width thin beam element is given in Chart 2.25. These curves are obtained by increasing the photoelastic K_{tn} values (Frocht 1935), which as in tension are known to be low, to agree with the semicircular notch mathematical values of Chart 2.24, which are assumed to be accurate. Photoelastic (Wilson and White 1973) and numerical results (Kitagawa and Nakade 1970) are in good agreement. Approximate K_{tn} values for extended r/d values are given in Charts 2.26 and 2.27.

2.6.4 V-Shaped Notches in a Flat Beam Element

The effect of notch angle on the stress concentration factors is presented in Chart 2.28 for a bar in bending with a V-shaped notch on one side (Leven and Frocht 1953). The K_{tn} value is for a U notch. $K_{t\alpha}$ is for a notch with inclined sides having an included angle α but with all other dimensions the same as for the corresponding U notch case. The curves of Chart 2.28 are based on data from specimens covering a H/d range up to 1.82. Any effect of H/d up to this value is sufficiently small that single α curves are adequate. For larger H/d values, the α curves may be lower (see Chart 2.7).

2.6.5 Notch on One Side of a Thin Beam

Chart 2.29 provides K_{tn} for bending of a semi-infinite thin element with a deep hyperbolic notch (Neuber 1958). In Chart 2.30a, K_{tn} curves based on photoelastic tests (Leven and Frocht 1953) are given. Corresponding Neuber K_{tn} factors obtained by use of Chart 2.29 and Eq. (2.1) are on the average 6% higher than the K_{tn} factors of Chart 2.30a.

The curve for the semicircular notch is obtained by noting that for this case $H/d = 1 + r/d$ and that $K_{tn} = 3.065$ at $r/d \to 0$.

In Chart 2.30b, finite-width correction factors are given for a bar with a notch on one side. The full curve represents a crack (Wilson 1970) and the dashed curved a semicircular notch (Leven and Frocht 1953). The correction factor for the crack is the ratio of the stress-intensity factors. In the small-radius, narrow-notch limit, the ratio is valid for stress concentration (Irwin 1960; Paris and Sih 1965). Note that the end points of the curves are 1.0 at $t/H = 0$ and $1/K_{t\infty}$ at $t/H = 1$. The $1/K_{t\infty}$ values at $t/H = 1$ for elliptical notches are obtained from $K_{t\infty}$ of Chart 2.2. These $1/K_t$ values at $t/H = 1$ are useful in sketching in approximate values for elliptical notches.

If $K_{tg}/K_{t\infty}$ (not shown in Chart 2.30b) is plotted, the curves start at 1.0 at $t/H = 0$, dip below 1.0, reach a minimum in the $t/H = 0.10$ to 0.15 range, and then turn upward to go to infinity at $t/H = 1.0$. This means that for bending the effect of the nominal stress gradient is to cause $\sigma_{\max}$ to decrease slightly as the notch is cut into the surface, but beyond a depth of $t/H = 0.25$ to 0.3 the maximum stress is greater than for the infinitely deep bar. The same effect, only of slight magnitude, was obtained by Isida (1953) for the bending case of a bar with opposite semicircular notches (Chart 2.24; K_{tg} not shown). In tension, since there is no nominal stress gradient, this effect is not obtained.

Chart 2.31 gives K_{tn} for various impact specimens.

2.6.6 Single or Multiple Notches with Semicircular or Semielliptical Notch Bottoms

From work on propellant grains, it is known (Tsao, Ching, and Okubo 1965) "that invariably the stress concentration factor for an optimized semielliptic notch is significantly lower than that for the more easily formed semicircular notch." Photoelastic tests (Tsao, Ching, and Okubo 1965; Nishioka and Hisemitsu 1962; Ching, Okabu, and Tsao 1968) were made on beams in bending, with variations of beam and notch depth, notch spacing, and semielliptical notch bottom shape. The ratio of beam depth to notch depth H/t (notch on one side only) varied from 2 to 10. Chart 2.32 provides results for $H/t = 5$.

For the single notch with $t/a = 2.666$, the ratio of K_{tn} for the semicircular bottom to the K_{tn} for the optimum semielliptical bottom, $a/b = 2.4$, is 1.25 (see Chart 2.32). In other words, a considerable stress reduction (20% in this case) results from using a semielliptical notch bottom instead of a semicircular notch bottom. As can be found from Chart 2.32, even larger stress reductions can be obtained for multiple notches.

Although these results are for a specific case of a beam in bending, it is reasonable to expect that, in general, a considerable stress reduction can be obtained by use of the semielliptic notch bottom. The optimum a/b of the semiellipse varies from 1.8 to greater than 3, with the single notch and the wider spaced multiple notches averaging at about 2 and the closer spaced notches increasing toward 3 and greater.

Other uses of the elliptical contour are found in Chart 4.59 for the slot end, where the optimum a/b is about 3 and in Chart 3.9, for the shoulder fillet.

2.6.7 Notches with Flat Bottoms

Chart 2.33 offers stress concentration factors K_{tn} for thin beams with opposite notches with flat bottoms (Neuber 1958; Sobey 1965). For a shaft with flat-bottom grooves in bending and/or tension, stress concentration factors K_{tn} are given in Chart 2.34 (Rubenchik 1975; ESDU 1981).

2.7 BENDING OF PLATES WITH NOTCHES

2.7.1 Various Edge Notches in an Infinite Plate in Transverse Bending

Stress concentration factors K_{tn} for opposite deep hyperbolic notches in a thin plate (Neuber 1958; Lee 1940) are given in Chart 2.35. The factors were obtained for transverse bending.

The bending of a semi-infinite plate with a V-shaped notch or a rectangular notch with rounded corners (Shioya 1959) is covered in Chart 2.36. At $r/t = 1$, both curves have the same K_{tn} value (semicircular notch). Note that the curve for the rectangular notch has a minimum K_{tn} at about $r/t = 1/2$.

In Chart 2.37, K_{tn} factors are given for the elliptical notch (Shioya 1960). For comparison, the corresponding curve for the tension case from Chart 2.2 is shown. This reveals that the tension K_{tn} factors are considerably higher.

In Chart 2.38 the K_t factor for an infinite row of semicircular notches is given as a function of the notch spacing (Shioya 1963). As the notch spacing increases, the K_t value for the single notch is approached asymptotically.

2.7.2 Notches in Finite-Width Plate in Transverse Bending

Approximate values have been obtained by the Neuber method (Neuber 1958; Peterson 1953) which makes use of the exact values for the deep hyperbolic notch (Lee 1940) and the shallow elliptical notch (Shioya 1960) in infinitely wide members and modifies these for finite-width members by using a second-power relation that has the correct end conditions. The results are shown for the thin plate in Chart 2.39.

No direct results are available for intermediate thicknesses. If we consider the tension case as representing maximum values for a thick plate in bending, we can use Chart 2.4 for $t/h \rightarrow 0$. For the thin plate ($t/h \rightarrow \infty$) use is made of Chart 2.39, as described in the preceding paragraph. (For intermediate thickness ratios, some guidance can be obtained from Chart 4.83; see also values on Chart 4.85 in the region of $b/a = 1$.)

2.8 BENDING OF SOLIDS WITH GROOVES

2.8.1 Deep Hyperbolic Groove in an Infinite Member

Stress concentration factors K_{tn} for Neuber's exact solution (Neuber 1958; Peterson 1953) are given in Chart 2.40 for the bending of an infinite three-dimensional solid with a deep hyperbolic groove. The net section on the groove plane is circular.

2.8.2 U-Shaped Circumferential Groove in a Bar of Circular Cross Section

The K_{tn} values of Chart 2.41 for a U-shaped circumferential groove in a bar of circular cross section are obtained by the method used in the tension case (see Section 2.5.2). Approximate K_{tn} factors for small r/d values are given in Chart 2.42 and for large r/d values (e.g., test specimens), in Chart 2.43.

Example 2.1 Design of a Shaft with a Circumferential Groove Suppose that we wish to estimate the bending fatigue strength of the shaft shown in Fig. 2.8 for two materials: an axle steel (normalized 0.40% C), and a heat-treated 3.5% nickel steel. These materials

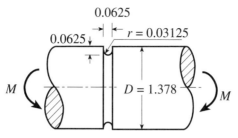

Figure 2.8 Grooved shaft.

will have fatigue strengths (endurance limits) of approximately 30,000 and 70,000 lb/in.^{2}, respectively, when tested in the conventional manner, with no stress concentration effects, in a rotating beam machine.

First we determine K_{tn}. From Fig. 2.8, $d = 1.378 - (2)0.0625 = 1.253$ in. and $r = 0.03125$ in. We calculate $D/d = 1.10$ and $r/d = 0.025$. From Chart 2.41 we find that $K_{tn} = 2.90$.

From Fig. 1.31 we obtain, for $r = 0.03125$ in., a q value of 0.76 for the axle steel and 0.93 for the heat-treated alloy steel.

Substituting in Eq. (1.53), for axle steel,

$$K_{tf} = 1 + 0.76(2.90 - 1) = 2.44 \tag{2.9}$$

$$\sigma_{tf} = \frac{30{,}000}{2.44} = 12{,}300 \text{ lb/in.}^2$$

for heat-treated alloy steel,

$$K_{tf} = 1 + 0.93(2.90 - 1) = 2.77 \tag{2.10}$$

$$\sigma_{tf} = \frac{70{,}000}{2.77} = 25{,}200 \text{ lb/in.}^2$$

This tells us that we can expect strength values of approximately 12,000 and 25,000 lb/in.^{2} under fatigue conditions for the shaft of Fig. 2.8 when the shaft is made of normalized axle steel and quenched-and-tempered alloy steel (as specified), respectively. These are not working stresses, since a factor of safety must be applied that depends on type of service, consequences of failure, and so on. Different factors of safety are used throughout industry depending on service and experience. The strength of a member, however, is not, in the same sense, a matter of opinion or judgment and should be estimated in accordance with the best methods available. Naturally a test of the member is desirable whenever possible. In any event, an initial calculation is made, and this should be done carefully and include all known factors.

2.8.3 Flat-Bottom Grooves in Bars of Circular Cross Section

Stress concentration factors for a bar of circular cross section with flat bottomed grooves are presented in Chart 2.44 (Peterson 1953; Sobey 1965; ESDU 1981).

2.9 DIRECT SHEAR AND TORSION

2.9.1 Deep Hyperbolic Notches in an Infinite Thin Element in Direct Shear

The stress concentration factors given in Chart 2.45 are from Neuber (1958). Shearing forces are applied to an infinite thin element with deep hyperbolic notches. These forces are parallel to the notch axis[1] as shown in Chart 2.45.

The location of $\sigma_{\max}$ is at

$$x = \frac{r}{\sqrt{1 + (2r/d)}} \qquad (2.11)$$

The location of $\tau_{\max}$ along the line corresponding to the minimum section is at

$$y = \frac{d}{2}\sqrt{\frac{(d/2r) - 2}{d/2r}} \qquad (2.12)$$

At the location of $\sigma_{\max}$, $K_{ts} = \tau_{\max}/\tau_{\text{nom}} = (\sigma_{\max}/2)/\tau_{\text{nom}} = K_t/2$, is greater than the K_{ts} value for the minimum section shown in Chart 2.45.

For combined shear and bending, Neuber (1958) shows that for large d/r values it is a good approximation to add the two K_t factors (Charts 2.35 and 2.45), even though the maxima do not occur at the same location along the notch surface. The case of a twisted sheet with hyperbolic notches has been analyzed by Lee (1940).

2.9.2 Deep Hyperbolic Groove in an Infinite Member

Stress concentration factors K_{ts} for Neuber's exact solution (Neuber 1958; Peterson 1953) are presented in Chart 2.46 for the torsion of an infinite three-dimensional solid with a deep hyperbolic groove. The net section is circular in the groove plane.

2.9.3 U-Shaped Circumferential Groove in a Bar of Circular Cross Section

Chart 2.47, for a U-shaped circumferential groove in a bar of circular cross section, is based on electric analog results (Rushton 1967), using a technique that has also provided results in agreement with the exact values for the hyperbolic notch in the parameter range of present interest. Mathematical results for semicircular grooves (Matthews and Hooke 1971; Hamada and Kitagawa 1968) are in reasonably good agreement with Chart 2.47. The K_{ts} values of Chart 2.47 are somewhat higher (average 4.5%) than the photoelastic values of Leven (1955). However, the photoelastic values are not in agreement with certain other published values (Okubo 1952, and 1953).

Chart 2.48 shows a leveling of the K_{ts} curve at a D/d value of about 2 or less for high r/d values. Approximate K_{ts} factors beyond the r/d range of Chart 2.47 are given for small r/d values in Chart 2.49 and for large r/d values (e.g., test specimens) in Chart 2.50.

Example 2.2 Analysis of a Circular Shaft with a U-Shaped Groove The circular shaft of Fig. 2.9 has a U-shaped groove, with $t = 10.5$ mm deep. The radius of the groove root

---

[1] For equilibrium, the shear force couple $2bV$ must be counterbalanced by an equal couple symmetrically applied remotely from the notch (Neuber 1958). To avoid possible confusion with the combined shear and bending case, the countercouple is not shown in Chart 2.45.

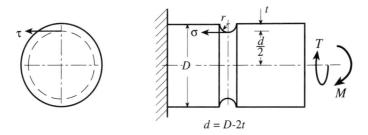

Figure 2.9 Shaft, with circumferential U-shaped groove, subject to torsion and bending.

is $r = 7$ mm, and the bar diameter away from the notch is $D = 70$ mm. The shaft is subjected to a bending moment of $M = 1.0$ kN · m and a torque of $T = 2.5$ kN · m. Find the maximum shear stress and the equivalent stress at the root of the notch.

The minimum radius of this shaft is

$$d = D - 2t = 70 - 2 \times 10.5 = 49 \text{ mm} \tag{1}$$

Then

$$\frac{r}{d} = \frac{7}{49} = 0.143 \quad \text{and} \quad \frac{D}{d} = \frac{70}{49} = 1.43 \tag{2}$$

From Chart 2.41 the stress concentration factor for bending is to be approximately

$$K_{tn} = 1.82 \tag{3}$$

Similarly from Chart 2.48 the stress concentration factor for torsion is

$$K_{ts} = 1.46 \tag{4}$$

As indicated in Chart 2.41, σ_{nom} is found as

$$\sigma_{\text{nom}} = \frac{32M}{\pi d^3} = \frac{32 \times 1.0 \times 10^3}{\pi \times (0.049)^3} = 86.58 \text{ MPa} \tag{5}$$

Thus the maximum tensile stress at the root of the groove is

$$\sigma_{\max} = K_{tn}\sigma_{\text{nom}} = 1.82 \times 86.58 = 157.6 \text{ MPa} \tag{6}$$

In the case of torsion, the shear stress τ_{nom} is found to be

$$\tau_{\text{nom}} = \frac{16T}{\pi d^3} = \frac{16 \times 2.5 \times 10^3}{\pi \times 0.049^3} = 108.2 \text{ MPa} \tag{7}$$

so that the maximum torsional shear stress at the bottom of the groove is

$$\tau_{\max} = K_{ts}\tau_{\text{nom}} = 1.46 \times 108.2 = 158.0 \text{ MPa} \tag{8}$$

76 NOTCHES AND GROOVES

The principal stresses are found to be (Pilkey 1994)

$$\sigma_1 = \frac{1}{2}\sigma_{max} + \frac{1}{2}\sqrt{\sigma_{max}^2 + 4\tau_{max}^2} \qquad (9)$$

$$= \frac{1}{2} \times 157.6 + \frac{1}{2}\sqrt{157.6^2 + 4 \times 158.0^2}$$

$$= 78.8 + 176.6 = 255.4 \text{ MPa}$$

$$\sigma_2 = \frac{1}{2}\sigma_{max} - \frac{1}{2}\sqrt{\sigma_{max}^2 + 4\tau_{max}^2} = 78.8 - 176.6 = -97.8 \text{ MPa}$$

Thus the corresponding maximum shear stress is

$$\frac{(\sigma_1 - \sigma_2)}{2} = 176.6 \text{ MPa} \qquad (10)$$

which, of course, differs from the maximum torsional shear stress of (8).

Finally, the equivalent stress (Eq. 1.35) becomes

$$\sigma_{eq} = \sqrt{\sigma_1^2 - \sigma_1\sigma_2 + \sigma_2^2} \qquad (11)$$

$$= \sqrt{255.4^2 - 255.4 \times (-97.8) + (-97.8)^2}$$

$$= 315.9 \text{ MPa}$$

2.9.4 V-Shaped Circumferential Groove in a Bar of Circular Cross Section

Chart 2.51 shows $K_{ts\alpha}$ for the V groove (Rushton 1967), with variable angle α, using the style of Charts 2.7 and 2.28. For $\alpha \leq 90°$ the curves are approximately independent of r/d. For $\alpha = 135°$ separate curves are needed for $r/d = 0.005, 0.015$, and 0.05. The effect of the V angle may be compared with Charts 2.7 and 2.28.

2.9.5 Shaft in Torsion with Grooves with Flat Bottoms

The stress concentration factors K_{ts} for flat bottom notches in a shaft of circular cross section under tension are given in Chart 2.52.

2.10 TEST SPECIMEN DESIGN FOR MAXIMUM K_t FOR A GIVEN r/D OR r/H

In designing a test piece, suppose that we have a given outside diameter (or width), D (or H).[2] For a particular notch bottom radius, r, we want to know the notch depth (the d/D or r/H ratio) that gives a maximum K_t.[3]

From the curves of Charts 2.5, 2.20, 2.26, 2.42, and 2.49, maximum K_t values are plotted in Chart 2.53 with r/H and d/H as variables for two-dimensional problems and

---
[2]The width D is often dictated by the available bar size.
[3]The minimum notch bottom radius is often dictated by the ability of the shop to produce accurate, smooth, small radius.

r/D and d/D for three-dimensional. Although these values are approximate, in that the Neuber approximation is involved (as detailed in the introductory remarks at the beginning of this chapter), the maximum region is quite flat, and therefore K_t is not highly sensitive to variations in d/D or d/H in the maximum region.

From Chart 2.53 it can be seen that a rough guide for obtaining maximum K_t in a specimen in the most used r/D or r/H range is to make the smaller diameter, or width, about three-fourths of the larger diameter, or width (assuming that one is working with a given r and D or H).

Another specimen design problem occurs when r and d are given. The smaller diameter d may, in some cases, be determined by the testing machine capacity. In this case K_t increases with increase of D/d, reaching a "knee" at a D/d value which depends on the r/d value (see Chart 2.48). For the smaller r/d values, a value of $d/D = 1/2$ where the "knee" is reached would be indicated, and for the larger r/d values, the value of $d/D = 3/4$ would be appropriate.

REFERENCES

Appl, F. J., and Koerner, D. R., 1968, "Numerical Analysis of Plane Elasticity Problems," *Proc. Am. Soc. Civil Eng.*, Vol. 94, p. 743.

Appl, F. J., and Koerner, D. R., 1969, "Stress Concentration Factors for U-Shaped, Hyperbolic and Rounded V-Shaped Notches," ASME Paper 69-DE-2, Eng. Soc. Library, United Eng. Center, New York.

Atsumi, A., 1958, "Stress Concentration in a Strip under Tension and Containing an Infinite Row of Semicircular Notches," *Q. J. Mech. & Appl. Math.*, Vol. 11, Part 4, p. 478.

Atsumi, A., 1967, "Stress Concentrations in a Strip Under Tension and Containing Two Pairs of Semicircular Notches Placed on the Edges Symmetrically," *Trans. ASME, Applied Mechanics Section*, Vol. 89, p. 565.

Barrata, F. I., and Neal, D. M., 1970, "Stress Concentration Factors in U-Shaped and Semi-Elliptical Shaped Edge Notches," *J. Strain Anal.*, Vol. 5, p. 121.

Barrata, F. I., 1972, "Comparison of Various Formulae and Experimental Stress-Concentration Factors for Symmetrical U-Notched Plates," *J. Strain Anal.*, Vol. 7, p. 84.

Bowie, O. L., 1963, "Rectangular Tensile Sheet with Symmetric Edge Cracks," Army Materials and Mechanics Research Center, AMRA TR 63-22.

Bowie, O. L., 1966, "Analysis of Edge Notches in a Semi-Infinite Region," Army Materials & Mechanics Research Center, AMRA TR 66-07.

Brown, W. F., and Strawley, J. E., 1966, "Plane Strain Crack Toughness Testing of High Strength Metallic Materials," STP 410, Amer. Soc. Testing Mtls., Philadelphia, Pa., p. 11.

Cheng, Y. F., 1970, "Stress at Notch Root of Shafts under Axially Symmetric Loading," *Exp. Mechanics*, Vol. 10, p. 534.

Ching, A., Okubo, S., and Tsao, C. H., 1968, "Stress Concentration Factors for Multiple Semi-Elliptical Notches in Beams Under Pure Bending," *Exp. Mechanics*, Vol. 8, p. 19N.

Cole, A. G., and Brown, A. F., 1958, "Photoelastic Determination of Stress Concentration Factors Caused by a Single U-Notch on One Side of a Plate in Tension," *J. Royal Aero. Soc.*, Vol. 62, p. 597.

Cowper, G. R., 1962, "Stress Concentrations Around Shallow Spherical Depressions in a Flat Plate," Aero Report LR-340, National Research Laboratories, Ottawa, Canada.

Denardo, B. P., 1968, "Projectile Shape Effects on Hypervelocity Impact Craters in Aluminum," NASA TN D-4953, Washington, D.C.

Dixon, J. R., 1962, "Stress Distribution Around Edge Slits in a Plate Loaded in Tension–The Effect of Finite Width of Plate," *J. Royal Aero. Soc.*, Vol. 66, p. 320.

Durelli, A. J., Lake, R. L., and Phillips, E., 1952, "Stress Concentrations Produced by Multiple Semi-Circular Notches in Infinite Plates under Uniaxial State of Stress," *Proc. Soc. Exp. Stress Analysis*, Vol. 10, No. 1, p. 53.

ESDU, (Engineering Science Data Unit), 1981, *Stress Concentrations*, London.

Eubanks, R. A., 1954, "Stress Concentration Due to a Hemispherical Pit at a Free Surface," *Trans. ASME, Applied Mechanics Section*, Vol. 76, p. 57.

Flynn, P. D., and Roll, A. A., 1966, "Re-examination of Stresses in a Tension Bar with Symmetrical U-Shaped Grooves," *Proc. Soc. Exp. Stress Analysis*, Vol. 23, Pt. 1, p. 93.

Flynn, P. D., and Roll, A. A., 1967, "A Comparison of Stress Concentration Factors in Hyperbolic and U-Shaped Grooves," *Proc. Soc. Exp. Stress Analysis*, Vol. 24, Pt. 1, p. 272.

Frocht, M. M., 1935, "Factors of Stress Concentration Photoelasticity Determined," *Trans. ASME, Applied Mechanics Section*, Vol. 57, p. A-67.

Frocht, M. M., and Landsberg, D., 1951, "Factors of Stress Concentration in Bars with Deep Sharp Grooves and Fillets in Torsion," *Trans. ASME, Applied Mechanics Section*, Vol. 73, p. 107.

Grayley, M. E., 1979, "Estimation of the Stress Concentration Factors at Rectangular Circumferential Grooves in Shafts under Torsion," Engineering Science Data Unit, London, ESDU Memorandum No. 33.

Hamada, M., and Kitagawa, H., 1968, "Elastic Torsion of Circumferentially Grooved Shafts," *Bull. Japan Soc. Mech. Eng.*, Vol. 11, p. 605.

Hartman, J. B., and Leven, M. M., 1951, "Factors of Stress Concentration in Bars with Deep Sharp Grooves and Fillets in Tension," *Proc. Soc. Exp. Stress Analysis*, Vol. 8, No. 2.

Hetényi, M., 1943, "The Distribution of Stress in Threaded Connections," *Proc. Soc. Exp. Stress Analysis*, Vol. 1, No. 1, p. 147.

Hetényi, M. and Liu, T. D., 1956, "Method for Calculating Stress Concentration Factors," *J. of Applied Mech.*, Vol. 23.

Heywood, R. B., 1952, *Designing by Photoelasticity*, Chapman and Hall, London, p. 163.

Hooke, C. J., 1968, "Numerical Solution of Plane Elastostatic Problems by Point Matching," *J. Strain Anal.*, Vol. 3, p. 109.

Inglis, C. E., 1913, "Stresses in a Plate due to the Presence of Cracks and Sharp Corners," *Engineering (London)*, Vol. 95, p. 415.

Irwin, G. R., 1958, *Fracture*, Vol. 6, Encyclopedia of Physics, Springer, Berlin, p. 565.

Irwin, G. R., 1960, "Fracture Mechanics," in *Structural Mechanics*, Pergamon, New York.

Isida, M., 1953, "On the Tension of the Strip with Semi-Circular Notches," *Trans. Japan Soc. Mech. Eng.*, Vol. 19, p. 5.

Isida, M., 1955, "On the Tension of a Strip with a Central Elliptic Hole," *Trans. Japan Soc. Mech. Eng.*, Vol. 21.

Kikukawa, M., 1962, "Factors of Stress Concentration for Notched Bars Under Tension and Bending," *Proc. 10th Intern. Cong. Appl. Mech.*, Elsevier, New York, p. 337.

Kitagawa, H., and Nakade, K., 1970, "Stress Concentrations in Notched Strip Subjected to In-Plane Bending," *Technology Reports of Osaka University*, Vol. 20, p. 751.

Koiter, W. T., 1965, "Note on the Stress Intensity Factors for Sheet Strips with Crack under Tensile Loads," Rpt. 314 of Laboratory of Engr. Mechanics, Technological University, Delft, Holland.

Lee, G. H., 1940, "The Influence of Hyperbolic Notches on the Transverse Flexure of Elastic Plates," Trans. *ASME, Applied Mechanics Section*, Vol. 62, p. A-53.

Leibowitz, H., Vandervelt, H., and Sanford, R. J., 1967, "Stress Concentrations Due to Sharp Notches," *Exper. Mech.*, Vol. 7, p. 513.

Leven, M. M., and Frocht, M. M., 1953, "Stress Concentration Factors for a Single Notch in a Flat Plate in Pure and Central Bending," *Proc. SESA*, Vol. 11, No. 2, p. 179.

Leven, M. M., 1955, "Quantitative Three-Dimensional Photoelasticity," *Proc. SESA*, Vol. 12, No. 2, p. 167.

Ling, Chi-Bing, 1967, "On Stress Concentration at Semicircular Notch," *Trans. ASME, Applied Mechanics Section*, Vol. 89, p. 522.

Ling, Chi-Bing, 1968, "On Stress Concentration Factor in a Notched Strip," *Trans. ASME, Applied Mechanics Section*, Vol. 90, p. 833.

Matthews, G. J., and Hooke, C. J., 1971, "Solution of Axisymmetric Torsion Problems by Point Matching," *J. Strain Anal.*, Vol. 6, p. 124.

Moore, R. R., 1926, "Effect of Grooves, Threads and Corrosion upon the Fatigue of Metals," *Proc. ASTM*, Vol. 26, Part 2, p. 255.

Neuber, H., 1946, *Theory of Notch Stresses: Principal for Exact Calculation of Strength with Reference To Structural Form and Material*, J. W. Edwards Co., Ann Arbor, Michigan.

Neuber, H., 1958, *Kerbspannungslehre*, 2nd ed., Springer, Berlin; Translation, 1961, *Theory of Notch Stresses*, Office of Technical Services, Dept. of Commerce, Washington, D.C.

Nishioka, K. and Hisamitsu, N., 1962, "On the Stress Concentration in Multiple Notches," *Trans. ASME, Applied Mechanics Section*, Vol. 84, p. 575.

Okubo, H., 1952, "Approximate Approach for Torsion Problem of a Shaft with a Circumferential Notch," *Trans. ASME, Applied Mechanics Section*, Vol. 54, p. 436.

Okubo, H., 1953, "Determination of Surface Stress by Means of Electroplating," *J. Appl. Physics*, Vol. 24, p. 1130.

Paris, P. C. and Sih, G. C., 1965, "Stress Analysis of Cracks," ASTM Special Tech. Publ. 381, p. 34.

Peterson, R. E., 1953, *Stress Concentration Design Factors*, John Wiley, New York.

Peterson, R. E., 1974, *Stress Concentration Factors*, John Wiley, New York.

Pilkey, W. D., 1994, *Formulas for Stress, Strain, and Structural Matrices*, Wiley, New York.

Reed, R. E., and Wilcox, P. R., 1970, "Stress Concentration Due to a Hyperboloid Cavity in a Thin Plate," NASA TN D-5955, Washington, D. C.

Rubenchik, V. Ya, 1975, "Stress Concentration Close to Grooves," *Vestnik Mashinostroeniya*, Vol. 55, No. 12, pp. 26-28.

Rushton, K. R., 1964, "Elastic Stress Concentrations for the Torsion of Hollow Shouldered Shafts Determined by an Electrical Analogue," *Aeronaut. Q.*, Vol. 15 p. 83.

Rushton, K. R., 1967, "Stress Concentrations Arising in the Torsion of Grooved Shafts," *J. Mech. Sci.*, Vol. 9, p. 697.

Schulz, K. J., 1941, "Over den Spannungstoestand in doorborde Platen," (On the State of Stress in Perforated Plates), Doctoral Thesis, Techn. Hochschule, Delft. (in Dutch).

Schulz, K. J., 1942, "On the State of Stress in Perforated Strips and Plates," Proc. Koninklÿke Nederlandsche Akademie van Wetenschappen (Netherlands Royal Academy of Science), Amsterdam, Vol. 45, p. 233, 341, 457, 524, (in English).

Schulz, K. J., 1943-1945, "On the State of Stress in Perforated Strips and Plates," Proc. Koninklÿke Nederlandsche Akademie van Wetenschappen (Netherlands Royal Academy of Science), Amsterdam, Vol. 46-48, p.282, 292 (in English).

Seika, M., 1960, "Stresses in a Semi-Infinite Plate Containing a U-Type Notch Under Uniform Tension," *Ingenieur-Archiv.*, Vol. 27, p. 20.

Shioya, S., 1959, "The Effect of Square and Triangular Notches with Fillets on the Transverse Flexure of Semi-Infinite Plates," *Z. angew. Math. u. Mech.*, Vol. 39, p. 300.

Shioya, S., 1960, "On the Transverse Flexure of a Semi-Infinite Plate with an Elliptic Notch," *Ingenieur-Archiv*, Vol. 29, p. 93.

Shioya, S., 1963, "The Effect of an Infinite Row of Semi-Circular Notches on the Transverse Flexure of a Semi-Infinite Plate," *Ingenieur-Archiv*, Vol. 32, p. 143.

Slot, T., 1972, *Stress Analysis of Thick Perforated Plates*, Technomic Publ. Co., Westport, Conn.

Sobey, A. J., 1965, "Stress Concentration Factors for Round Rectangular Holes in Infinite Sheets," ARC R&M 3407.

Suzuki, S. I., 1967, "Stress Analysis of a Semi-Infinite Plate Containing a Reinforced Notch Under Uniform Tension," *Intern. J. Solids and Structures*, Vol. 3, p. 649.

Tsao, C. H., Ching, A., and Okubo, S., 1965, "Stress Concentration Factors for Semi-elliptical Notches in Beams Under Pure Bending," *Exp. Mechanics*, Vol. 5, p. 19A.

Weber, C., 1942, "Halbebene mit periodisch gewelltem Rand," *Z. angew. Math. u. Mech.*, Vol. 22, p. 29.

Westergaard, H. M., 1939, "Bearing Pressures and Cracks," *Trans. ASME, Applied Mechanics Section*, Vol. 61, p. A-49.

Wilson, I. H., and White, D. J., 1973, "Stress Concentration Factors for Shoulder Fillets and Grooves in Plates," *J. Strain Anal.*, Vol. 8, p. 43.

Wilson, W. K., 1970, "Stress Intensity Factors for Deep Cracks in Bending and Compact Tension Specimens," *Engineering Fracture Mechanics*, Vol. 2, Pergamon, London.

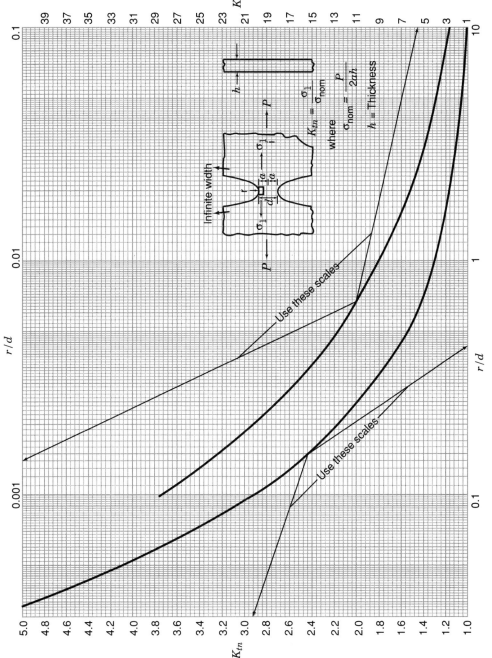

Chart 2.1 Stress concentration factors K_{tn} for opposite deep hyperbolic notches in an infinitely wide thin element in tension (Neuber, 1958).

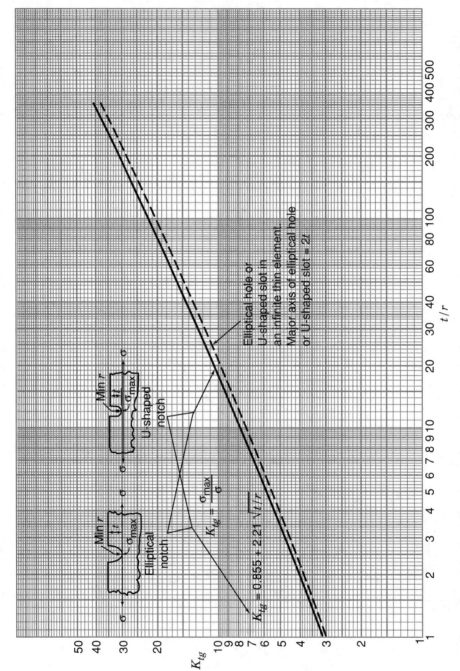

Chart 2.2 Stress concentration factors K_{tg} for an elliptical or U-shaped notch in a semi-infinite thin element in tension (Seika 1960; Bowie 1966; Baratta and Neal 1970).

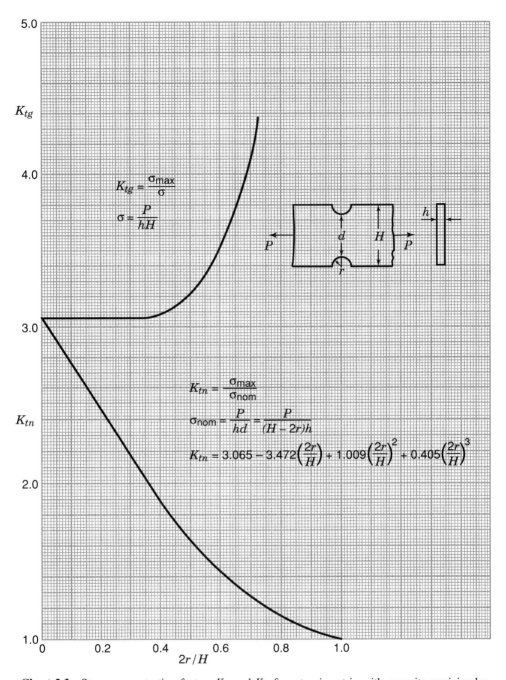

Chart 2.3 Stress concentration factors K_{tg} and K_{tn} for a tension strip with opposite semicircular edge notches (Isida 1953; Ling, 1968).

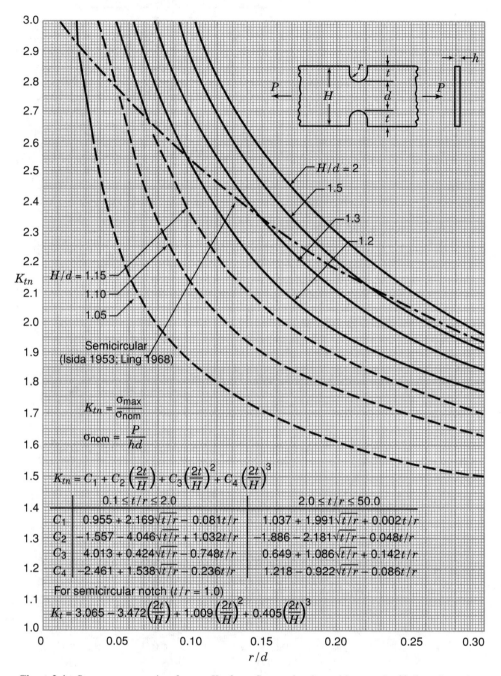

Chart 2.4 Stress concentration factors K_{tn} for a flat tension bar with opposite U-shaped notches (from data of Flynn and Roll 1966; Appl and Koerner 1969; Isida 1953; Ling 1968).

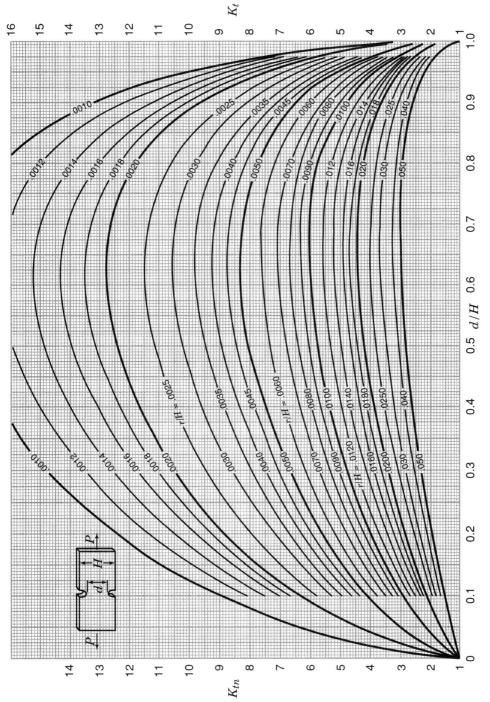

Chart 2.5 Stress concentration factors K_{tn} for a flat tension bar with opposite U-shaped notches (calculated using Neuber 1958 theory, Eq. 2.1), r/H from 0.001 to 0.05.

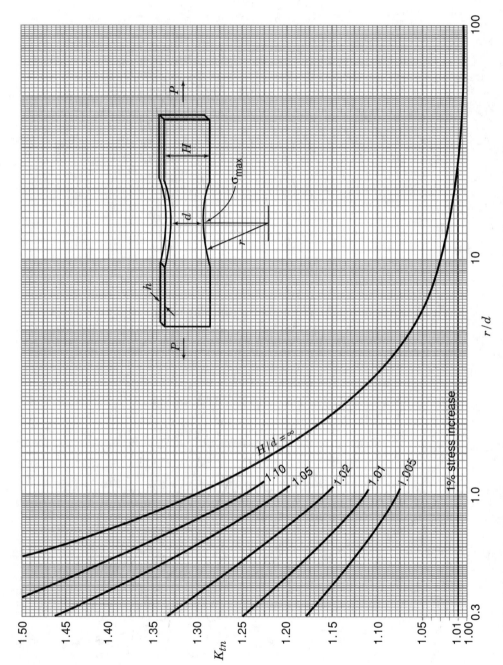

Chart 2.6 Stress concentration factors K_{tn} for a flat test specimen with opposite shallow U-shaped notches in tension (calculated using Neuber 1958 theory, Eq. 2.1).

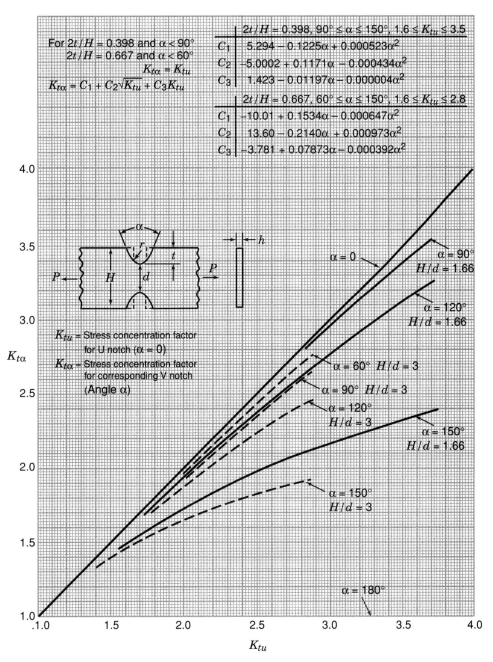

Chart 2.7 Stress concentration factors $K_{t\alpha}$ for a flat tension bar with opposite V-shaped notches (from data of Appl and Koerner 1969).

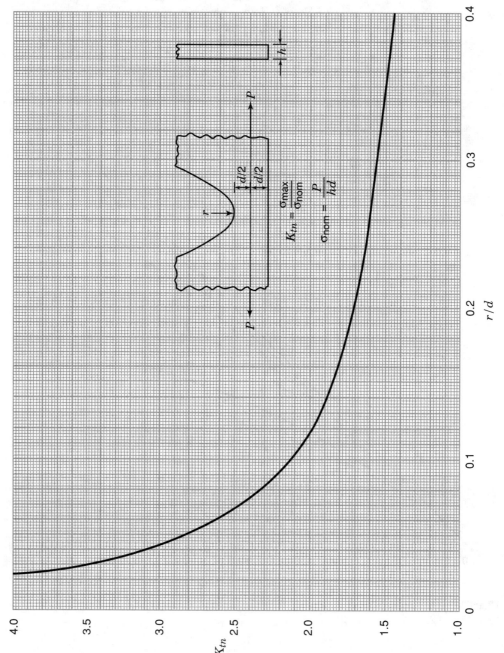

Chart 2.8 Stress concentration factors K_{tn} for tension loading of a semi-infinite thin element with a deep hyperbolic notch, tension loading in line with middle of minimum section (approximate values; Neuber 1958).

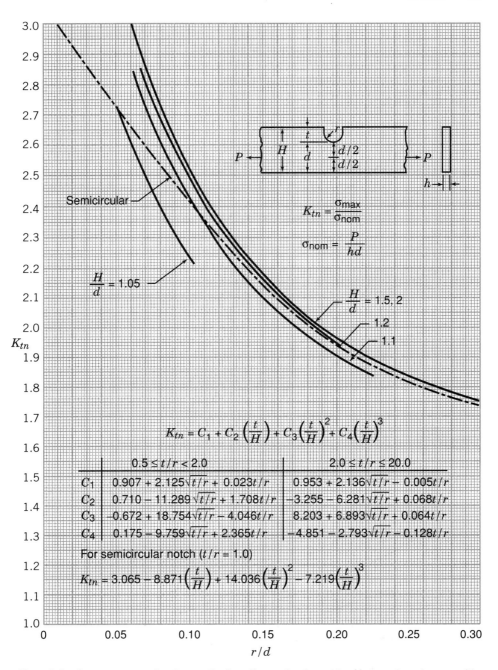

Chart 2.9 Stress concentration factors K_{tn} for a flat tension bar with a U-shaped notch at one side. (from photoelastic data of Cole and Brown 1958). Tension loading in line with middle of minimum section.

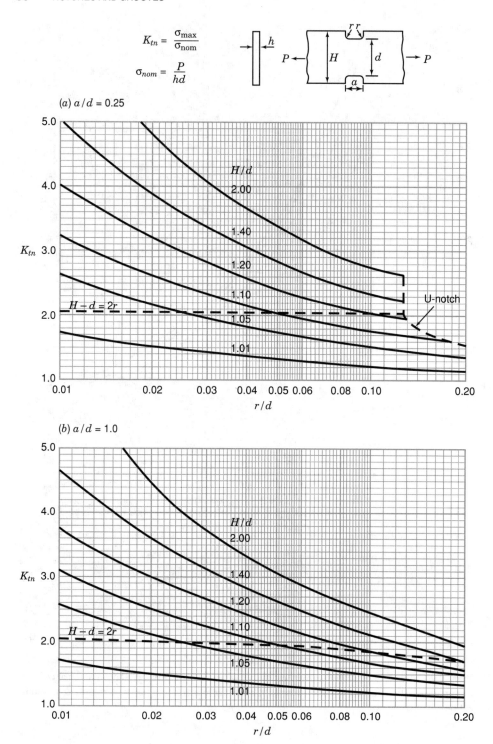

Chart 2.10 Stress concentration factors K_{tn} for opposing notches with flat bottoms in finite-width flat elements in tension (Neuber 1958; Hetenyi and Liu 1956; Sobey 1965; ESDU 1981). (a) $a/d = 0.25$; (b) $a/d = 1.0$.

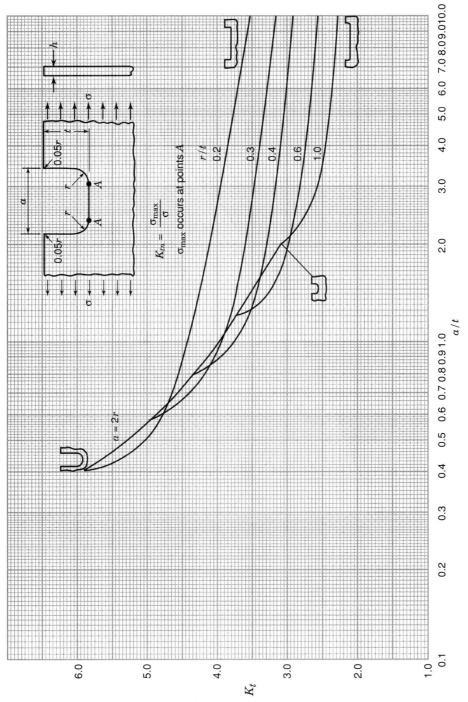

Chart 2.11 Stress concentration factors K_t for notches with flat bottoms in semi-infinite flat elements in tension (Rubenchik 1975; ESDU 1981).

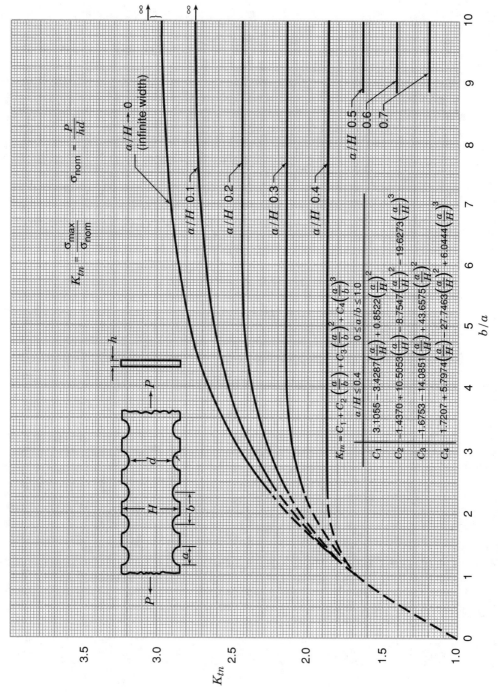

Chart 2.12 Stress concentration factors K_{tn} for a tension bar with infinite rows of semicircular edge notches (from data of Atsumi 1958).

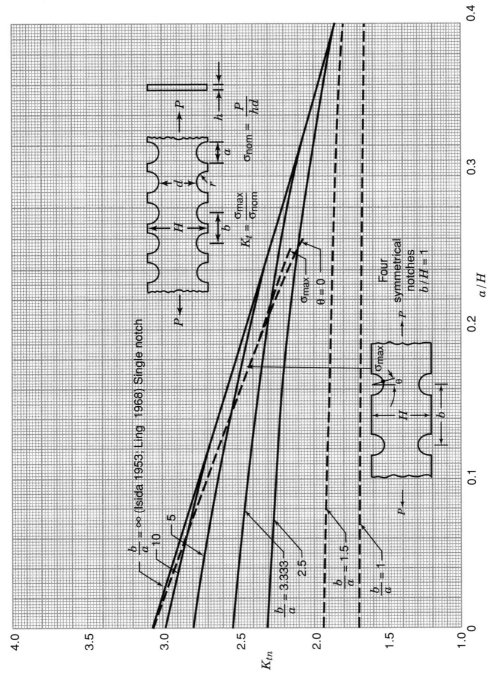

Chart 2.13 Stress concentration factors K_{tn} for tension bar with infinite rows of semicircular edge notches (from data of Atsumi 1958).

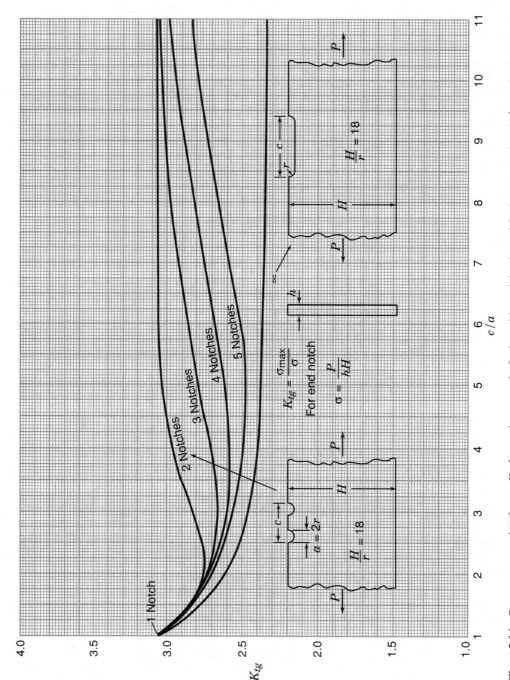

Chart 2.14 Stress concentration factors K_{tg} for tension case of a flat bar with semicircular and flat bottom notches, $H/r = 18$ (photoelastic tests by Durelli, Lake, and Phillips 1952).

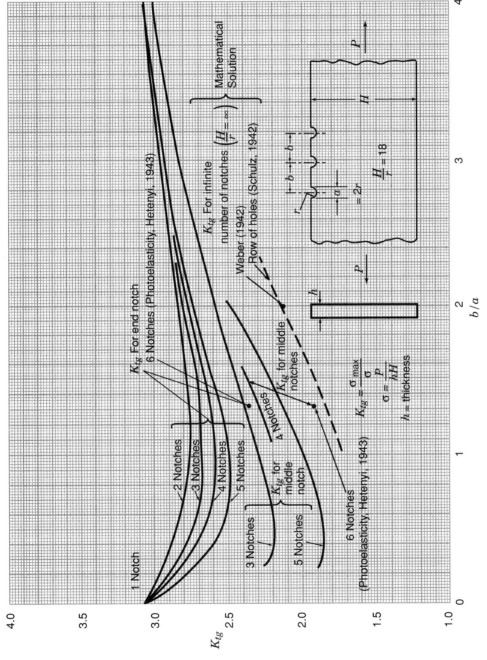

Chart 2.15 Stress concentration factors K_{tg} for tension case of a flat bar with semicircular notches, $H/r = 18$. (photoelastic tests by Durelli, Lake, and Phillips 1952).

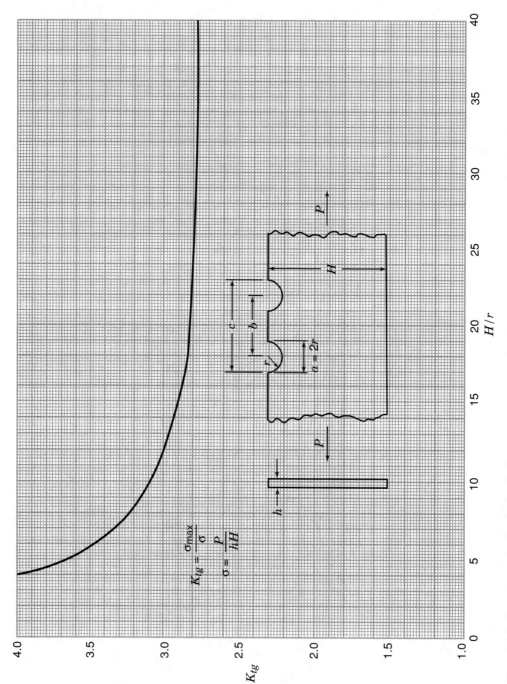

Chart 2.16 Stress concentration factors K_{tg} for tension case of a flat bar with two semicircular notches, $b/a = 2$, $c/a = 3$ (from photoelastic tests by Durelli, Lake, and Phillips 1952).

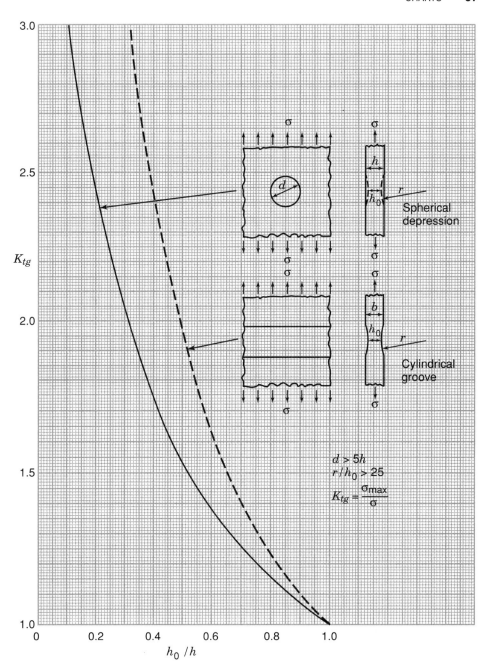

Chart 2.17 Stress concentration factors K_{tg} for a uniaxially stressed infinite thin element with opposite shallow depressions (dimples) (Cowper 1962).

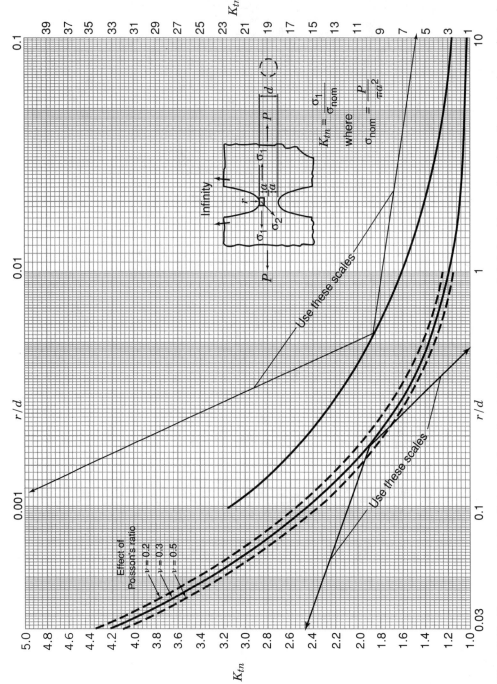

Chart 2.18 Stress concentration factors K_{tn} for a deep hyperbolic groove in an infinitely wide member, three dimensional case, in tension (Neuber 1958 solution).

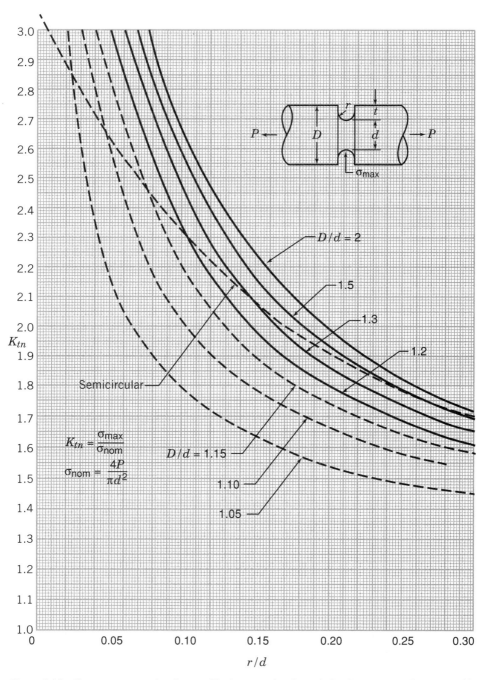

Chart 2.19 Stress concentration factors K_{tn} for a tension bar of circular cross section with a U-shaped groove. Values are approximate.

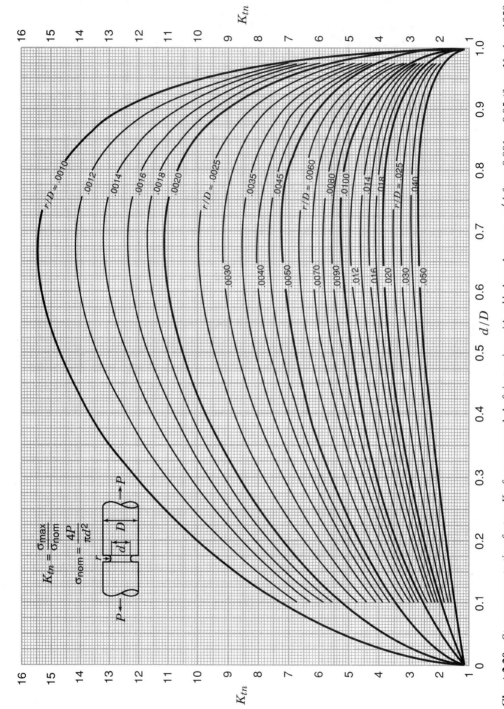

Chart 2.20 Stress concentration factors K_{tn} for a grooved shaft in tension with a U-shaped groove, r/d from 0.001 to 0.05 (from Neuber 1958 formulas).

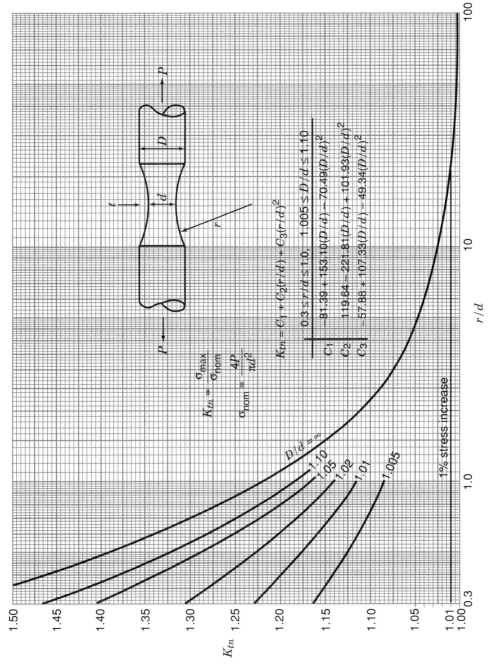

Chart 2.21 Stress concentration factors K_{tn} for a test specimen of circular cross section in tension with a U-shaped groove (curves represent calculated values using Neuber 1958 theory).

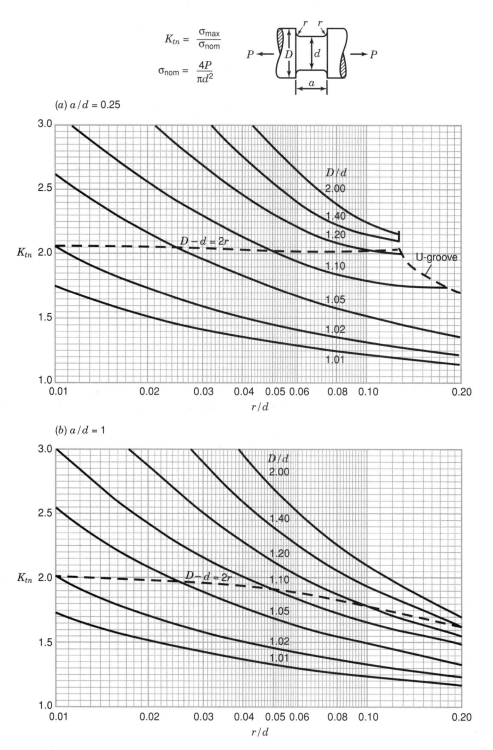

Chart 2.22 Stress concentration factors K_{tn} for flat bottom grooves in tension (Neuber 1958 formulas; ESDU 1981). (*a*) $a/d = 0.25$; (*b*) $a/d = 1$.

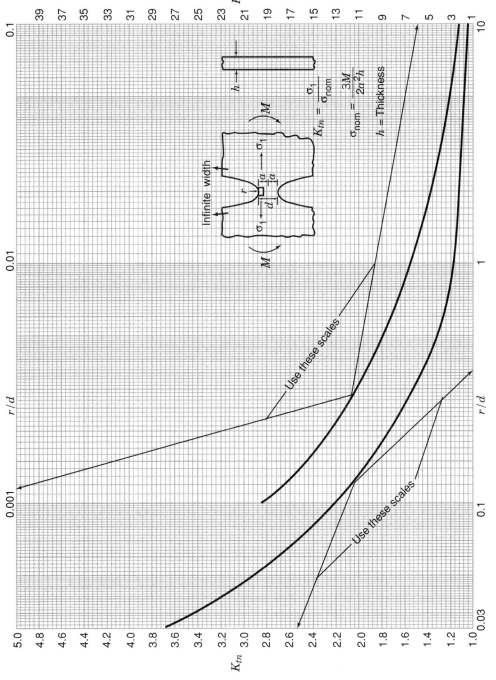

Chart 2.23 Stress concentration factors K_{tn} for opposite deep hyperbolic notches in an infinitely wide thin element, two-dimensional case, subject to in-plane moments (Neuber 1958 solution).

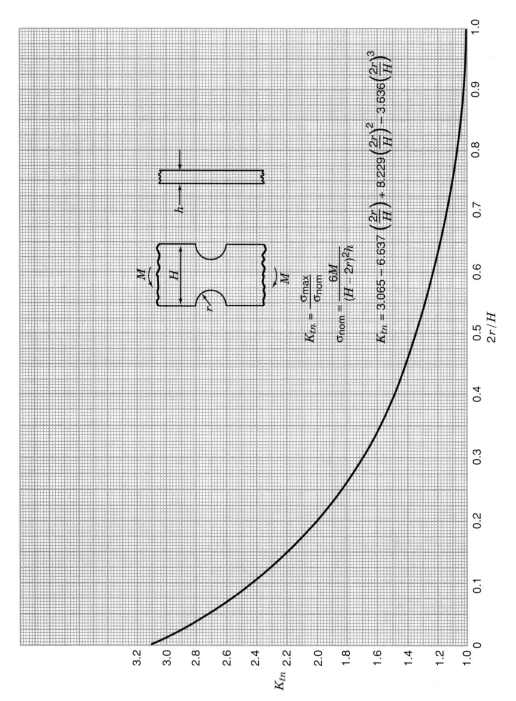

Chart 2.24 Stress concentration factors K_{tn} for bending of a flat beam with semicircular edge notches (Isida 1953; Ling 1967).

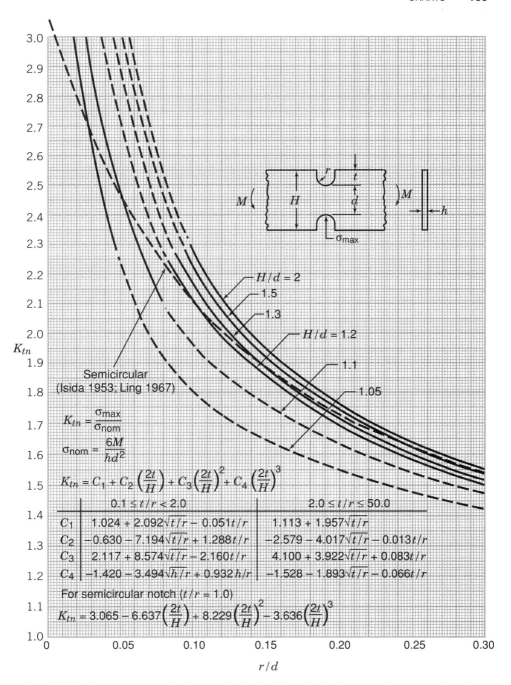

Chart 2.25 Stress concentration factors K_{tn} for bending of a flat beam with opposite U notches. (from data of Frocht 1935; Isida 1953; Ling 1967).

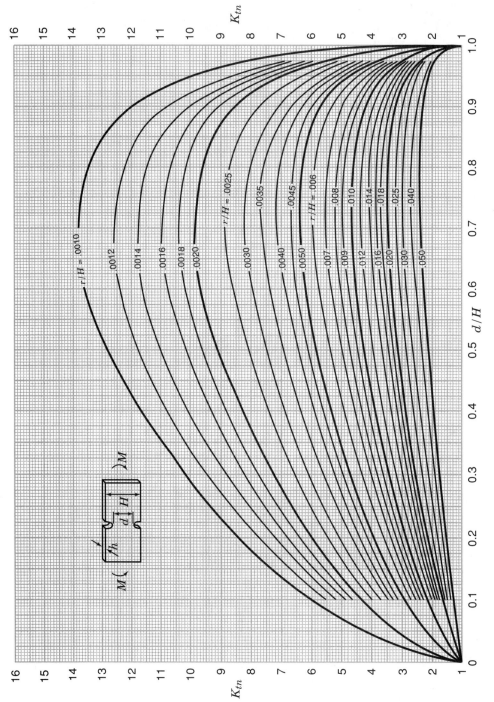

Chart 2.26 Stress concentration factors K_{tn} for bending of flat beam with opposite U notches, r/H from 0.001 to 0.05 (from Neuber 1958 formulas).

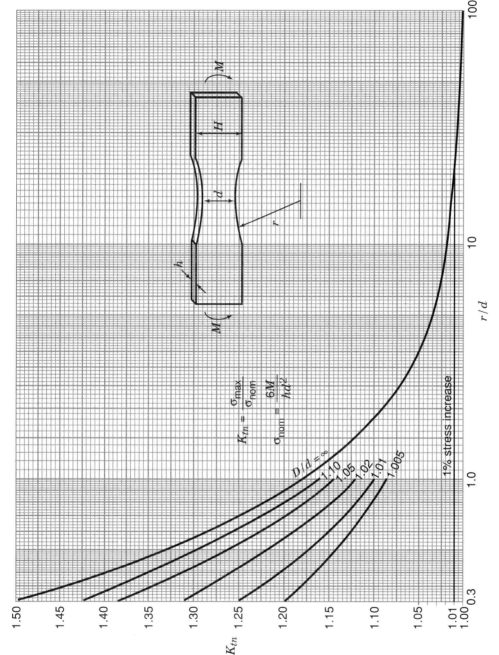

Chart 2.27 Stress concentration factors K_{tn} for bending of a flat beam with opposite shallow U notches (curves represent calculated values using Neuber 1958 theory).

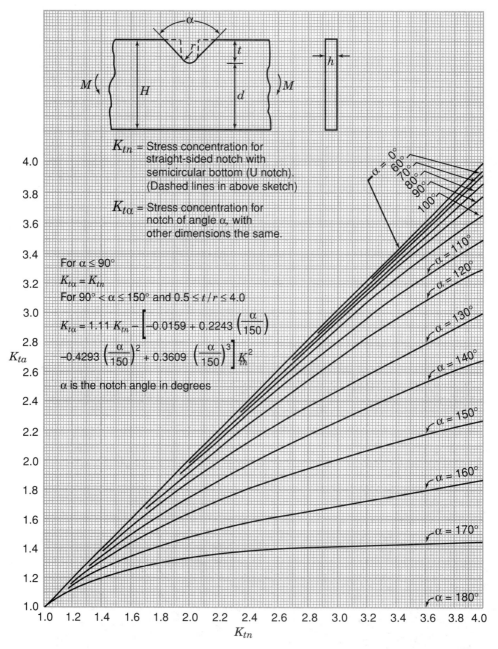

Chart 2.28 Effect of notch angle on stress concentration factors for a thin beam element in bending with a V-shaped notch on one side (Leven and Frocht 1953).

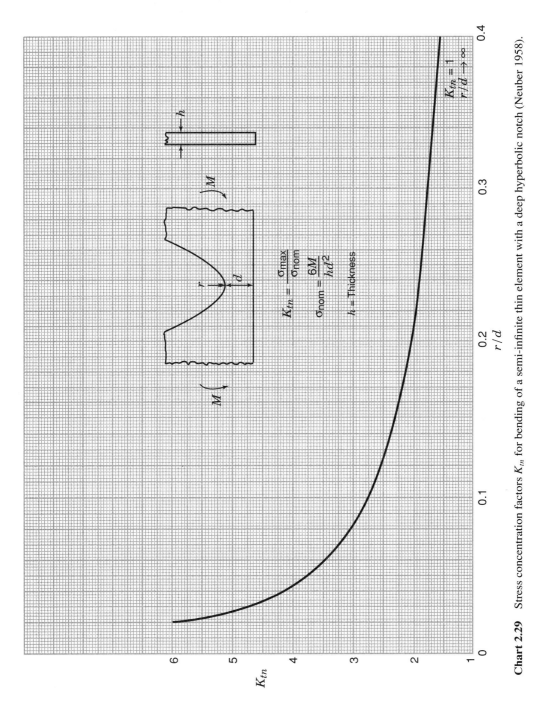

Chart 2.29 Stress concentration factors K_{tn} for bending of a semi-infinite thin element with a deep hyperbolic notch (Neuber 1958).

110 NOTCHES AND GROOVES

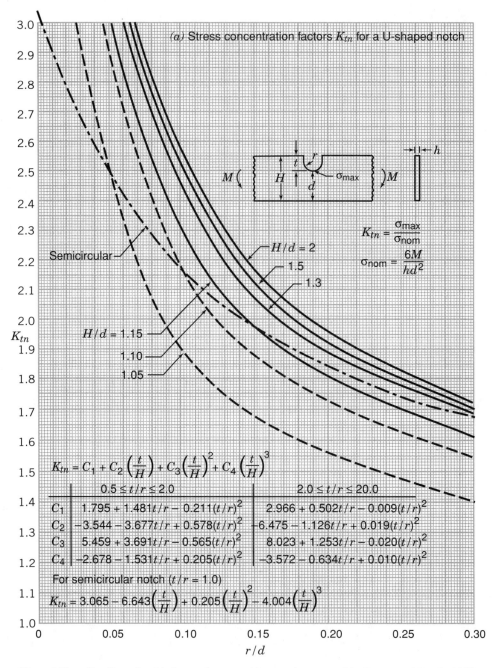

Chart 2.30a Bending of a thin beam element with a notch on one side (Leven and Frocht 1953). (*a*) Stress concentration factors K_{tn} for a U-shaped notch.

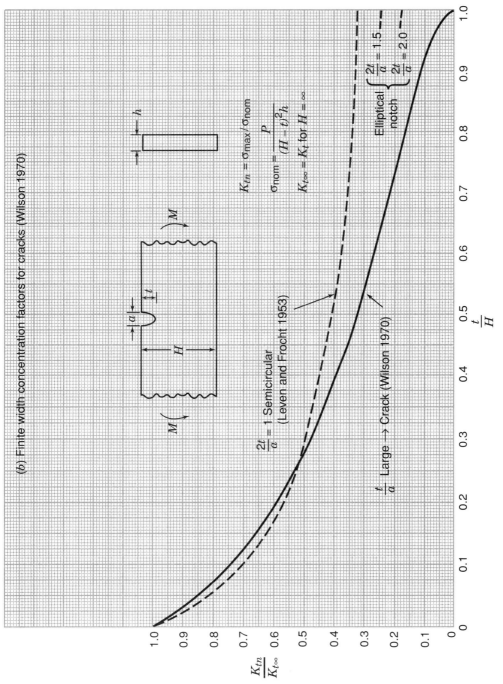

Chart 2.30b Bending of a thin beam element with a notch on one side (Leven and Frocht 1953). (b) Finite width concentration factors for cracks (Wilson 1970).

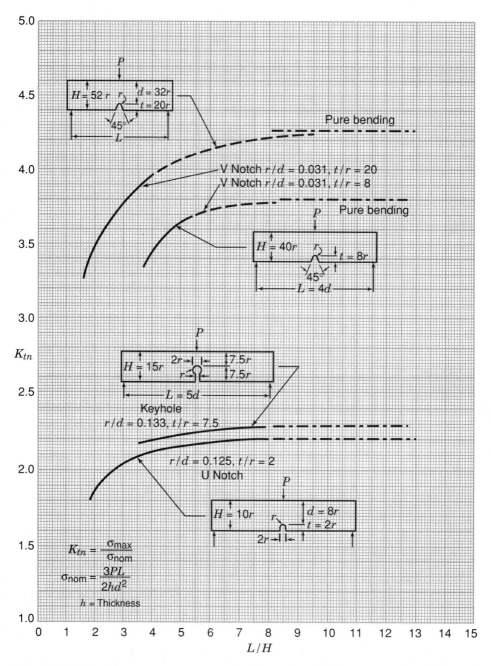

Chart 2.31 Effect of span on stress concentration factors for various impact test pieces (Leven and Frocht 1953).

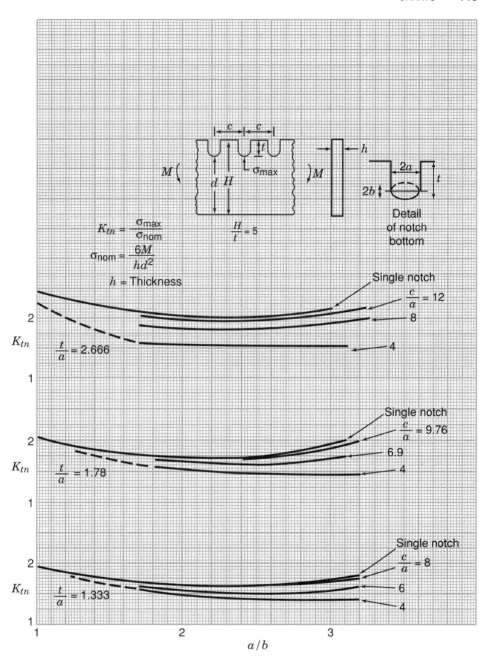

Chart 2.32 Stress concentration factors K_{tn} for bending of a thin beam having single or multiple notches with a semielliptical bottom (Ching, Okubo, and Tsao 1968).

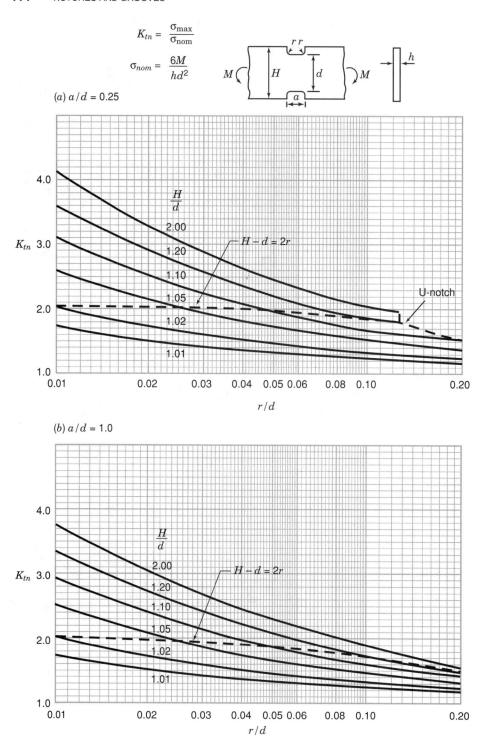

Chart 2.33 Stress concentration factors K_{tn} for thin beam in bending with opposite notches with flat bottoms (Neuber 1958; Sobey 1965; ESDU 1981). (*a*) $a/d = 0.25$; (*b*) $a/d = 1.0$.

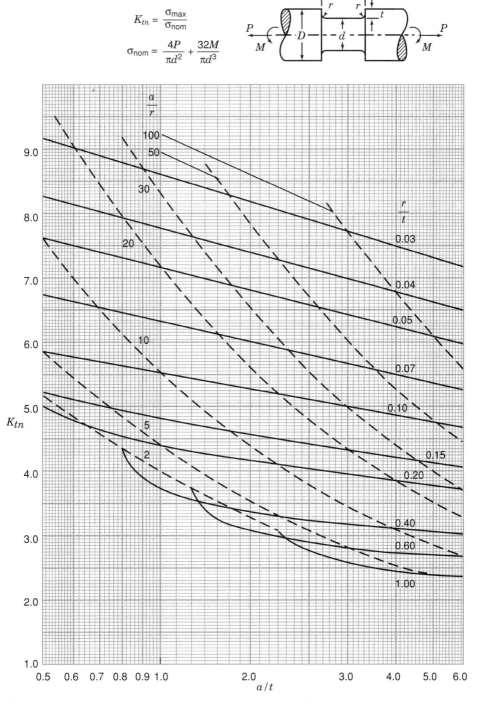

Chart 2.34 Stress concentration factors K_{tn} for a shaft in bending and/or tension with flat-bottom groove (Rubenchik 1975; ESDU 1981).

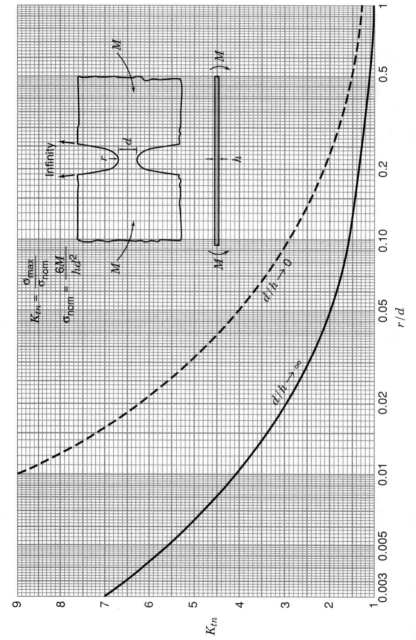

Chart 2.35 Stress concentration factors K_{tn} for a deep hyperbolic notch in an infinitely wide thin plate in transverse bending, $\nu = 0.3$ (Neuber 1958; Lee 1940).

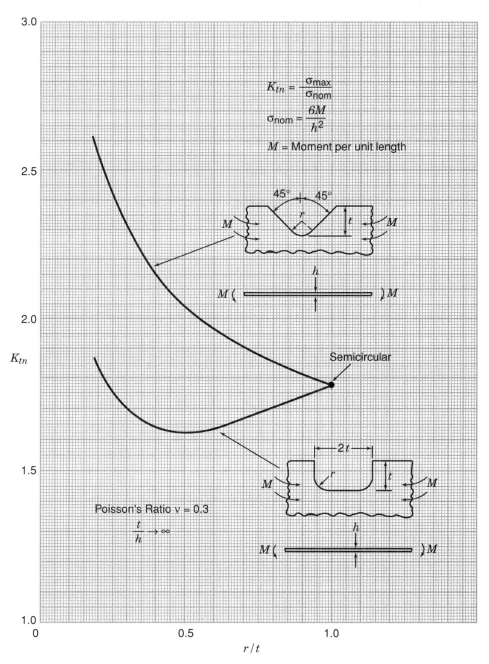

Chart 2.36 Stress concentration factors K_{tn} for rounded triangular or rectangular notches in semi-infinite plate in transverse bending (Shioya 1959).

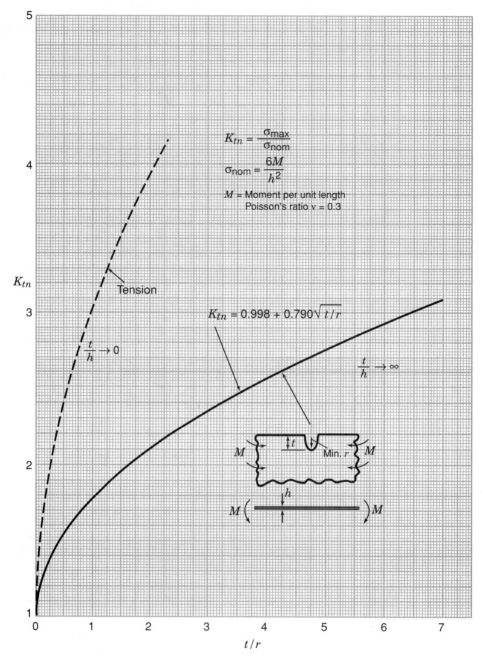

Chart 2.37 Stress concentration factors K_{tn} for an elliptical notch in a semi-infinite plate in transverse bending (from data of Shioya 1960).

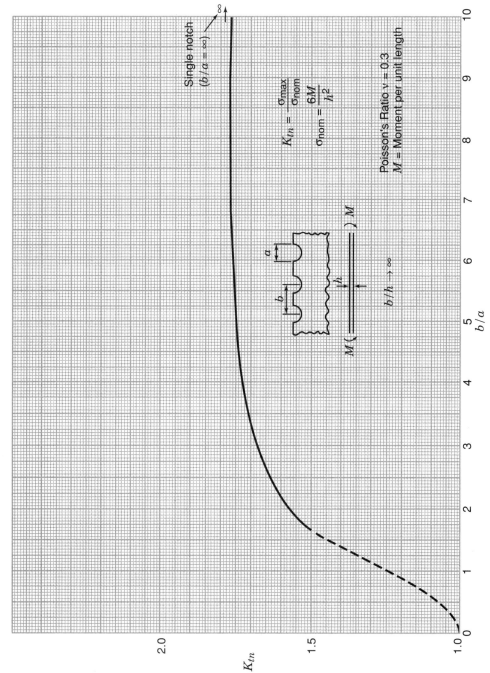

Chart 2.38 Stress concentration factors K_{tn} for infinite row of semicircular notches in a semi-infinite plate in transverse bending (from data of Shioya 1963).

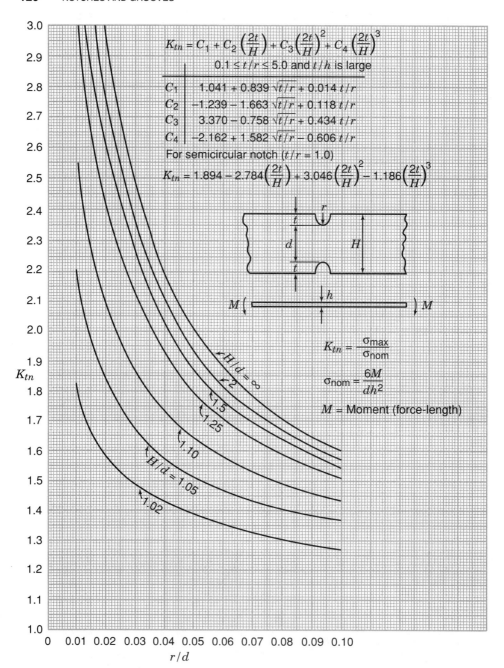

Chart 2.39 Stress concentration factors K_{tn} for a thin notched plate in transverse bending, t/h large (based on mathematical analyses of Shioya 1960; Lee 1940; Neuber 1958).

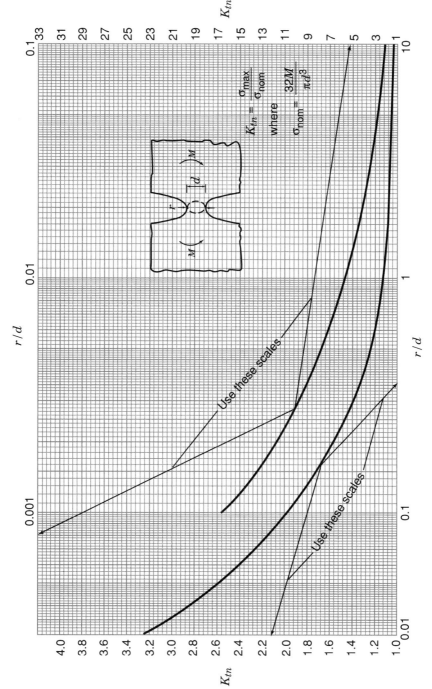

Chart 2.40 Stress concentration factors K_{tn} for a deep hyperbolic groove in an infinite member, three-dimensional case, subject to moments (Neuber solution 1958). The net cross section is circular.

122 NOTCHES AND GROOVES

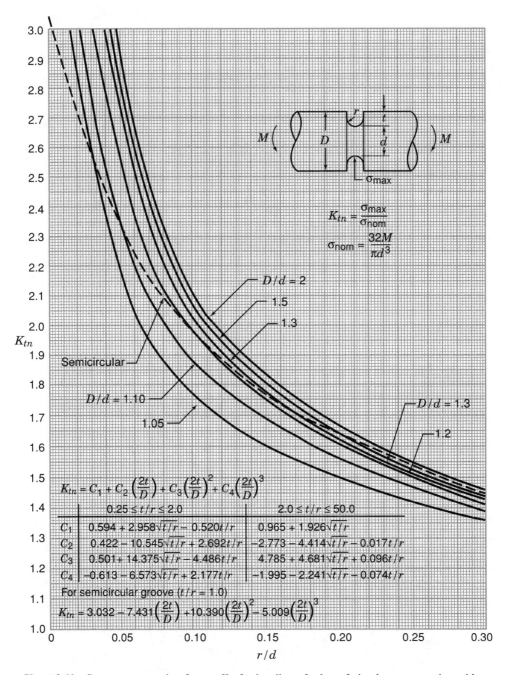

Chart 2.41 Stress concentration factors K_{tn} for bending of a bar of circular cross section with a U-shaped groove. K_t values are approximate.

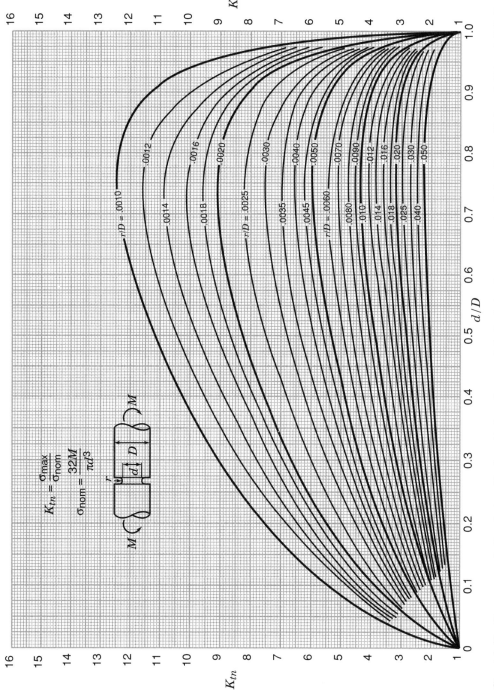

Chart 2.42 Stress concentration factors K_{tn} for a U-shaped grooved shaft of circular cross section in bending, r/D from 0.001 to 0.050 (from Neuber 1958 formulas).

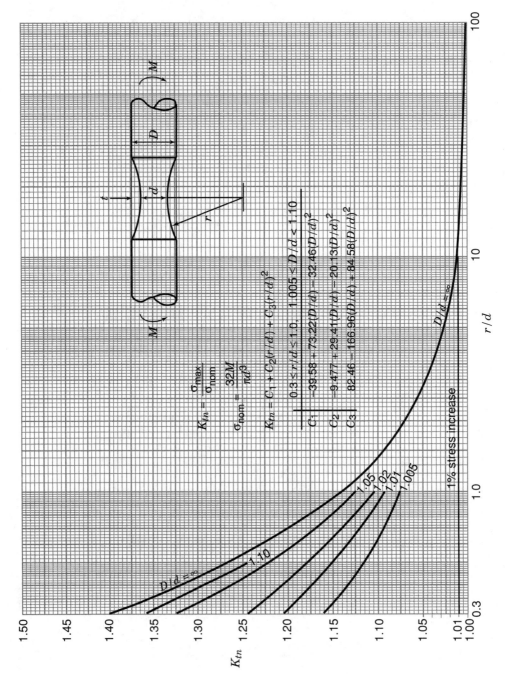

Chart 2.43 Stress concentration factors K_{tn} for bending of a bar of circular cross section with a shallow U-shaped groove (curves represent calculated values using Neuber 1958 theory).

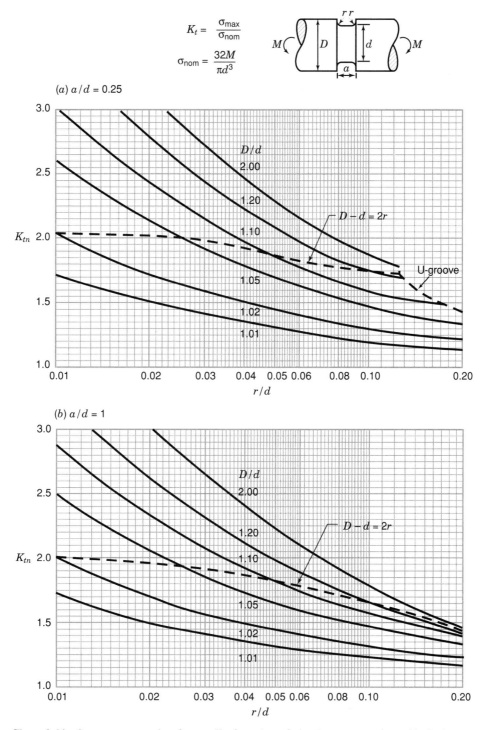

Chart 2.44 Stress concentration factors K_{tn} for a bar of circular cross section with flat-bottom grooves (from Peterson 1953; ESDU 1981). (*a*) $a/d = 0.25$; (*b*) $a/d = 1.0$.

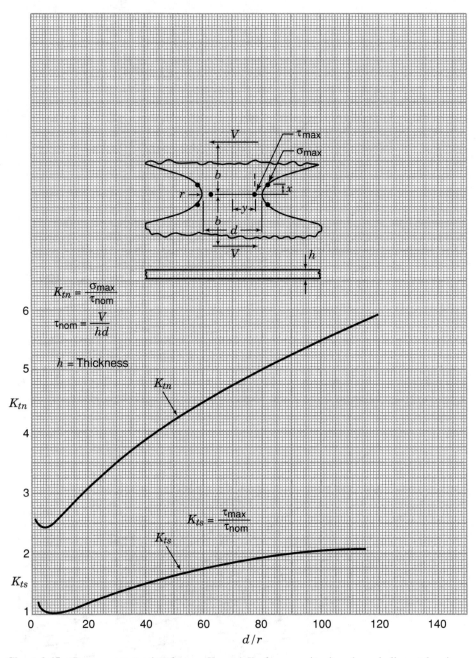

Chart 2.45 Stress concentration factors K_{tn} and K_{ts} for opposite deep hyperbolic notches in an infinite thin element in shear (Neuber 1958).

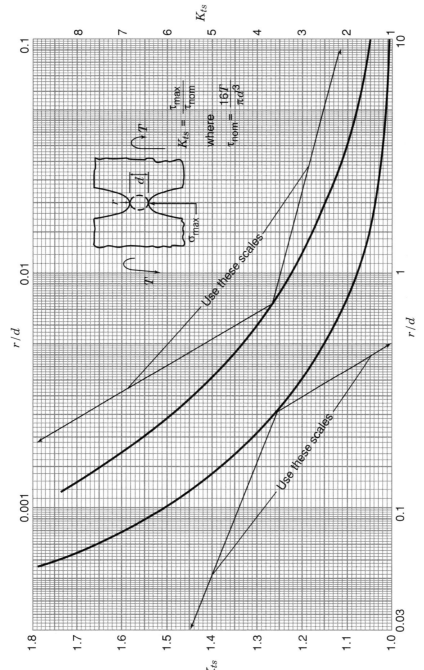

Chart 2.46 Stress concentration factors K_{ts} for a deep hyperbolic groove in an infinite member, torsion (Neuber 1958 solution). The net cross section is circular.

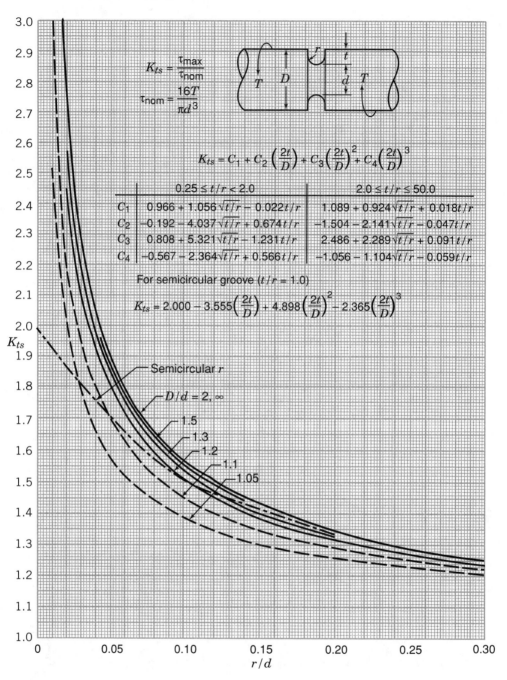

Chart 2.47 Stress concentration factors K_{ts} for torsion of a bar of circular cross section with a U-shaped groove (from electrical analog data of Rushton 1967).

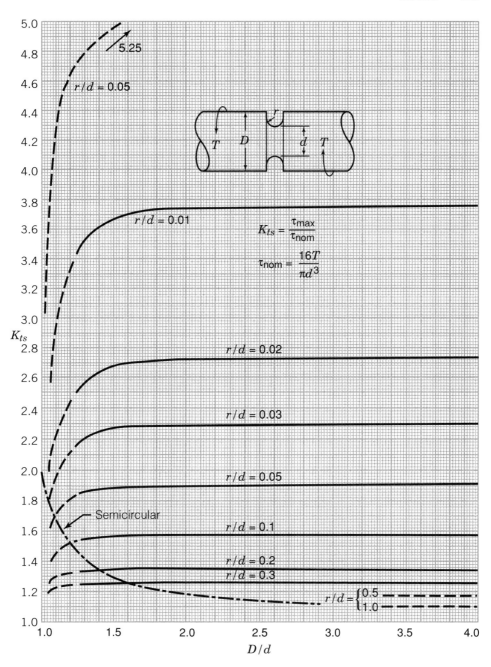

Chart 2.48 Stress concentration factors K_{ts} for torsion of a bar of circular cross section with a U-shaped groove (from electrical analog data of Rushton 1967).

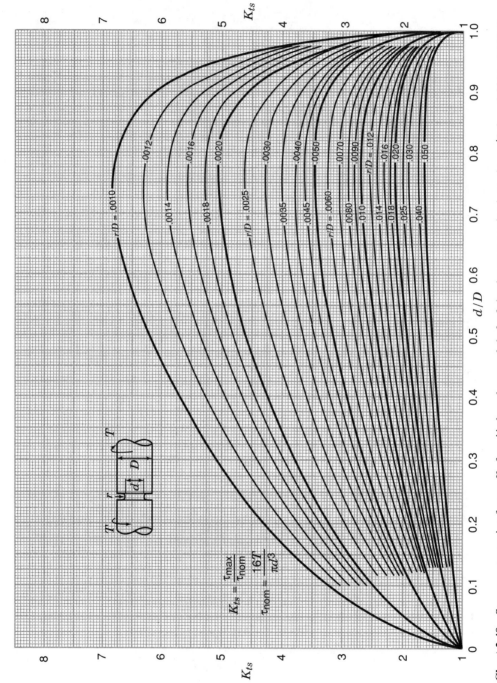

Chart 2.49 Stress concentration factors K_{ts} for a U-shaped grooved shaft of circular cross section in torsion, r/D from 0.001 to 0.050 (from Neuber 1958 formulas).

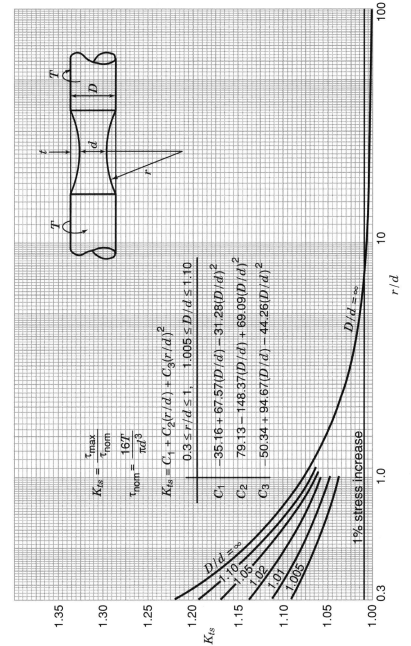

Chart 2.50 Stress concentration factors K_{ts} for the torsion of a bar of circular cross section with a shallow U-shaped groove (curves represent calculated values using Neuber 1958 theory).

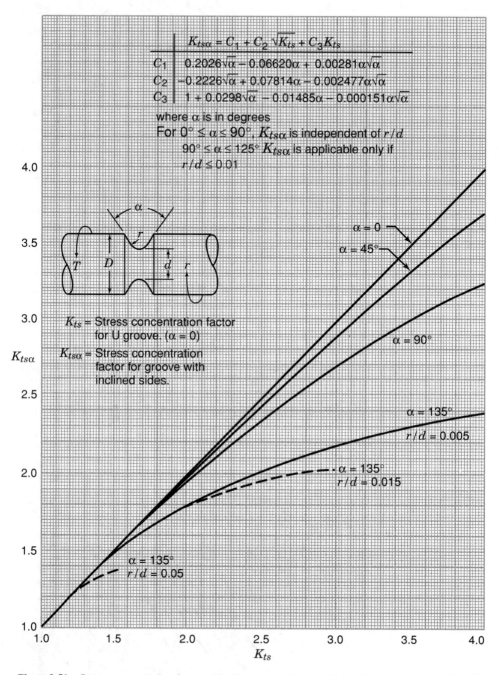

Chart 2.51 Stress concentration factors $K_{ts\alpha}$ for torsion of a bar of circular cross section with a V groove (Rushton 1967).

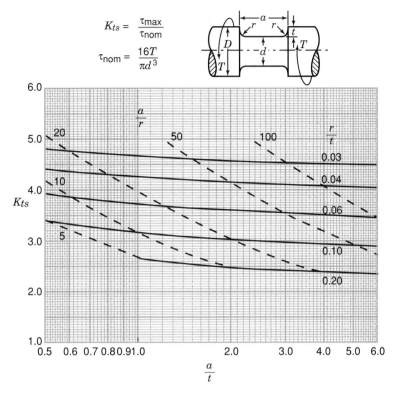

Chart 2.52 Stress concentration factors K_{ts} for a shaft in torsion with flat bottom groove (Rubenchik 1975; Grayley 1979; ESDU 1981).

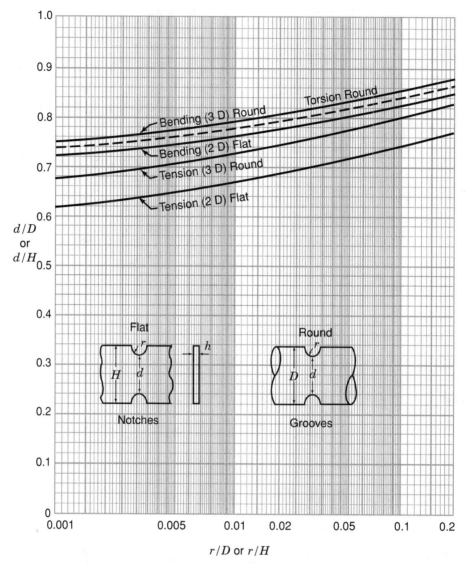

Chart 2.53 Approximate geometric relations for maximum stress concentration for notched and grooved specimens (based on Neuber 1958 relations).

CHAPTER 3

SHOULDER FILLETS

The shoulder fillet (Fig. 3.1) is the type of stress concentration that is more frequently encountered in machine design practice than any other. Shafts, axles, spindles, rotors, and so forth, usually involve a number of diameters connected by shoulders with rounded fillets replacing the sharp corners that were often used in former years.

3.1 NOTATION

Definition:
 Panel. A thin flat element with in-plane loading
Symbols:

a = semimajor axis of an ellipse

b = semiminor axis of an ellipse

D = larger diameter of circular bar

d = smaller diameter of circular bar; smaller width of thin flat element

d_f = middle diameter or width of streamline fillet

d_i = diameter of central (axial) hole

H = larger width (depth) of thin flat element

H_x = depth of equivalent wide shoulder element

h = thickness of a thin flat element

K_t = stress concentration factor

K_{tI}, K_{tII} = stress concentration factors at I, II

L = length or shoulder width

136 SHOULDER FILLETS

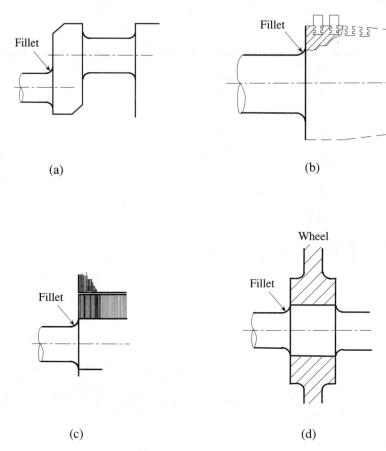

Figure 3.1 Examples of filleted members: (*a*) Engine crankshaft; (*b*) turbine rotor; (*c*) motor shaft; (*d*) railway axle.

L_x = radial height of fillet
L_y = axial length of fillet
M = bending moment
P = applied tension force
r = fillet radius
r_1 = fillet radius at end of compound fillet that merges into shoulder fillet
r_2 = fillet radius at end of compound fillet that merges into shaft
T = torque
t = fillet height
σ = stress
σ_{nom} = nominal stress
σ_{max} = maximum stress
τ_{max} = maximum shear stress
τ_{nom} = nominal shear stress
θ = angle

TENSION (AXIAL LOADING) **137**

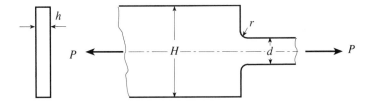

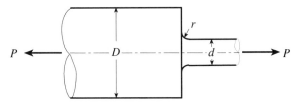

Figure 3.2 Fillets in a thin element and a circular bar.

3.2 STRESS CONCENTRATION FACTORS

Unless otherwise specified, the stress concentration factor K_t is based on the smaller width or diameter, d. In tension (Fig. 3.2) $K_t = \sigma_{max}/\sigma_{nom}$, where $\sigma_{nom} = P/hd$ for a thin flat element of thickness h and $\sigma_{nom} = 4P/\pi d^2$ for a circular bar.

The fillet factors for tension and bending are based on photoelastic values. For torsion the fillet factors are from a mathematical analysis. A method was given in Peterson (1953) for obtaining approximate K_t values for smaller r/d values where r is the fillet radius. The charts in this book extend well into the small r/d range, owing to use of lately published results.

The K_t factors for the thin flat members considered in this chapter are for two-dimensional states of stress (plane stress) and apply only to very thin sheets or, more strictly, to where $h/r \rightarrow 0$. As h/r increases, a state of plane strain is approached. The stress at the fillet surface at the middle of the plate thickness increases, and the stress at the plate surface decreases.

Some of the stress concentration cases of Chapter 5 on miscellaneous design elements are related to fillets.

3.3 TENSION (AXIAL LOADING)

3.3.1 Opposite Shoulder Fillets in a Flat Bar

Chart 3.1 presents stress concentration factors K_t for a stepped flat tension bar. These curves are modifications of the K_t factors determined through photoelastic tests (Frocht 1935) whose values have been found to be too low, owing probably to the small size of the models and to possible edge effects. The curves in the r/d range of 0.03 to 0.3 have been obtained as follows:

$$K_t(\text{Chart 3.1}) = K_t(\text{Fig. 57, Peterson 1953})$$
$$\times \left[\frac{K_t \text{ (Chart 2.4)}}{K_t \text{ (notch, Frocht 1935)}} \right] \quad (3.1)$$

The r/d range has been extended to lower values by photoelastic tests (Wilson and White 1973). These data fit well with the above results from Eq. (3.1) for $H/d > 1.1$.

Other photoelastic tests (Fessler et al. 1969) give K_t values that agree reasonably well with the $H/d = 1.5$ and two curves of Chart 3.1.

3.3.2 Effect of Shoulder Geometry in a Flat Member

The factors of Chart 3.1 are for the case where the large width H extends back from the shoulder a relatively great distance. Frequently one encounters a case in design where this shoulder width L (Fig. 3.3) is relatively narrow.

In one of the early investigations in the photoelasticity field, Baud (1928) noted that in the case of a narrow shoulder the outer part is unstressed, and he proposed the formula

$$H_x = d + 0.3L \quad (3.2)$$

where H_x is the depth (Fig. 3.3) of a wide shoulder member that has the same K_t factor.

The same result can be obtained graphically by drawing intersecting lines at an angle θ of 17° (Fig. 3.3). Sometimes a larger angle θ is used, up to 30°. The Baud (1934) rule, which was proposed as a rough approximation, has been quite useful.

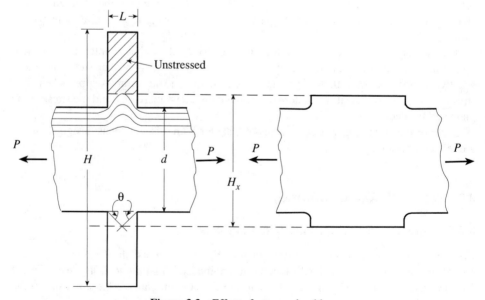

Figure 3.3 Effect of narrow shoulder.

Although the K_t factors for bending of flat elements with narrow shoulders (Charts 3.7 to 3.9) were published (Leven and Hartman) in 1951, it was not until 1968, that the tension case was systematically investigated (Kumagai and Shimada) (Chart 3.2). Referring to Chart 3.2c and d, note that at $L/d = 0$ a cusp remains. Also $K_t = 1$ at $L/d = -2r/d$ (see the dashed lines in Charts 3.2c and d for extrapolation to $K_t = 1$). The extrapolation formula gives the exact L/d value for $K_t = 1$ for $H/d = 1.8$ (Chart 3.2c) when $r/d \le 0.4$, and for $H/d = 5$ (Chart 3.2d) when $r/d \le 2$. Kumagai and Shimada (1968) state that their results are consistent with previous data (Scheutzel and Gross 1966; Spangenberg 1960) obtained for somewhat different geometries. Empirical formulas were developed by Kumagai and Schimada (1968) to cover their results.

Round bar values are not available. It is suggested that Eq. (1.15) be used.

3.3.3 Effect of a Trapezoidal Protuberance on the Edge of a Flat Bar

A weld bead can sometimes be adequately approximated as a trapezoidal protuberance. The geometrical configuration is shown in the sketch in Chart 3.3. A finite difference method was used to find the stress concentration factor (Derecho and Munse 1968). The resulting K_t factors for $\theta = 30°$ and $60°$ are given in Chart 3.3. The dashed curve corresponds to a protuberance height where the radius is exactly tangent to the angular side; that is, below the dashed curve there are no straight sides, only segments of a circle. See sketch in Chart 3.3.

A comparison (Derecho and Munse 1968) of K_t factors with corresponding (large L/t) factors, obtained from Figs. 36 and 62 of Peterson (1953) for filleted members with angle correction, showed the latter to be about 7% higher on the average, with variations from 2 to 15%. A similar comparison by Peterson, using the increased K_t fillet values of Chart 3.1, showed these values (corrected for angle) to be about 17% higher (varying between 14 and 22%) than the Derecho and Munse values.

Strain gage measurements (Derecho and Munse 1968) resulted in K_t factors 32%, 23%, and 31% higher, with one value (for the lowest K_t) 2.3% lower than the computed values. They comment: "the above comparisons suggest that the values [in Chart 3.3] ... may be slightly lower than they should be. It may be noted here that had a further refinement of the spacing been possible in the previously discussed finite-difference solution, slightly higher values of the stress concentration factor could have been obtained." It is possible that the factors may be more than slightly lower.

A typical weld bead would correspond to a geometry of small t/L, with H/d near 1.0. Referring, for example, to Chart 3.3a, we see that for $t/L = 0.1$ and $r/L = 0.1$, K_t is surprisingly low, 1.55. Even if we increase this by 17%, to be on the safe side in design, we still have a relatively low factor, $K_t = 1.8$.

3.3.4 Fillet of Noncircular Contour in a Flat Stepped Bar

Circular fillets are usually used for simplicity in drafting and machining. The circular fillet does not correspond to minimum stress concentration.

The variable radius fillet is often found in old machinery (using many cast-iron parts) where the designer or builder apparently produced the result intuitively. Sometimes the variable radius fillet is approximated by two radii, resulting in the compound fillet illustrated in Fig. 3.4.

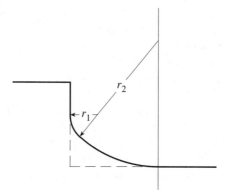

Figure 3.4 Compound fillet.

Baud (1934) proposed a fillet form with the same contour as that given mathematically for an ideal, frictionless liquid flowing by gravity from an opening in the bottom of a tank (Fig. 3.5):

$$x = 2\frac{d}{\pi} \sin^2 \frac{\theta}{2} \qquad (3.3)$$

$$y = \frac{d}{\pi} \left[\log \tan \left(\frac{\theta}{2} + \frac{\pi}{4} \right) - \sin \theta \right] \qquad (3.4)$$

Baud noted that in this case the liquid at the boundary has constant velocity and reasoned that the same boundary may also be the contour of constant stress for a tension member. By means of a photoelastic test in tension, Baud observed that no appreciable stress concentration occurred with a fillet of streamline form.

For bending and torsion Thum and Bautz (1934) applied a correction in accordance with the cube of the diameter, resulting in a shorter fillet than for tension. This correction led to Table 3.1. Thum and Bautz also demonstrated by means of fatigue tests in bending and in torsion that, with fillets having the proportions of Table 3.1, no appreciable stress concentration effect was obtained.

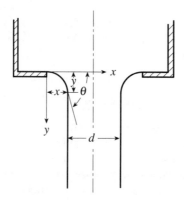

Figure 3.5 Ideal frictionless liquid flow from opening in bottom of tank.

TENSION (AXIAL LOADING)   141

TABLE 3.1 Proportions for streamline fillet[a]

y/d	d_f/d for Tension	d_f/d for Bending or Torsion	y/d	d_f/d for Tension	d_f/d for Bending or Torsion
0.0	1.636	1.475	0.3	1.187	1.052
0.002	1.610	1.420	0.4	1.134	1.035
0.005	1.594	1.377	0.5	1.096	1.026
0.01	1.572	1.336	0.6	1.070	1.021
0.02	1.537	1.287	0.7	1.051	1.018
0.04	1.483	1.230	0.8	1.037	1.015
0.06	1.440	1.193	0.9	1.027	1.012
0.08	1.405	1.166	1.0	1.019	1.010
0.10	1.374	1.145	1.3	1.007	1.005
0.15	1.310	1.107	1.6	1.004	1.003
0.2	1.260	1.082	∞	1.000	1.000

[a]See Fig. 3.6 for notation.

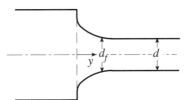

Figure 3.6 Notation for Table 3.1.

To reduce the length of the streamline fillet Deutler and Harvers (Lurenbaum 1937) suggested a special elliptical form based on theoretical considerations of Föppl (Timoshenko and Goodier 1970).

Grodzinski (1941) mentions fillets of parabolic form. He also gives a simple graphical method, which may be useful in making a template or a pattern for a cast part (Fig. 3.7). Dimensions a and b are usually dictated by space or design considerations. Divide each distance into the same number of parts and number in the order shown. Connect points hav-

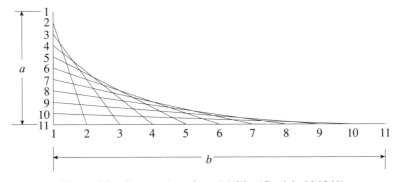

Figure 3.7 Construction of special fillet (Grodzinski 1941).

142   SHOULDER FILLETS

ing the same numbers by straight lines. This results in an envelope of gradually increasing radius, as shown in Fig. 3.7.

For heavy shafts or rolls, Morgenbrod (1939) has suggested a tapered fillet with radii at the ends, the included angle of the tapered portion being between 15° and 20° (Fig. 3.8). This is similar to the basis of the tapered cantilever fatigue specimen of McAdam (1923), which has been shown (Peterson 1930) to have a stress variation of less than 1% over a 2 in. length, with a nominal diameter of 1 in. This conical surface is tangent to the constant-stress cubic solid of revolution.

Photoelastic tests have provided values for a range of useful elliptical fillets for bending (Section 3.4.3). The degree of improvement obtained may be useful in considering a tension case. Clock (1952) has approximated an elliptical fillet by using an equivalent segment of a circle and has provided corresponding K_t values.

An excellent treatment of optimum transition shapes has been provided by Heywood (1969). His discussion includes some interesting observations about shapes found in nature (tree trunks and branches, thorns, animal bones).

3.3.5   Stepped Bar of Circular Cross Section with a Circumferential Shoulder Fillet

Consider a stepped bar of circular cross section with a circumferential shoulder fillet. The K_t values for this case (Chart 3.4) were obtained by ratioing the K_t values of Chart 3.1 in accordance with the three- to two-dimensional notch values, as explained in Section 2.5.2. Chart 3.4 is labeled "approximate" in view of the procedure.

For d/D values considered valid for comparison (0.6, 0.7, 0.9), photoelastic results for round bars (Allison 1962) are somewhat lower than the values of Chart 3.4. Photoelastic tests (Fessler et al. 1969) give K_t values for $D/d = 1.5$ that are in good agreement with Chart 3.4.

3.3.6   Tubes

Stress concentration factors K_t are given in Chart 3.5 for thin-walled tubes with fillets. The data are based on the work of Lee and Ades (1956). In Chart 3.5, K_t is shown versus t/r for various values of t/h for a tube subject to tension. The plot holds only when $(d_i/h + d_i/t) > 28$. For $d_i < 28ht/(t + h)$, K_t will be smaller.

For solid shafts ($d_i = 0$), K_t is reduced as t/h increases. The location of $\sigma_{\max}$ is in the fillet near the smaller tube.

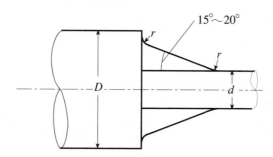

Figure 3.8   Tapered fillet suggested by Morgenbrod (1939).

3.3.7 Stepped Pressure Vessel Wall with Shoulder Fillets

Chart 3.6 is for a pressure vessel with a stepped wall with shoulder fillets. The K_t curve is based on calculated values of Griffin and Thurman (1967). A direct comparison (Griffin and Kellogg 1967) with a specific photoelastic test by Leven (1965) shows good agreement. The strain gage results of Heifetz and Berman (1967) are in reasonably good agreement with Chart 3.6. Lower values have been obtained in the finite element analysis of Gwaltney et al. (1971).

For comparison the model shown in Chart 3.1 may be considered to be split in half axially. The corresponding K_t curves have the same shape as in Chart 3.6, but they are somewhat higher. However, the cases are not strictly comparable and, furthermore, Chart 3.1 is approximate.

3.4 BENDING

3.4.1 Opposite Shoulder Fillets in a Flat Bar

Stress concentration factors for the in-plane bending of a thin element with opposing shoulder fillets are displayed in Chart 3.7. Photoelastic values of Leven and Hartman (1951) cover the r/d range from 0.03 to 0.3, whereas the photoelastic tests of Wilson and White (1973) cover r/d values in the 0.003 to 0.03 range. These results blend together reasonably well and form the basis of Chart 3.7.

3.4.2 Effect of Shoulder Geometry in a Flat Thin Member

In Chart 3.8, K_t factors are given for various shoulder parameters for a fillet bar in bending (Leven and Hartman 1951). For $L/H = 0$ a cusp remains. For $H/d = 1.25$ (Chart 3.8a) and $r/d \leq 1/8$, $K_t = 1$ when $L/H = -1.6r/d$. For $H/d = 2$ (Chart 3.8b) and $r/d \leq 1/2$, $K_t = 1$ when $L/H = -r/d$. For $H/d = 3$ (Chart 3.8c) and $r/d \leq 1$, $K_t = 1$ when $L/H = -(2/3)(r/d)$. The dashed lines in Chart 3.8 show extrapolations to $K_t = 1$. Only limited information on bars of circular cross section is available. It is suggested that the designer obtain an adjusted value by ratioing in accordance with the corresponding Neuber three- to two-dimensional notch values (Peterson 1953, p. 61), or Eq. (1.15).

3.4.3 Elliptical Shoulder Fillet in a Flat Member

Photoelastic tests by Berkey (1944) have provided K_t factors for the flat element with in-plane bending (Chart 3.9). The corresponding factors for a round shaft should be somewhat lower. An estimate can be made by comparing the corresponding Neuber three- to two-dimensional notch factors, as discussed in Section 2.5.2. Equation (1.15) can also be used.

3.4.4 Stepped Bar of Circular Cross Section with a Circumferential Shoulder Fillet

Photoelastic tests (Leven and Hartman 1951) have been made of stepped bars of circular cross section in the r/d range of 0.03 to 0.3. By use of the plane bending tests of Wilson and White (1973) reasonable extensions of curves have been made in the r/d range below 0.03. The results are presented in Chart 3.10.

In comparison with other round bar photoelastic tests (Allison 1961) for the d/D ratios considered valid for comparison, there is reasonably good agreement for $d/D = 0.6, 0.8$. However, for $d/D = 0.9$ the results are lower.

In the design of machinery shafts, where bending and torsion are the primary loadings of concern, small steps (D/d near 1.0) are often used. Since for this region Chart 3.10 is not very suitable, Chart 3.11 has been provided, wherein the curves go to $K_t = 1.0$ at $D/d = 1.0$.

3.5 TORSION

3.5.1 Stepped Bar of Circular Cross Section with a Circumferential Shoulder Fillet

Investigations of the filleted shaft in torsion have been made by use of photoelasticity (Fessler et al. 1969; Allison 1961), with strain gages (Weigland 1943), by use of the electrical analog (Jacobsen 1925; Rushton 1964), and computationally (Matthews and Hooke 1971). The computational approach, using a numerical technique based on elasticity equations and a point-matching method for approximately satisfying boundary conditions, is believed to be of satisfactory accuracy. Currently most computational stress concentration studies are performed using finite element based general purpose structural analysis software. The Matthews and Hooke method provides K_{ts} values (Chart 3.12) lower than those used previously (Peterson 1953), and in the lower r/d range it provides higher values than from Rushton's electrical analog.

An empirical relation (Fessler et al. 1969) based on published data including two photoelastic tests by the authors is in satisfactory agreement with the values of Chart 3.12 in the area covered by their tests.

In design of machinery shafts, where bending and torsion are the main cases of interest, small steps (D/d near 1.0) are often used. For this region Chart 3.12 is not very suitable, and Chart 3.13 has been provided, wherein the curves go to $K_{ts} = 1.0$ at $D/d = 1.0$.

3.5.2 Stepped Bar of Circular Cross Section with a Circumferential Shoulder Fillet and a Central Axial Hole

Central (axial) holes are used in large forgings for inspection purposes and in shafts for cooling or fluid transmission purposes.

For the hollow shaft, a reasonable design procedure is to find the ratios of stress concentration factors from Chart 3.14, that have been obtained from the electrical analog values (Rushton 1964) and then, using the K_{ts} values of Charts 3.12 and 3.13, to find the stress concentration factor K_{ts} of the hollow shaft. Chart 3.14 provides the ratios of the K_{ts} values for the hollow shaft to the $K_{ts} = K_{tso}$ values for the solid shaft of Charts 3.12 and 3.13. These ratios, $(K_{ts} - 1)/(K_{tso} - 1)$, are plotted against the ratio d_i/d. Chart 3.15 gives K_{ts} for hollow shafts, plotted, in contrast to the previous table, versus r/d. Both Charts 3.14 and 3.15 are based on the data from Rushton (1964).

The strength/width ratio of the small-diameter portion of the shaft increases with increasing hollowness ratio. However, this is usually not of substantial benefit in practical designs because of the relatively larger weight of the large-diameter portion of the shaft. An exception may occur when the diameters are close together ($D/d = 1.2$ or less).

3.5.3 Compound Fillet

For a shouldered shaft in torsion, the stress concentration factor can by controlled by adjusting the size of a single radius fillet. Specifically the stress concentration factor is reduced by increasing the radius of the fillet. However, the increase in radius may not be possible due to practical constraints on the axial length (L_x) and radial height (L_y). See Fig. 3.9. Occasionally the lowest single radius fillet stress concentration factor K_{ts} (e.g., from Charts 3.12 and 3.13) that fits within the restrictions on L_y and L_x can be improved somewhat (up to about 20%) by using a double radius fillet.

For a double radius fillet, two distinct maximum stress concentrations occur. One is on circumferential line II, which is located close to where radii r_1 and r_2 are tangential to each other. The other occurs where r_2 is first parallel to d on circumferential line I. For cases that satisfy constraints on L_y and L_x, the lowest maximum shear stress occurs for the largest fillet for which K_{tI} equals K_{tII}. Care should be taken to ensure that the two fillets fit well at their intersection. Small changes in r_1 can lead to corresponding changes in the shear stress at II due to stress concentration.

For $K_{tI} = K_{tII}$, Chart 3.16 provides plots of r_2/d versus L_x/d and L_y/d for $r_2/r_1 = 3$ and 6 (Battenbo and Baines 1974; ESDU 1981). The corresponding reduction in $K_{tI}/K_t = K_{tII}/K_t$ versus r_2/r_1 is given in Chart 3.17.

Example 3.1 Design of a Fillet for a Shaft in Torsion Suppose that a fillet with a stress concentration factor of less than 1.26 is to be chosen for a shaft in torsion. In the notation of Fig. 3.9, $d = 4$ in., $D = 8$ in. There is a spacing washer carrying the maximum allowable 45° chamfer of 0.5 in. to accommodate the fillet.

For a single radius (r) fillet, let $r = L_y$ and calculate

$$\frac{D}{d} = \frac{8}{4} = 2, \quad \frac{r}{d} = \frac{0.5}{4} = 0.125 \tag{1}$$

From Chart 3.12, $K_t = 1.35$. This value exceeds the desired $K_t = 1.26$.

If a double radius fillet is employed, then for a stress concentration factor of less than 1.26

$$\frac{K_{tI}}{K_t} = \frac{K_{tII}}{K_t} = \frac{1.26}{1.35} = 0.93 \tag{2}$$

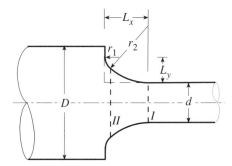

Figure 3.9 Double radius fillet.

146 SHOULDER FILLETS

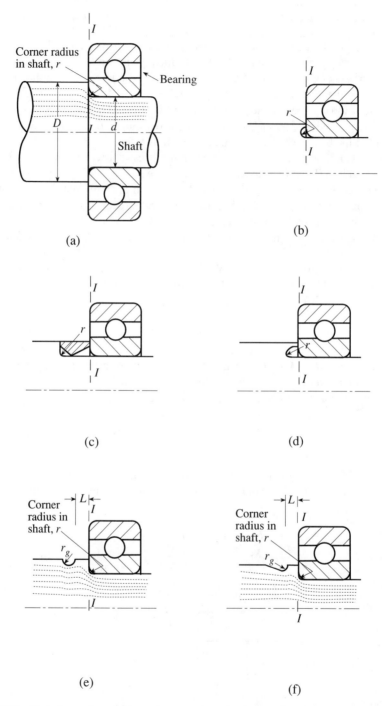

Figure 3.10 Techniques for reducing stress concentration in stepped shaft with bearing: (*a*) With corner radius only; (*b*) undercut; (*c*) inserted ring; (*d*) undercut to simulate a ring; (*e*) relief groove; (*f*) relief groove.

Both L_x and L_y must be less than 0.5 in. For a double radius fillet, $L_x > L_y$ so that $L_y = 0.5$ in. is the active constraint. From the upper curve in Chart 3.17, $r_2/r_1 = 3$ can satisfy this constraint. Use $L_x = 0.5$ in. in Chart 3.16a, and observe that for $L_y/d = 0.5/4 = 0.125$

$$\frac{r_2}{d} = 0.19 \tag{3}$$

for which L_x/d is 0.08. Finally, the double radius fillet would have the properties

$$r_2 = 0.19 \times 4 = 0.76 \text{ in.}$$
$$L_y = 0.08 \times 4 = 0.32 \text{ in.} \tag{4}$$
$$r_1 = 0.76/3 = 0.2533 \text{ in.}$$

3.6 METHODS OF REDUCING STRESS CONCENTRATION AT A SHOULDER

One of the problems occurring in the design of shafting, rotors, and so forth, is the reduction of stress concentration at a shoulder fillet (Fig. 3.10a) while maintaining the positioning line I–I and dimensions D and d. This can be done in a number of ways, some of which are illustrated in Fig. 3.10b, c, d, e, and f. By cutting into the shoulder, a larger fillet radius can be obtained (Fig. 3.10b) without developing interference with the fitted member. A ring insert could be used as at Fig. 3.10c, but this represents an additional part. A similar result could be obtained as shown in Fig. 3.10d, except that a smooth fillet surface is more difficult to realize.

Sometimes the methods of Fig. 3.10b, c, and d are not helpful because the shoulder height $(D - d)/2$ is too small. A relief groove (Fig. 3.10e, f) may be used provided that this does not conflict with the location of a seal or other shaft requirements. Fatigue tests (Oschatz 1933; Thum and Bruder 1938) show a considerable gain in strength due to relief grooving.

It should be mentioned that in the case at hand there is also a combined stress concentration and fretting corrosion problem at the bearing fit. (See Section 5.5.) The gain due to fillet improvement in this case might be limited by failure at the fitted surface. However, fatigue tests (Thum and Bruder 1938) showed that at least for the specific proportions tested a gain in strength was realized by the use of relief grooves.

REFERENCES

Allison, I. M., 1961a, "The Elastic Stress Concentration Factors in Shouldered Shafts," *Aeronautical Q.*, Vol. 12, p. 189.

Allison, I. M., 1961b, "The Elastic Concentration Factors in Shouldered Shafts, Part II: Shafts Subjected to Bending," *Aeronautical Q.*, Vol. 12, p. 219.

Allison, I. M., 1962, "The Elastic Concentration Factors in Shouldered Shafts, Part III: Shafts Subjected to Axial Load," *Aeronautical Q.*, Vol. 13, p. 129.

Appl, F. J., and Koerner, D. R., 1969, "Stress Concentration Factors for U-Shaped, Hyperbolic and Rounded V-Shaped Notches," ASME Paper 69-DE-2.

Battenbo, H., and Baines, B. H., 1974, "Numerical Stress Concentrations for Stepped Shafts in Torsion with Circular and Stepped Fillets," *J. Strain Anal.*, Vol. 2, pp. 90–101.

Baud, R. V., 1928, "Study of Stresses by Means of Polarized Light and Transparencies," *Proc. Engrs. Soc. West. Penn.*, Vol. 44, p. 199.

Baud, R. V., 1934, "Beiträge zur Kenntnis der Spannungsverteilung in Prismatischen und Keilförmigen Konstruktionselementen mit Querschnittsübergängen," *Eidgenöss. Materialprüf. Ber 83*, Zurich. See also *Product Eng.*, 1934, Vol. 5, p.133.

Berkey, D. C., 1944, "Reducing Stress Concentration with Elliptical Fillets," *Proc. Soc. Exp. Stress Analysis*, Vol. 1, No. 2, p. 56.

Clock, L. S., 1952, "Reducing Stress Concentration with an elliptical Fillet," *Design News*, Rogers Publishing Co., Detroit, MI (May 15).

Derecho, A. T., and Munse, W. H., 1968, "Stress Concentration at External Notches in Members Subjected to Axial Loading," *Univ. Illinois Eng. Expt. Eng. Sta. Bull. No. 494*.

ESDU (Engineering Science Data Unit), 1981, "Stress Concentrations," London.

Fessler, H., Rogers, C. C., and Stanley, P., 1969, "Shouldered Plates and Shafts in Tension and Torsion," *J. Strain Anal.*, Vol. 4, p. 169.

Frocht, M. M., 1935, "Factors of Stress Concentration Photoelastically Determined," *Trans. ASME, Applied Mechanics Section*, Vol. 57, p. A-67.

Griffin, D. S., and Thurman, A. L., 1967, "Comparison of DUZ Solution with Experimental Results for Uniaxially and Biaxially Loaded Fillets and Grooves," WAPD TM-654, Clearinghouse for Scientific and Technical Information, Springfield, VA.

Griffin, D. S., and Kellogg, R. B., 1967, "A Numerical Solution for Axially Symmetrical and Plane Elasticity Problems," *Intern. J. Solids and Structures*, Vol. 3, p. 781.

Grodzinski, P., 1941, "Investigation on Shaft Fillets," *Engineering (London)*, Vol. 152, p. 321.

Gwaltney, R. C., Corum, J. M., and Greenstreet, W. L., 1971, "Effect of Fillets on Stress Concentration in Cylindrical Shells with Step Changes in Outside Diameter," *Trans. ASME, J. Eng. for Industry*, Vol. 93, p. 986.

Heifetz, J. H., and Berman, I., 1967, "Measurements of Stress Concentration Factors in the External Fillets of a Cylindrical Pressure Vessel," *Expt. Mechanics.* Vol. 7, p. 518.

Heywood, R. B., 1969, *Photoelasticity for Designers*, Pergamon, New York, Chapter 11.

Jacobsen, L. S., 1925, "Torsional Stress Concentrations in Shafts of Circular Cross Section and Variable Diameter," *Trans. ASME, Applied Mechanics Section*, Vol. 47, p. 619.

Kumagai, K., and Shimada, H., 1968, "The Stress Concentration Produced by a Projection under Tensile Load," *Bull. Japan Soc. Mech. Eng.*, Vol. 11, p. 739.

Lee, L. H. N., and Ades, C. S., 1956, "Stress Concentration Factors for Circular Fillets in Stepped Walled Cylinders Subject to Axial Tension," *Proc. Soc. Expt. Stress Analysis*, Vol. 14, No. 1.

Leven, M. M., 1965, "Stress Distribution in a Cylinder with an External Circumferential Fillet Subjected to Internal Pressure," *Res. Memo* 65-9D7-520-M1, Westinghouse Research Lab.

Leven, M. M., and Hartman, J. B., 1951, "Factors of Stress Concentration for Flat Bars with Centrally Enlarged Section," *Proc. SESA*, Vol. 19, No. 1, p. 53.

Lurenbaum, K., 1937, *Ges. Vortrage der Hauptvers. der Lilienthal Gesell.*, p. 296.

Matthews, G. J., and Hooke, C. J., 1971, "Solution of Axisymmetric Torsion Problems by Point Matching," *J. Strain Anal.*, Vol. 6, p. 124.

McAdam, D. J., 1923, "Endurance Properties of Steel," *Proc. ASTM*, Vol. 23, Part II, p. 68.

Morgenbrod, W., 1939, "Die Gestaltftestigkeit von Walzen und Achsen mit Hohlkehlen," *Stahl u. Eisen*, Vol. 59, p. 511.

Oschatz, H., 1933, "Gesetzmässigkeiten des Dauerbruches und Wege zur Steigerung der Dauerhaltbarkeit," *Mitt. der Materialprüfungsanstalt an der Technischen Hochschule Darmstadt*, Vol. 2.

Peterson, R. E., 1930, "Fatigue Tests of Small Specimens with Particular Reference to Size Effect," *Proc. Am. Soc. Steel Treatment*, Vol. 18, p. 1041.

Peterson, R. E., 1953, *Stress Concentration Design Factors*, Wiley, New York.

Rushton, K. R., 1964, "Elastic Stress Concentration for the Torsion of Hollow Shouldered Shafts Determined by an Electrical Analogue," *Aeronautical Q.*, Vol. 15, p. 83.

Scheutzel, B., and Gross, D., 1966, *Konstruktion*, Vol. 18, p. 284.

Spangenberg, D., 1960, *Konstruktion*, Vol. 12, p. 278.

Thum, A., and Bautz, W., 1934, "Der Entlastungsübergang—Günstigste Ausbildung des Überganges an abgesetzten Wellen u. dg.," *Forsch. Ingwes.*, Vol. 6, p. 269.

Thum, A., and Bruder, E., 1938, "Dauerbruchgefahr an Hohlkehlen von Wellen und Achsen und ihre Minderung," *Deutsche Kraftfahrtforschung im Auftrag des Reichs-Verkehrsministeriums*, No. 11, VDI Verlag, Berlin.

Timoshenko, S., and Goodier, J. N., 1970, *Theory of Elasticity*, 3rd Ed., McGraw Hill, New York, p. 398.

Weigand, A., 1943, "Ermittlung der Formziffer der auf Verdrehung beanspruchten abgesetzen Welle mit Hilfe von Feindehnungsmessungen," *Luftfahrt Forsch.*, Vol. 20, p. 217.

Wilson, I. H., and White, D. J., 1973, "Stress Concentration Factors for Shoulder Fillets and Grooves in Plates," *J. Strain Anal.*, Vol. 8, p. 43.

150 SHOULDER FILLETS

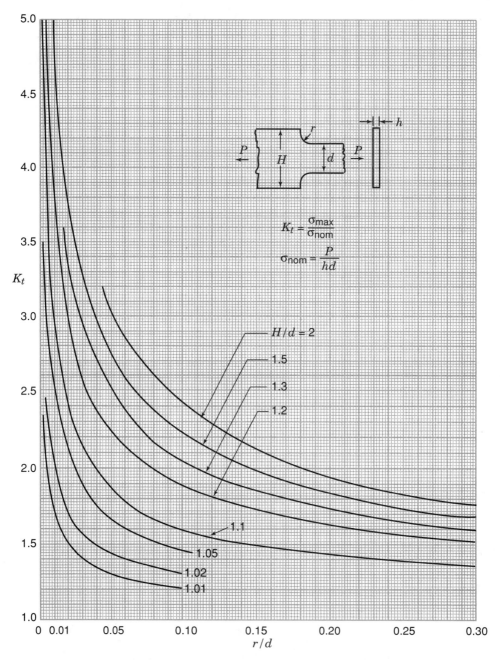

Chart 3.1 Stress concentration factors K_t for a stepped flat tension bar with shoulder fillets (based on data of Frocht 1935; Appl and Koerner 1969; Wilson and White 1973).

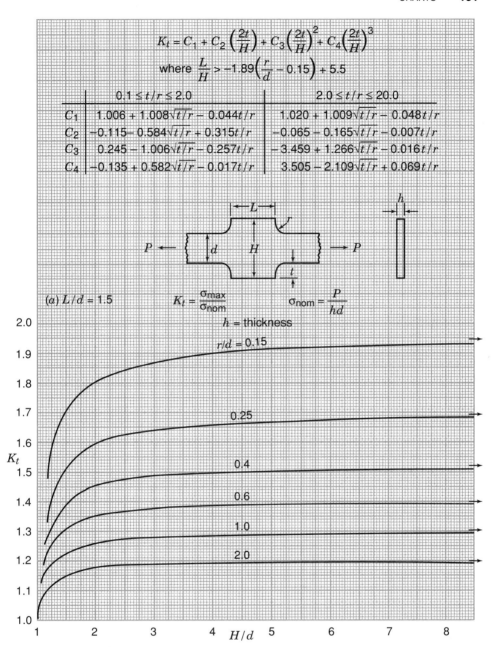

Chart 3.2a Stress concentration factors K_t for a stepped flat tension bar with shoulder fillets (Kumagai and Shimada, 1968). (a) $L/d = 1.5$.

152 SHOULDER FILLETS

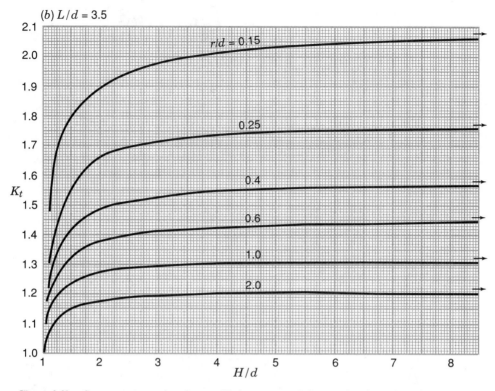

Chart 3.2b Stress concentration factors K_t for a stepped flat tension bar with shoulder fillets (Kumagai and Shimada, 1968). (b) $L/d = 3.5$.

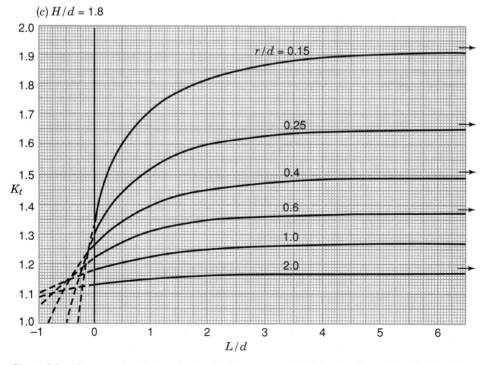

Chart 3.2c Stress concentration factors K_t for a stepped flat tension bar with shoulder fillets (Kumagai and Shimada, 1968). (c) $H/d = 1.8$.

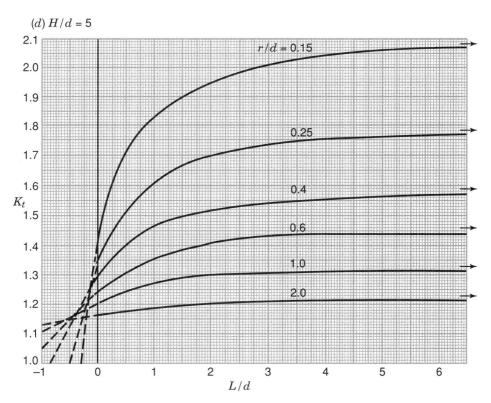

Chart 3.2d Stress concentration factors K_t for a stepped flat tension bar with shoulder fillets (Kumagai and Shimada, 1968). (*d*) $H/d = 5$.

154 SHOULDER FILLETS

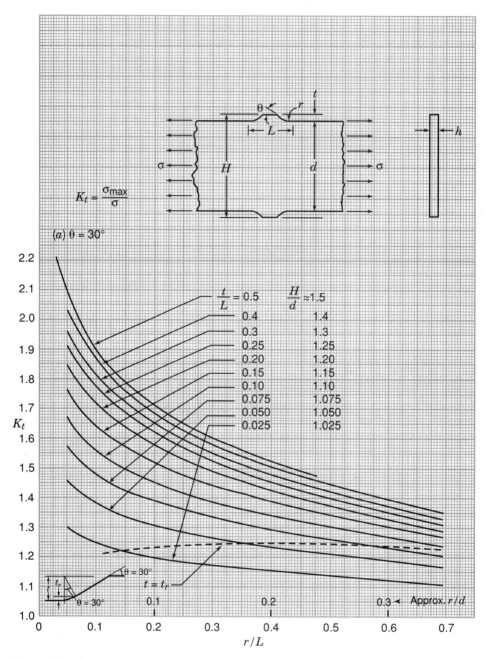

Chart 3.3a Stress concentration factors K_t for a trapezoidal protuberance on a tension member $L/(d/2) = 1.05$ (Derecho and Munse 1968). (a) $\theta = 30°$.

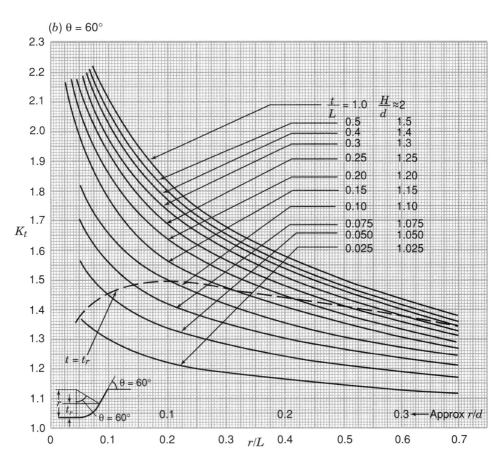

Chart 3.3b Stress concentration factors K_t for a trapezoidal protuberance on a tension member $L/(d/2) = 1.05$ (Derecho and Munse 1968). (b) $\theta = 60°$.

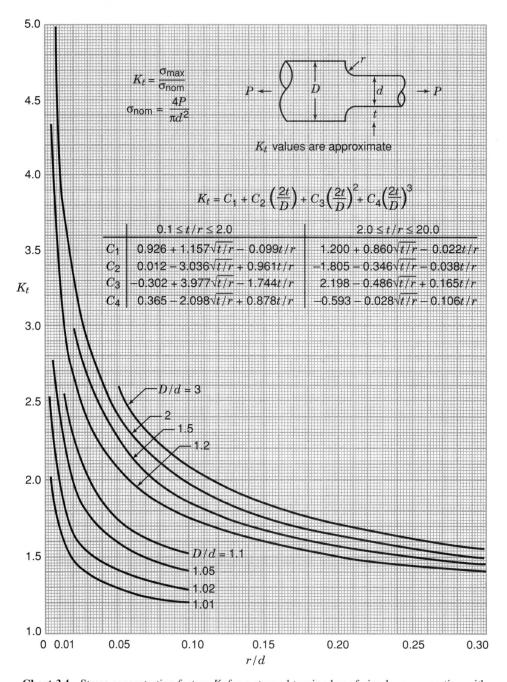

Chart 3.4 Stress concentration factors K_t for a stepped tension bar of circular cross section with shoulder fillet.

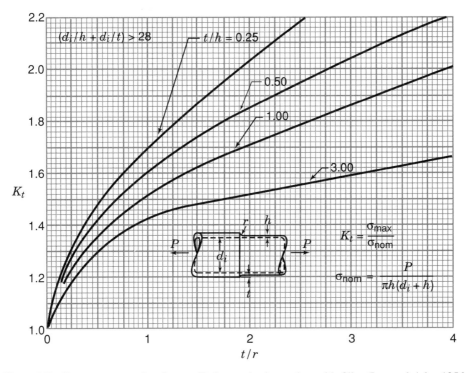

Chart 3.5 Stress concentration factors K_t for a tube in tension with fillet (Lee and Ades 1956; ESDU 1981).

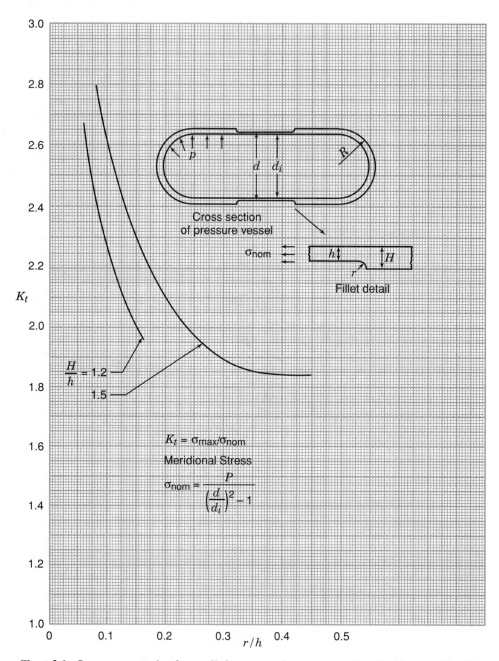

Chart 3.6 Stress concentration factors K_t for a stepped pressure vessel wall with a shoulder fillet $R/H \approx 10$ (Griffin and Thurman 1967).

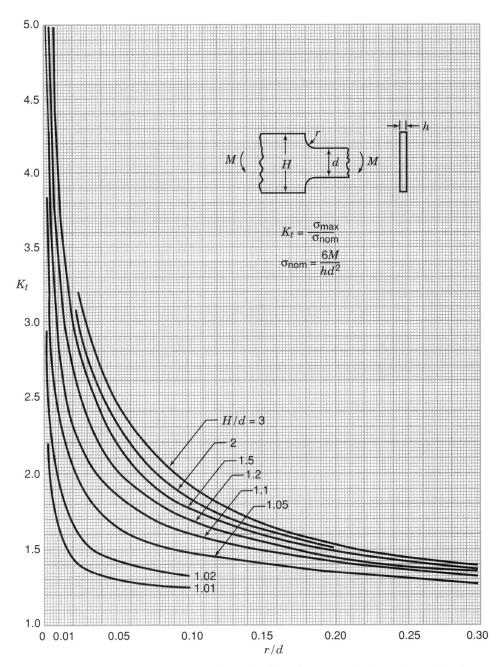

Chart 3.7 Stress concentration factors K_t for bending of a stepped flat bar with shoulder fillets (based on photoelastic tests of Leven and Hartman 1951; Wilson and White 1973).

160 SHOULDER FILLETS

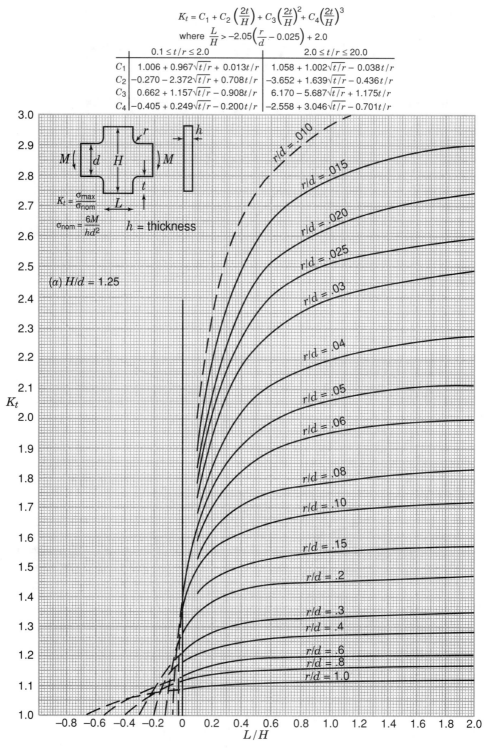

Chart 3.8a Effect of shoulder width L on stress concentration factors K_t for filleted bars in bending (based on photoelastic data by Leven and Hartman 1951). (a) $H/d = 1.25$.

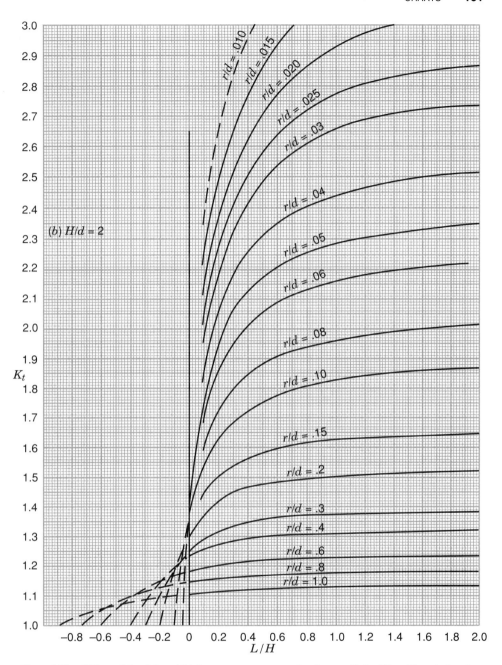

Chart 3.8b Effect of shoulder width L on stress concentration factors K_t for filleted bars in bending (based on photoelastic data by Leven and Hartman 1951). (b) $H/d = 2$.

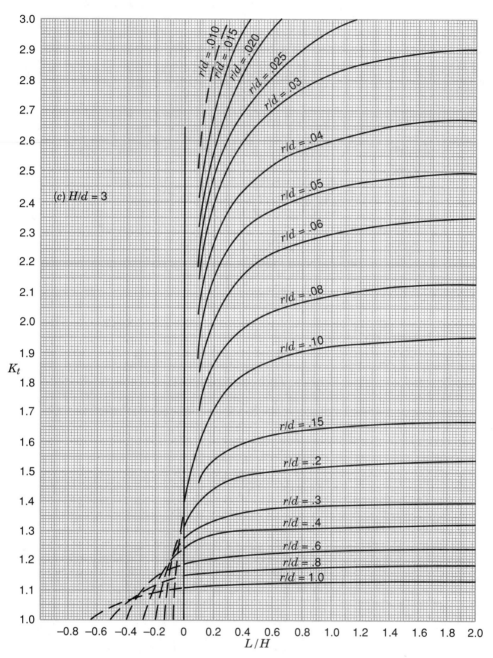

Chart 3.8c Effect of shoulder width L on stress concentration factors K_t for filleted bars in bending (based on photoelastic data by Leven and Hartman 1951). (c) $H/d = 3$.

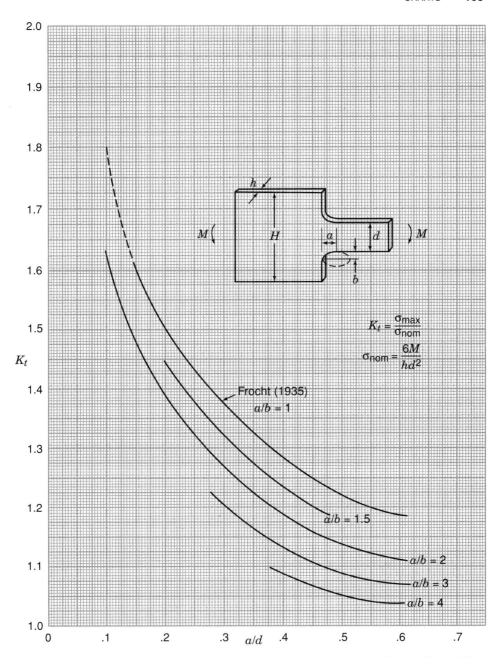

Chart 3.9 Stress concentration factors K_t for the bending case of a flat bar with an elliptical fillet. $H/d \approx 3$ (photoelastic tests of Berkey 1944).

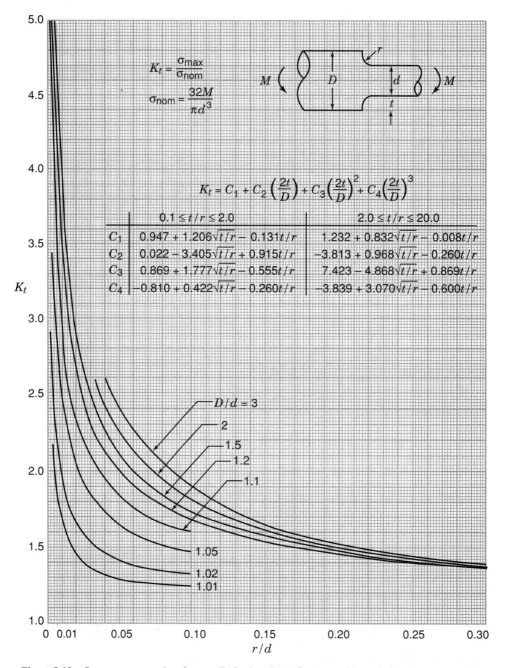

Chart 3.10 Stress concentration factors K_t for bending of a stepped bar of circular cross section with a shoulder fillet (based on photoelastic tests of Leven and Hartman 1951; Wilson and White 1973).

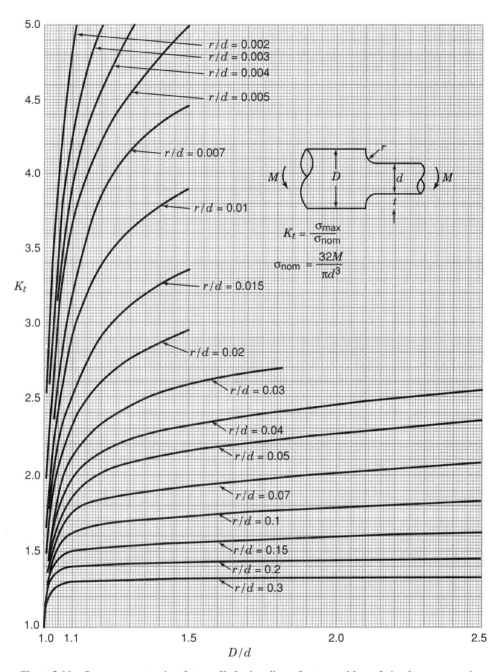

Chart 3.11 Stress concentration factors K_t for bending of a stepped bar of circular cross section with a shoulder fillet. This chart serves to supplement Chart 3.10 (based on photoelastic tests of Leven and Hartman 1951; Wilson and White 1973).

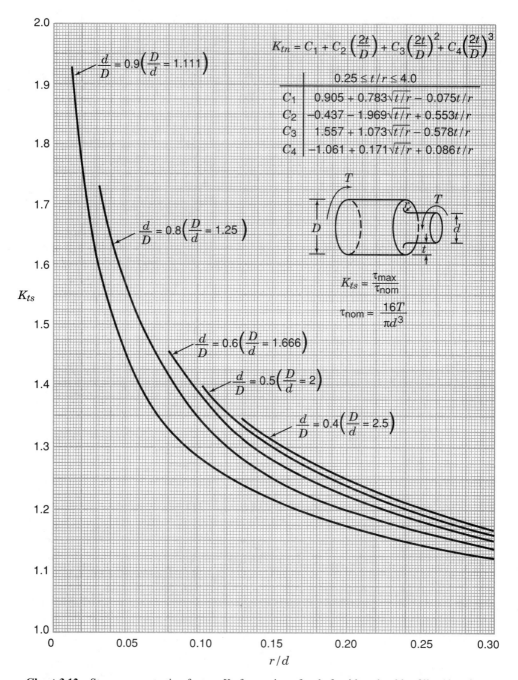

Chart 3.12 Stress concentration factors K_{ts} for torsion of a shaft with a shoulder fillet (data from Matthews and Hooke 1971).

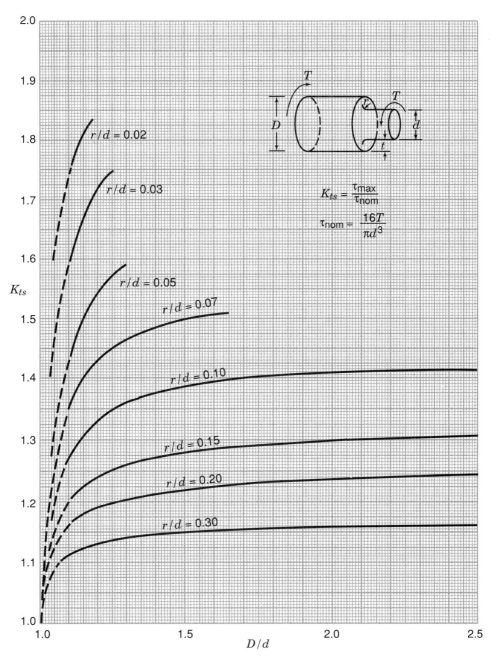

Chart 3.13 Stress concentration factors K_{ts} for torsion of a shaft with a shoulder fillet (data from Matthews and Hooke 1971). This chart serves to supplement Chart 3.12.

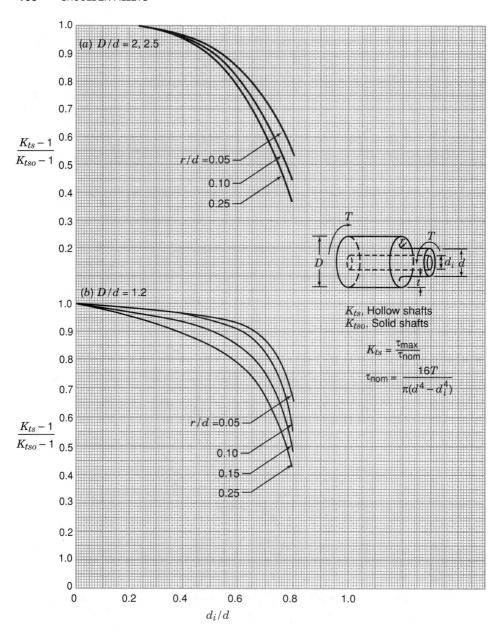

Chart 3.14 Effect of axial hole on stress concentration factors of a torsion shaft with a shoulder fillet (from data of Rushton 1964). (a) $D/d = 2, 2.5$; (b) $D/d = 1.2$.

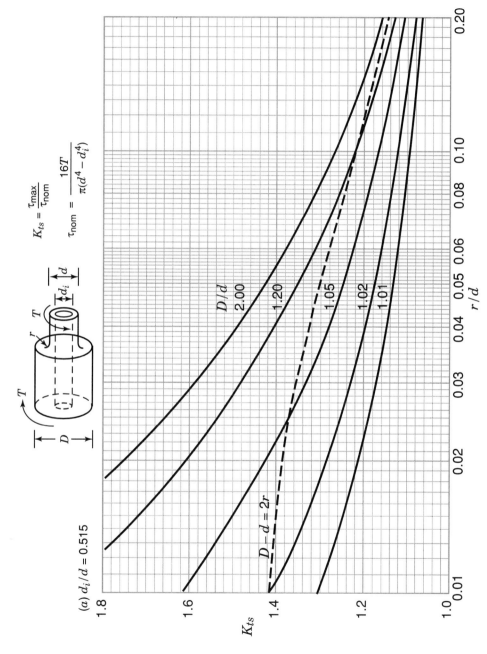

Chart 3.15a Stress concentration factors of a torsion tube with a shoulder fillet (Rushton 1964; ESDU 1981). (a) $d_i/d = 0.515$.

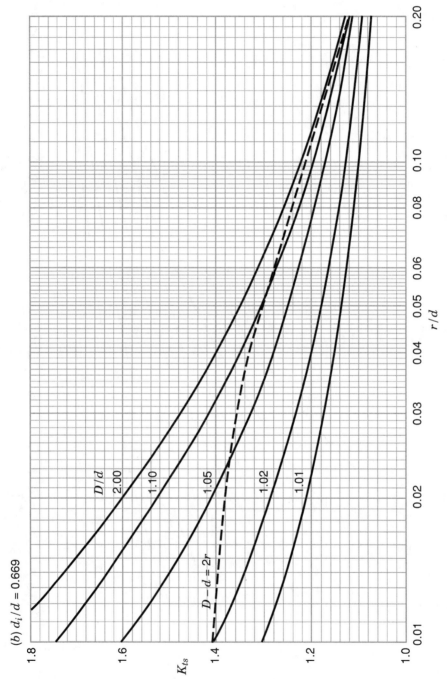

Chart 3.15b Stress concentration factors of a torsion tube with a shoulder fillet (Rushton 1964; ESDU 1981). (*b*) $d_i/d = 0.669$.

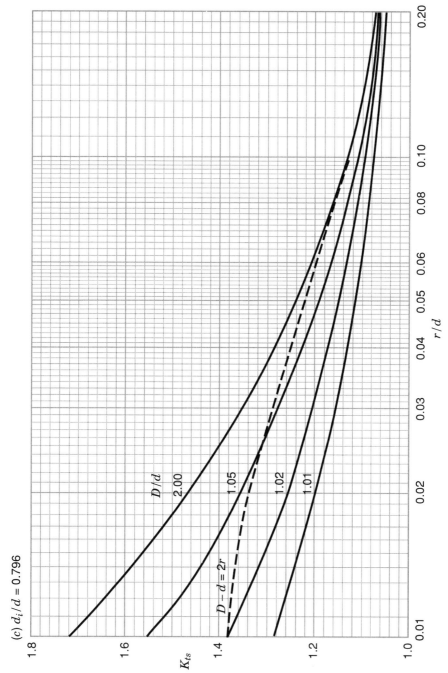

Chart 3.15c Stress concentration factors of a torsion tube with a shoulder fillet (Rushton 1964; ESDU 1981). (c) $d_i/d = 0.796$.

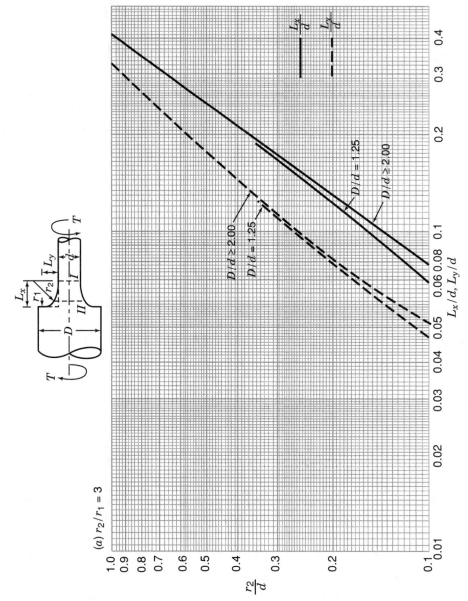

Chart 3.16a Radius of compound fillet for shoulder shaft in torsion. $K_{ts} = K_{tII}$ (Battenbo and Baines 1974; ESDU 1981). (a) $r_2/r_1 = 3$.

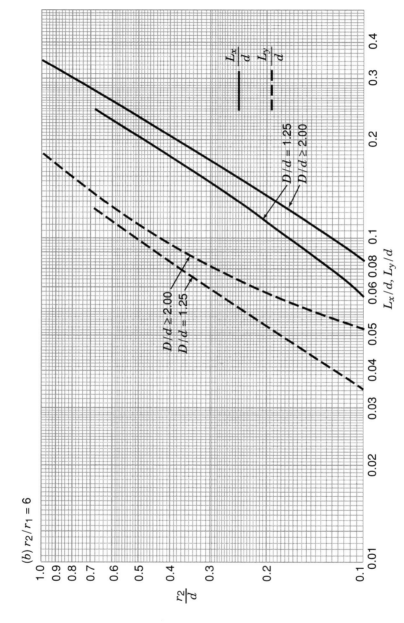

Chart 3.16b Radius of compound fillet for shoulder shaft in torsion. $K_{tl} = K_{tll}$ (Battenbo and Baines 1974; ESDU 1981). (b) $r_2/r_1 = 6$.

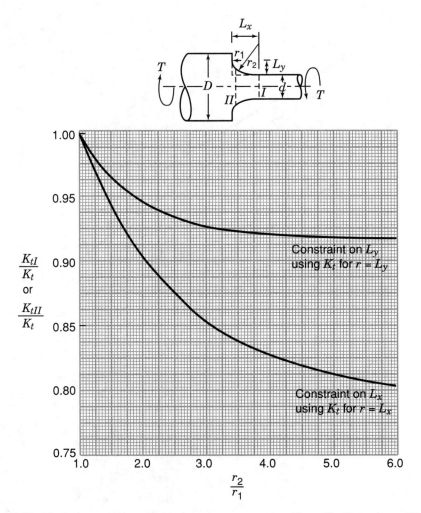

Chart 3.17 Maximum possible reduction in stress concentration. $K_{tI} = K_{tII}$ (Battenbo and Baines 1974; ESDU 1981).

CHAPTER 4

HOLES

Some structural members with transverse holes are shown in Fig. 4.1. The formulas and figures of the stress concentration factors are arranged in this chapter according to the loading (tension, torsion, bending, etc.), the shape of the hole (circular, elliptical, rectangular, etc.), single and multiple holes, two- and three-dimensional cases. In addition to "empty holes," various shaped inclusions are treated.

4.1 NOTATION

Definitions:
 Panel. A thin flat element with in-plane loading. This is a plane sheet that is sometimes referred to as a *membrane* or *diaphragm*.
 Plate. A thin flat element with transverse loading. This element is characterized by transverse displacements (i.e., deflections).
Symbols:

A = area (or point)
A_r = effective cross-sectional area of reinforcement
a = radius of hole
a = major axis of ellipse
a = half crack length
b = minor axis of ellipse
c = distance from center of hole to the nearest edge of element
C_A = reinforcement efficiency factor
C_s = shape factor

176 HOLES

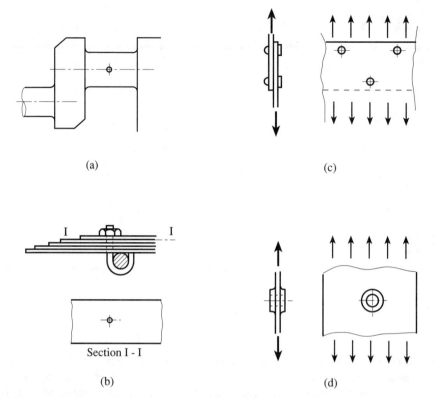

Figure 4.1 Examples of parts with transverse holes: (*a*) Oil hole in crankshaft (bending and torsion); (*b*) clamped leaf spring (bending); (*c*) riveted flat elements; (*d*) hole with reinforcing bead.

d = diameter of hole

D = outer diameter of reinforcement surrounding a hole

e = distance from center of hole to the furthest edge of the element

E = modulus of elasticity

E' = modulus of elasticity of inclusion material

h = thickness

h_r = thickness of reinforcement

h_t = total thickness, including reinforcement ($h_t = h + h_r$ or $h + 2h_r$)

H = height or width of element

K_t = theoretical stress concentration factor for normal stress

K_{te} = stress concentration factor at edge of the hole based on von Mises stress

K_{tf} = estimated fatigue notch factor for normal stress

K_{tg} = stress concentration factor with the nominal stress based on gross area

K_{tn} = stress concentation factor with the nominal stress based on net area

l = pitch, spacing between notches or holes

L = length of element

p = pressure

P = load
q = notch sensitivity factor
r = radius of hole, arc, notch
r, θ = polar coordinates
r, θ, x = cylindrical coordinates
R = radius of thin cylinder or sphere
s = distance between the edges of two adjacent holes
x, y, z = rectangular coordinates
α = material constant for evaluating notch sensitivity factor
ν = Poisson's ratio
σ = normal stress, typically the normal stress on gross section
σ_n = normal stress based on net area
σ_{eq} = equivalent stress
$\sigma_{\max}$ = maximum normal stress or maximum equivalent stress
σ_{nom} = nominal or reference normal stress
σ_{tf} = estimated fatigue strength
σ_1, σ_2 = biaxial in-plane normal stresses
$\sigma_1, \sigma_2, \sigma_3$ = principal stresses
τ = shear stress

4.2 STRESS CONCENTRATION FACTORS

As discussed in Section 1.2, the stress concentration factor is defined as the ratio of the peak stress in the body to a reference stress. Usually the stress concentration factor is K_{tg}, for which the reference stress is based on the gross cross-sectional area, or K_{tn}, for which the reference stress is based on the net cross-sectional area. For a two-dimensional element with a single hole (Fig. 1.4a), the formulas for these stress concentration factors are

$$K_{tg} = \frac{\sigma_{\max}}{\sigma} \tag{4.1}$$

where K_{tg} is the stress concentration factor based on gross stress, $\sigma_{\max}$ is the maximum stress, at the edge of the hole, σ is the stress on gross section far from the hole, and

$$K_{tn} = \frac{\sigma_{\max}}{\sigma_n} \tag{4.2}$$

where K_{tn} is the stress concentration factor based on net (nominal) stress and σ_n is the net stress $\sigma/(1 - d/H)$, with d the hole diameter and H the width of element (Fig. 1.4a). From the foregoing,

$$K_{tn} = K_{tg}\left(1 - \frac{d}{H}\right) = K_{tg}\frac{\sigma}{\sigma_n} \tag{4.3}$$

178 HOLES

or

$$K_{tg} = \frac{\sigma_n}{\sigma} K_{tn} = \frac{\sigma_n}{\sigma} \cdot \frac{\sigma_{max}}{\sigma_n} \qquad (4.4)$$

The significance of K_{tg} and K_{tn} can be seen by referring to Chart 4.1. The factor K_{tg} takes into account the two effects: (1) increased stress due to loss of section (term σ_n/σ in Eq. 4.4); and (2) increased stress due to geometry (term σ_{max}/σ_n). As the element becomes narrower (the hole becomes larger), $d/H \to 1$, $K_{tg} \to \infty$. However, K_{tn} takes account of only one effect; increased stress due to geometry. As the hole becomes larger, $d/H \to 1$, the element becomes in the limit a uniform tension member, with $K_{tn} = 1$. Either Eq. (4.1) or (4.2) can be used to evaluate σ_{max}. Usually the simplest procedure is to use K_{tg}. If the stress gradient is of concern as in certain fatigue problems, the proper factor to use is K_{tn}. See Section 1.2 for more discussion on the use of K_{tg} and K_{tn}.

Example 4.1 Fatigue Stress of Element with Square Pattern of Holes Consider a thin, infinite element with four holes arranged in a square pattern subjected to uniaxial tension with $s/l = 0.1$ (Fig. 4.2). From Chart 4.2, it can be found that $K_{tg} = 10.8$ and $K_{tn} = (s/l)K_{tg} = 1.08$, the difference being that K_{tg} takes account of the loss of section. The material is low carbon steel and the hole diameter is 0.048 in. Assuming a fatigue strength (specimen without stress concentration) $\sigma_f = 30,000$ lb/in.^{2}, we want to find the estimated fatigue strength of the member with holes.

From Eq. (1.49), expressed in terms of estimated values, the estimated fatigue stress σ_{tf} of the member with holes is given by

$$\sigma_{tf} = \frac{\sigma_f}{K_{tf}} \qquad (1)$$

where K_{tf} is the estimated fatigue notch factor of the member with holes. From Eq. (1.53),

$$K_{tf} = q(K_t - 1) + 1 \qquad (2)$$

where K_t is the theoretical stress concentration factor of the member and q is the notch sensitivity of the member and from Eq. (1.55),

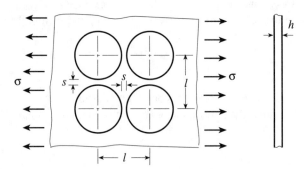

Figure 4.2 A thin infinite element with four holes.

$$q = \frac{1}{1 + \alpha/r} \tag{3}$$

where α is a material constant and r is the notch radius.

For annealed or normalized steel, it is found that $\alpha = 0.01$ (Section 1.9) and for a hole of radius $r = 0.024$, the notch sensitivity is given by

$$q = \frac{1}{1 + 0.01/0.024} \approx 0.7 \tag{4}$$

If the factor K_{tn} is used, (2) becomes

$$K_{tf} = q(K_{tn} - 1) + 1 = 0.7(1.08 - 1) + 1 = 1.056 \tag{5}$$

Substitute K_{tf} into (1)

$$\sigma_{tf} = \frac{30,000}{1.056} = 28,400 \text{ lb/in.}^2 \tag{6}$$

This means that if the effect of stress concentration is considered, the estimated fatigue stress on the net section is 28,400 lb/in.² Since $s/l = 0.1$, the area of the gross section $(l \times h)$ is ten times that of the net section $(s \times h) = (s/l)(l \times h) = 0.1(l \times h)$. Since the total applied loading remains unchanged, the estimated fatigue stress applied on the gross section should be 2840 lb/in.²

If the estimate is obtained by use of the factor K_{tg},

$$K'_{tf} = q(K_{tg} - 1) + 1 = 0.7(10.8 - 1) + 1 = 7.86 \tag{7}$$

$$\sigma'_{tgf} = \frac{\sigma_f}{K'_{tf}} = \frac{30,000}{7.86} = 3,820 \text{ lb/in.}^2 \tag{8}$$

Thus, if K_{tg} is used, the estimated fatigue stress on the gross section is 3820 lb/in.², and the corresponding estimated fatigue stress on the net section is

$$\sigma'_{tf} = \frac{\sigma'_{tgf}}{s/l} = 38,200 \text{ lb/in.}^2 \tag{9}$$

The result of (9) is erroneous, since it means that the fatigue limit of a specimen with holes (σ'_{tf}) is larger than the fatigue limit of a specimen without holes (σ_f). When q is applied, it is necessary to use K_{tn}. Note that 28,400 lb/in.² is close to the full fatigue strength of 30,000 lb/in.² This is because an element between two adjacent holes is like a tension specimen, with small stress concentration due to the relatively large holes.

In an element of arbitrary thickness with a transverse hole (Fig. 4.3), the maximum stress on the surface of the hole varies across the thickness of the element, being lower at the surface (point A) and somewhat higher in the interior (point A'). For a hole of radius r in an element of thickness $(3/2)r$, subjected to uniaxial tension and with $\nu = 0.3$, the maximum stress at the surface was 7% less than the two-dimensional stress concentration factor of 3.0

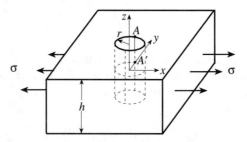

Figure 4.3 An element with a transverse hole.

(Section 4.3), whereas the stress at the midplane was less than 3% higher (Sternberg and Sadowsky 1949). They mention "the general assertion that factors of stress concentration based on two-dimensional analysis sensibly apply to elements of arbitrary thickness ratio."

In a later analysis (Youngdahl and Sternberg 1966) of an infinitely thick solid (semi-infinite body, mathematically) subjected to shear (or biaxial stress $\sigma_2 = -\sigma_1$), and with $\nu = 0.3$, the maximum stress at the surface of the hole was found to be 23% lower than the value normally utilized for a thin element (Eqs. 4.17 and 4.18), and the corresponding stress at a depth of the hole radius was 3% higher.

In summarizing the foregoing discussion of stress variation in the thickness direction of elements with a hole it can be said that the usual two-dimensional stress concentration factors are sufficiently accurate for design application to elements of arbitrary thickness. This is of interest in the mechanics of materials and failure analysis, since failure would be expected to start down in the hole rather than at the surface, in the absence of other factors such as those due to processing or manufacturing.

4.3 CIRCULAR HOLES WITH IN-PLANE STRESSES

4.3.1 Single Circular Hole in an Infinite Thin Element in Uniaxial Tension

A fundamental case of stress concentration is the study of the stress distribution around a circular hole in an infinite thin element (panel), which is subjected to uniaxial in-plane tension stress (Fig. 4.4). In polar coordinates, with applied stress σ the stresses are given as (Timoshenko and Goodier 1970)

$$\sigma_r = \frac{1}{2}\sigma\left(1 - \frac{a^2}{r^2}\right) + \frac{1}{2}\sigma\left(1 - \frac{4a^2}{r^2} + \frac{3a^4}{r^4}\right)\cos 2\theta$$

$$\sigma_\theta = \frac{1}{2}\sigma\left(1 + \frac{a^2}{r^2}\right) - \frac{1}{2}\sigma\left(1 + \frac{3a^4}{r^4}\right)\cos 2\theta \qquad (4.5)$$

$$\tau_{r\theta} = -\frac{1}{2}\sigma\left(1 + \frac{2a^2}{r^2} - \frac{3a^4}{r^4}\right)\sin 2\theta$$

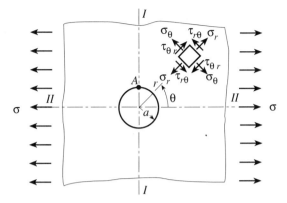

Figure 4.4 An infinite thin element with hole under tensile load.

where a is the radius of the hole, r and θ are the polar coordinates of a point in the element as shown in Fig. 4.4. At the edge of the hole with $r = a$,

$$\sigma_r = 0$$
$$\sigma_\theta = \sigma(1 - 2\cos 2\theta) \tag{4.6}$$
$$\tau_{r\theta} = 0$$

At point A, $\theta = \pi/2$ (or $3\pi/2$) and

$$\sigma_{\theta A} = 3\sigma$$

This is the maximum stress around the circle, so the stress concentration factor for this case is 3.

The distribution of σ_θ at the edge of the hole is shown in Fig. 4.5. At point B, with $\theta = 0$, Eq. (4.6) gives

$$\sigma_{\theta B} = -\sigma$$

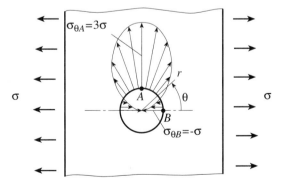

Figure 4.5 Circumferential stress distribution on the edge of a circular hole in an infinite thin element.

182 HOLES

When $\theta = \pm\pi/6$ (or $\pm 5\pi/6$)

$$\sigma_\theta = 0$$

Consider section I–I, which passes through the center of the hole and point A, as shown in Fig. 4.4. For the points on section I–I, $\theta = \pi/2$ (or $3\pi/2$) and Eq. (4.5) becomes

$$\sigma_r = \frac{3}{2}\sigma\left(\frac{a^2}{r^2} - \frac{a^4}{r^4}\right)$$

$$\sigma_\theta = \frac{1}{2}\sigma\left(2 + \frac{a^2}{r^2} + \frac{3a^4}{r^4}\right) \qquad (4.7)$$

$$\tau_{r\theta} = 0$$

From Eq. (4.7), it can be observed that on cross section I–I, when $r = a$, $\sigma_\theta = 3\sigma$, and as r increases, σ_θ decreases. Eventually, when r is large enough, $\sigma_\theta = \sigma$, and the stress distribution recovers to a uniform state. Also, it follows from Eq. (4.7) that the stress concentration caused by a single hole is localized. When, for example, $r = 5.0a$, σ_θ decreases to 1.02σ. Thus, after $5a$ distance from the center, the stress is very close to a uniform distribution.

The stress distribution over cross section II–II of Fig. 4.4 can be obtained using similar reasoning. Thus, from Eq. (4.5) with $\theta = 0$ (or $\theta = \pi$),

$$\sigma_r = \frac{1}{2}\sigma\left(2 - \frac{5a^2}{r^2} + \frac{3a^4}{r^4}\right)$$

$$\sigma_\theta = \frac{1}{2}\sigma\left(\frac{a^2}{r^2} - \frac{3a^4}{r^4}\right) \qquad (4.8)$$

$$\tau_{r\theta} = 0$$

Figure 4.6 shows the σ_θ distribution on section I–I and the σ_r distribution over section II–II. Note that on cross section II–II, $\sigma_r \leq \sigma$, although it finally reaches σ. The stress

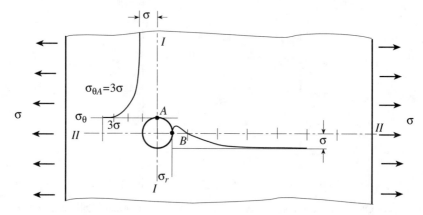

Figure 4.6 The distribution of σ_θ on section I–I and σ_r on section II–II.

gradient on section *II–II* is less than that on section *I–I*. For example, on section *II–II* when $r = 11.0a$, $\sigma_r = 0.98\sigma$ or $\sigma - \sigma_r = 2\%$. In contrast, on section *I–I*, when $r = 5.0a$, σ_θ reaches σ within the 2% deviation.

For the tension case of a finite-width thin element with a circular hole, K_t values are given in Chart 4.1 for $d/H \leq 0.5$ (Howland 1929–30). Photoelastic values (Wahl and Beeuwkes 1934) and analytical results (Isida 1953; Christiansen 1968) are in good agreement. For a row of holes in the longitudinal direction with a hole to hole center distance/hole diameter of 3, and with $d/H = 1/2$, Slot (1972) obtained good agreement with the Howland K_t value (Chart 4.1) for the single hole with $d/H = 1/2$.

In a photoelastic test (Coker and Filon 1931), it was noticed that as d/H approached unity the stress σ_θ on the outside edges of the panel approached ∞, which would correspond to $K_{tn} = 2$. Many other researchers also indicate that $K_{tn} = 2$ for $d/H \to 1$ (Wahl and Beeuwkes 1934; Koiter 1957; Heywood 1952). Wahl and Beeuwkes observed that when the hole diameter so closely approaches the width of the panel that the minimum section between the edge of the element and the hole becomes an infinitely thin filament; for any finite deformation they noted that "this filament may move inward toward the center of the hole sufficiently to allow for a uniform stress distribution, thus giving $K_{tn} = 1$. For infinitely small deformations relative to the thickness of this filament, however, K_{tn} may still be equal to 2." They found with a steel model test that the curve does not drop down to unity as fast as would appear from certain photoelastic tests (Hennig 1933). Since the inward movement varies with σ and E, the K_{tn} would not drop to 1.0 as rapidly as with a plastic model. The case of $d/H \to 1$, does not have much significance from a design standpoint. Further discussion is provided in Belie and Appl (1972).

An empirical formula for K_{tn} was proposed to cover the entire d/H range (Heywood 1952)

$$K_{tn} = 2 + \left(1 - \frac{d}{H}\right)^3 \tag{4.9}$$

The formula is in good agreement with the results of Howland (Heywood 1952) for $d/H < 0.3$ and is only about 1.5% lower at $d/H = 1/2$ ($K_{tn} = 2.125$ versus $K_{tn} = 2.16$ for Howland). The Heywood formula of Eq. (4.9) is satisfactory for many design applications, since in most cases d/H is less than 1/3. Note that the formula gives $K_{tn} = 2$ as $d/H \to 1$, which seems reasonable.

The Heywood formula, when expressed as K_{tg}, becomes

$$K_{tg} = \frac{2 + (1 - d/H)^3}{1 - (d/H)} \tag{4.10}$$

Factors for a circular hole near the edge of a semi-infinite element in tension are shown in Chart 4.2 (Mindlin 1948; Udoguti 1947; Isida 1955a). The load carried by the section between the hole and the edge of the panel is (Mindlin 1948)

$$P = \sigma c h \sqrt{1 - (a/c)^2} \tag{4.11}$$

where σ is the stress applied to semi-infinite panel, c is the distance from center of hole to edge of panel, a is the radius of hole, and h is the thickness of panel.

In Chart 4.2 the upper curve gives values of $K_{tg} = \sigma_B/\sigma$, where σ_B is the maximum stress at the edge of the hole nearest the edge of the thin tensile element. Although the factor

184 HOLES

K_{tg} may be used directly in design, it was thought desirable to also compute K_{tn} based on the load carried by the minimum net section. The K_{tn} factor will be comparable with the stress concentration factors for other cases (Example 4.1). Based on the actual load carried by the minimum net section (Eq. 4.11), the average stress on the net section A–B is

$$\sigma_{\text{net A-B}} = \frac{\sigma ch\sqrt{1 - (a/c)^2}}{(c - a)h} = \frac{\sigma\sqrt{1 - (a/c)^2}}{1 - a/c}$$

$$K_{tn} = \frac{\sigma_B}{\sigma_{\text{net A-B}}} = \frac{\sigma_B(1 - a/c)}{\sigma\sqrt{1 - (a/c)^2}} \tag{4.12}$$

The symbols σ, c, a, h have the same meaning as those in Eq. (4.11).

The case of a tension bar of finite width having an eccentrically located hole has been solved analytically by Sjöström (1950). The semi-infinite strip values are in agreement with Chart 4.2. Also the special case of the centrally located hole is in agreement with the Howland solution (Chart 4.1). The results of the Sjöström analysis are given as $K_{tg} = \sigma_{\max}/\sigma$ values in the upper part of Chart 4.3. These values may be used directly in design. An attempt will be made in the following to arrive at approximate K_{tn} factors based on the net section. When the hole is centrally located ($e/c = 1$ in Chart 4.3), the load carried by section A–B is σch. As e/c is increased to infinity, the load carried by section A–B is, from Eq. (4.11), $\sigma ch\sqrt{1 - (a/c)^2}$. Assuming a linear relation between the foregoing end conditions, that is, $e/c = 1$ and $e/c = \infty$, results in the following expression for the load carried by section A–B:

$$P_{A-B} = \frac{\sigma ch\sqrt{1 - (a/c)^2}}{1 - (c/e)(1 - \sqrt{1 - (a/c)^2})} \tag{4.13}$$

The stress on the net section A–B is

$$\sigma_{\text{net A-B}} = \frac{\sigma ch\sqrt{1 - (a/c)^2}}{h(c - a)[1 - (c/e)(1 - \sqrt{1 - (a/c)^2})]}$$

so that

$$K_{tn} = \frac{\sigma_{\max}}{\sigma_{\text{net}}} = \frac{\sigma_{\max}(1 - a/c)}{\sigma\sqrt{1 - (a/c)^2}}[1 - (c/e)(1 - \sqrt{1 - (a/c)^2})] \tag{4.14}$$

It is seen from the lower part of Chart 4.3 that this relation brings all the K_{tn} curves rather closely together. For all practical purposes, then, the curve for the centrally located hole ($e/c = 1$) is, under the assumptions of Chart 4.3, a reasonable approximation for all eccentricities.

4.3.2 Single Circular Hole in an Infinite Thin Element under Biaxial In-plane Stresses

If a thin infinite element is subjected to biaxial in-plane tensile stresses σ_1 and σ_2 as shown in Fig. 4.7, the stress concentration factor may be derived by superposition. Equation (4.5) is the solution for the uniaxial problem of Fig. 4.4. At the edge of the hole for the biaxial

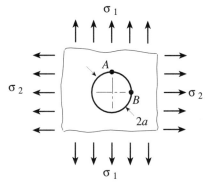

Figure 4.7 An infinite thin element under biaxial tensile in-plane loading.

case of Fig. 4.7, the stresses caused by σ_1 are calculated by setting $r = a$, $\sigma = \sigma_1$, $\theta = \theta + \pi/2$ in Eq. (4.6):

$$\sigma_r = 0$$
$$\sigma_\theta = \sigma_1(1 + 2\cos 2\theta) \quad (4.15)$$
$$\tau_{r\theta} = 0$$

Superimpose Eq. (4.15) and Eq. (4.6) with σ replaced by σ_2, which represents the stresses under uniaxial tension σ_2:

$$\sigma_r = 0$$
$$\sigma_\theta = (\sigma_2 + \sigma_1) - 2(\sigma_2 - \sigma_1)\cos 2\theta \quad (4.16)$$
$$\tau_{r\theta} = 0$$

Let $\sigma_2/\sigma_1 = \alpha$ so that

$$\sigma_\theta = \sigma_1(1 + \alpha) + 2\sigma_1(1 - \alpha)\cos 2\theta$$

Assume $\alpha \le 1$. Then

$$\sigma_{\theta\,max} = \sigma_{\theta B} = \sigma_1(3 - \alpha)$$
$$\sigma_{\theta\,min} = \sigma_{\theta A} = \sigma_1(3\alpha - 1)$$

If σ_1 is taken as the reference stress, the stress concentration factors at points A and B are

$$K_{tA} = \frac{\sigma_{\theta\,min}}{\sigma_1} = 3\alpha - 1 \quad (4.17)$$

$$K_{tB} = \frac{\sigma_{\theta\,max}}{\sigma_1} = 3 - \alpha \quad (4.18)$$

It is interesting to note that if σ_1 and σ_2 are both of the same sign (positive or negative), the stress concentration factor is less than 3, which is the stress concentration factor caused by uniaxial stress. For equal biaxial stresses, $\sigma_1 = \sigma_2$, the stresses at A and B are $\sigma_A = \sigma_B = 2\sigma_1$ or $K_t = 2$ ($h_r/h = 0, D/d = 1$ in Chart 4.13a). When σ_1 and σ_2 have same magnitude but are of opposite sign (the state of pure shear), $K_t = 4$ ($K_{tA} = -4$, $K_{tB} = 4$). This is equivalent to shear stresses $\tau = \sigma_1$ at $45°$ ($a/b = 1$ in Chart 4.88).

4.3.3 Single Circular Hole in a Cylindrical Shell with Tension or Internal Pressure

Considerable analytical work has been done on the stress in a cylindrical shell having a circular hole (Lekkerkerker 1964; Eringen et al. 1965; Van Dyke 1965). Stress concentration factors are given for tension in Chart 4.4 and for internal pressure in Chart 4.5. In both charts, factors for membrane (tension) and for total stresses (membrane plus bending) are given. The torsion case is given in Section 4.7, Chart 4.98.

For pressure loading the analysis assumes that the force representing the total pressure corresponding to the area of the hole is carried as a perpendicular shear force distributed around the edge of the hole. This is shown schematically in Chart 4.5. Results are given as a function of dimensionless parameter β:

$$\beta = \frac{\sqrt[4]{3(1-\nu^2)}}{2}\left(\frac{a}{\sqrt{Rh}}\right) \qquad (4.19)$$

where R is the mean radius of shell, h is the thickness of shell, a is the radius of hole, and ν is Poisson's ratio. In Charts 4.4 and 4.5, and Fig. 4.8 where $\nu = 1/3$,

$$\beta = 0.639 \frac{a}{\sqrt{Rh}} \qquad (4.20)$$

The analysis assumes a shallow, thin shell. Shallowness means a small curvature effect over the circumferential coordinate of the hole, which means a small a/R. Thinness of course implies a small h/R. The region of validity is shown in Fig. 4.8.

The physical significance of β can be evaluated by rearranging Eq. (4.20):

$$\beta = 0.639 \frac{(a/R)}{\sqrt{h/R}} \qquad (4.21)$$

For example, by solving Eq. (4.21) for h, a 10-in. diameter cylinder with a 1-in. hole would have a thickness of 0.082 in. for $\beta = 1/2$, a thickness of 0.02 in. for $\beta = 1$, a thickness of 0.005 in. for $\beta = 2$, and a thickness of 0.0013 in. for $\beta = 4$. Although $\beta = 4$ represents a very thin shell, large values of β often occur in aerospace structures. A formula is available (Lind 1968) for the pressurized shell where β is large compared to unity.

The K_t factors in Charts 4.4 and 4.5 are quite large for the larger values of β, corresponding to very thin shells. Referring to Fig. 4.8,

β	h/R
4	< 0.003
2	< 0.007
1	< 0.015
1/2	< 0.025

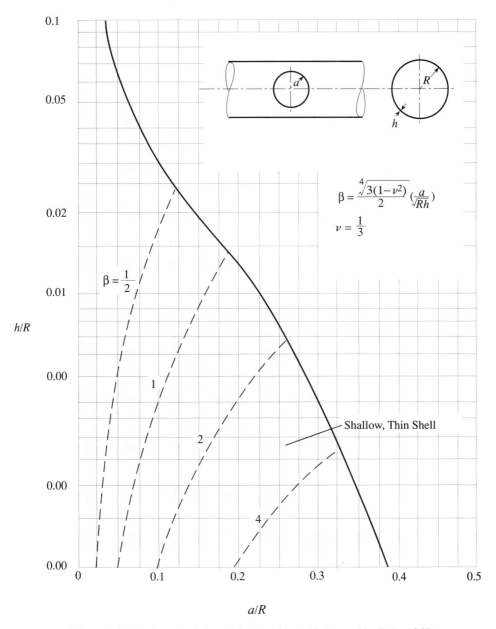

Figure 4.8 Region of validity of shallow, thin shell theory (Van Dyke 1965).

In the region of $\beta = 1/2$, the K_t factors are not unusually large.

The theoretical results (Lekkerkerker 1964; Eringen et al. 1965; Van Dyke 1965) are, with one exception, in good agreement. Experiments have been made by Houghton and Rothwell (1962) and by Lekkerkerker (1964). Comparisons made by Van Dyke (1965) showed reasonably good agreement for pressure loading (Houghton and Rothwell 1962).

Poor agreement was obtained for the tension loading (Houghton and Rothwell 1962). Referring to tests on tubular members (Jessop et al. 1959), the results for $d_i/D = 0.9$ are in good agreement for tension loadings (Chart 4.66). Photoelastic tests (Durelli et al. 1967) were made for the pressurized loading. Strain gage results (Pierce and Chou 1973) have been obtained for values of β up to 2 and agree reasonably well with Chart 4.4.

The case of two circular holes has been analyzed by Hanzawa et al. (1972) and Hamada et al. (1972). It was found that the interference effect is similar to that in an infinite thin element, although the stress concentration factors are higher for the shell. The membrane and bending stresses for the single hole (Hamada et al. 1972) are in good agreement with the results by Van Dyke (1965) on which Charts 4.4 and 4.5 are based.

Stress concentration factors have been obtained for the special case of a pressurized ribbed shell with a reinforced circular hole interrupting a rib (Durelli et al. 1971). Stresses around an elliptical hole in a cylindrical shell in tension have been determined by Murthy (1969), Murthy and Bopu Rao (1970), and Tingleff (1971).

4.3.4 Circular or Elliptical Hole in a Spherical Shell with Internal Pressure

Consider holes in the wall of a thin spherical shell subject to internal pressure. Chart 4.6 based on K_t factors determined analytically (Leckie et al. 1967) covers openings varying from a circle to an ellipse with $b/a = 2$. Referring to Chart 4.6, the K_t values for the four b/a values in an infinite flat element biaxially stressed are shown along the left-hand edge of the chart. The curves show the increase due to bending and shell curvature in relation to the flat element values. Experimental results (Leckie et al. 1967) are in good agreement. Application to the case of an oblique nozzle is discussed in the same article.

4.3.5 Reinforced Hole Near the Edge of a Semi-infinite Element in Uniaxial Tension

Consider a semi-infinite thin element subject to uniaxial tension. A circular hole with integral reinforcement of the same material is located near the edge of the element. Stress concentration factors are shown in Chart 4.7 (Davies 1963; Wittrick 1959; Mansfield 1955; ESDU 1981). High stresses would be expected to occur at points A and B. In the chart the values of K_{tgA} and K_{tgB} are plotted versus $A_e/(2ah)$ for a series of values of c/a. The quantity A_e is called the effective cross-sectional area of reinforcement,

$$A_e = C_A A_r \qquad (4.22)$$

where A_r is the cross-sectional area of the reinforcement (constant around hole), C_A is the reinforcement efficiency factor. Some values of C_A are given in Chart 4.7.

For point A, which is at the element edge, the gross stress concentration factor is defined as the ratio of the maximum stress acting along the edge and the tensile stress σ:

$$K_{tgA} = \frac{\sigma_{\max}}{\sigma} \qquad (4.23)$$

where $\sigma_{\max}$ is the maximum stress at point A along the edge.

At the junction (B) of the element and the reinforcement, the three-dimensional stress fields are complicated. It is reasonable to use the equivalent stress σ_{eq} (Section 1.8) at B as the basis to define the stress concentration factor. Define the gross stress concentration factor K_{tgB} as

$$K_{tgB} = \frac{\sigma_{eq}}{\sigma} \qquad (4.24)$$

As shown in the Figure in Chart 4.7, the two points B are symmetrically located with respect to the minimum crosssection I–I. For $A_e/(2ah) < 0.1$, the two points B coincide for any value of c/a. If $A_e/(2ah) > 0.1$, the two points B move further away as either c/a or $A_e/(2ah)$ increases. Similarly the two edge stress points A are also symmetrical relative to the minimum cross-section I–I and spread apart with an increase in c/a. For $c/a = 1.2$ the distance between two points A is equal to a. When $c/a = 5$, the distance is $6a$.

If the distance between element edges of a finite-width element and the center of the hole is greater than $4a$ and the reference stress is based on the gross cross section, the data from Chart 4.7 will provide a reasonable approximation.

The value of C_A depends on the geometry of the reinforcement and the manner in which it is mounted. If the reinforcement is symmetrical about the mid-plane of the thin element and if the reinforcement is connected to the thin element without defect, then the change in stress across the junction can be ignored and the reinforcement efficiency factor is equal to 1 ($C_A = 1$ and $A_e = A_r$). If the reinforcement is nonsymmetric and lies only to one side of the element, the following approximation is available:

$$C_A = 1 - \frac{A_r \bar{y}^2}{I} \qquad (4.25)$$

where $\bar{y}$ is the distance of the centroid of the reinforcement from the mid-plane of the element (e.g., see Fig. 4.9), I is the moment of inertia of the reinforcement about the mid-plane of the element. If the reinforcement is not symmetric, bending stress will be induced in the element. The data in Chart 4.7 ignore the effect of this bending.

Example 4.2   **L *Section Reinforcement***   Find the maximum stresses in a thin element with a 4.1 in. radius hole, whose center is 5.5 in. from the elements edge. The thickness of the element is 0.04 in. The hole is reinforced with an L section as shown in Fig. 4.9. A uniaxial in-plane tension stress of $\sigma = 6900$ lb/in^2. is applied to the thin element. For the reinforcement, with the dimensions of Fig. 4.9, $A_r = 0.0550$ in.^{2}, $\bar{y} = 0.0927$ in, and $I = 0.000928$ in.^{4}, where A_r is the cross-sectional area of the L section reinforcement, $\bar{y}$ is the distance of the centroid of the reinforcement from the mid-plane of the element, and I is the moment of inertia of the reinforcement about the mid-plane of the element (Fig. 4.9).

Begin by calculating the reinforcement efficiency factor C_A using Eq. (4.25)

$$C_A = 1 - \frac{A_r \bar{y}^2}{I} = 1 - \frac{0.0550 \cdot 0.0927^2}{0.000928} = 0.490 \qquad (1)$$

The effective cross-sectional area is given by (Eq. 4.22)

$$A_e = C_A A_r = 0.490 \cdot 0.0550 = 0.0270 \text{ in.}^2 \qquad (2)$$

Thus

$$\frac{A_e}{2ah} = \frac{0.0270}{2 \cdot 4.1 \cdot 0.04} = 0.0822 \quad \text{and} \quad \frac{c}{a} = \frac{5.5}{4.1} = 1.34 \qquad (3)$$

190 HOLES

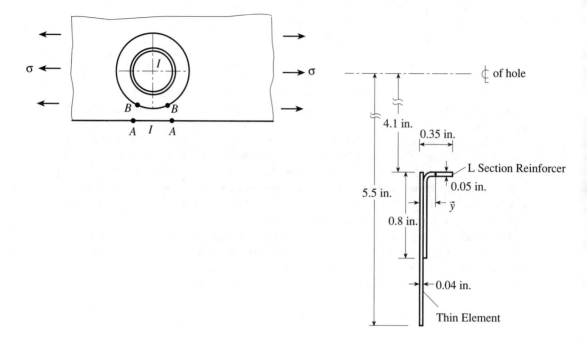

I-I Section Enlargement

Figure 4.9 A hole with L section reinforcement.

From the curves of Chart 4.7 for $A_e/(2ah) = 0.0822$, when $c/a = 1.3$, $K_{tgB} = 2.92$, $K_{tgA} = 2.40$, and when $c/a = 1.5$, $K_{tgB} = 2.65$, $K_{tgA} = 1.98$. The stress concentration factor at $c/a = 1.34$ can be derived by interpolation

$$K_{tgB} = 2.92 + \frac{2.65 - 2.92}{1.5 - 1.3} \cdot (1.34 - 1.3) = 2.86 \tag{4}$$

$$K_{tgA} = 2.40 + \frac{1.98 - 2.40}{1.5 - 1.3} \cdot (1.34 - 1.3) = 2.32 \tag{5}$$

The stresses at point A and B are (Eqs. 4.23 and 4.24)

$$\sigma_A = 2.32 \cdot 6900 = 16{,}008 \text{ lb/in.}^2 \tag{6}$$

$$\sigma_B = 2.86 \cdot 6900 = 19{,}734 \text{ lb/in.}^2 \tag{7}$$

where σ_B is the equivalent stress at point B.

4.3.6 Symmetrically Reinforced Hole in Finite-Width Element in Uniaxial Tension

For a symmetrically reinforced hole in a thin element of prescribed width, experimental results of interest for design application are the photoelastic test values of Seika et al. (1964,

1967). These tests used an element 6 mm thick, with a hole 30 mm in diameter. Cemented symmetrically into the hole was a stiffening ring of various thicknesses containing various diameters d of the central hole. The width of the element was also varied. A constant in all tests was $D/h =$ diameter of ring/thickness of element $= 5$.

Chart 4.8 presents $K_{tg} = \sigma_{max}/\sigma$ values, where $\sigma =$ gross stress, for various width ratios $H/D =$ width of element/diameter of ring. In all cases σ_{max} is located on the hole surface at 90° to the applied uniaxial tension. Only in the case of $H/D = 4$ was the effect of fillet radius investigated (Chart 4.8c).

For $H/D = 4$ and $D/h = 5$, Chart 4.9 shows the net stress concentration factor, defined as follows:

$$P = \sigma A = \sigma_{net} A_{net}$$

where P is the total applied force

$$K_{tn} = \frac{\sigma_{max}}{\sigma_{net}} = \frac{\sigma_{max} A_{net}}{\sigma A} = \frac{K_{tg} A_{net}}{A}$$

$$= K_{tg} \frac{(H-D)h + (D-d)h_t + (4-\pi)r^2}{Hh}$$

$$= K_{tg} \frac{[(H/d) - 1] + [1 - (d/D)](h_t/h) + (4-\pi)r^2/(Dh)}{H/D} \quad (4.26)$$

where d is the diameter of the hole, D is the outside diameter of the reinforcement, H is the width of the element, h is the thickness of the element, h_t is the thickness of the reinforcement, and r is the fillet radius at the junction of the element and the reinforcement.

Note from Chart 4.9 that the K_{tn} values are grouped closer together than the K_{tg} values of Chart 4.8c. Also note that the minimum K_{tn} occurs at $h_t/h \approx 3$ when $r > 0$. Thus for efficient section use the h_t/h should be set at about 3.

The $H/D = 4$ values are particularly useful in that they can be used without serious error for wide element problems. This can be demonstrated by using Eq. (4.26) to replot the K_{tn} curve in terms of $d/H =$ diameter of hole/width of element and extrapolating for $d/H = 0$, equivalent to an infinite element (see Chart 4.10).

It will be noted from Chart 4.8c that the lowest K_{tg} factor achieved by the reinforcements used in this series of tests was approximately 1.1, with $h_t/h \geq 4$, $d/D = 0.3$, and $r/h = 0.83$. By decreasing d/D, that is, by increasing D relative to d, the K_{tg} factor can be brought to 1.0. For a wide element without reinforcement, $K_{tg} = 3$; to reduce this to 1, it is evident that h_t/h should be 3 or somewhat greater.

An approximate solution was proposed by Timoshenko (1924), based on curved bar theory. A comparison curve is shown in Chart 4.8c.

4.3.7 Nonsymmetrically Reinforced Hole in Finite-Width Element in Uniaxial Tension

For an asymmetrically reinforced hole in a finite-width element in tension as shown in Chart 4.11, photoelastic tests were made with $d/h = 1.833$ (Lingaiah et al. 1966). Except for one series of tests, the volume of the reinforcement (V_R) was made equal to the volume of the hole (V_H). In Chart 4.11 the effect of varying the ring height (and corresponding ring diameter) is shown for various d/H ratios. A minimum K_t value is reached at about $h_t/h = 1.45$ and $D/d = 1.8$.

A shape factor is defined as

$$C_s = \frac{D/2}{h_t - h} \qquad (4.27)$$

For the photoelastic tests with $d/h = 1.833$ and $V_R/V_H = 1$, the shape factor C_s is chosen to be 3.666. This is shown in Fig. 4.10. If one wishes to lower K_t by increasing V_R/V_H, the shape factor $C_s = 3.666$ should be maintained as an interim procedure.

In Chart 4.12, where the abscissa scale is d/H, extrapolation is shown to $d/H = 0$. This provides intermediate values for relatively wide elements.

The curves shown are for a zero fillet radius. A fillet radius r of 0.7 of the element thickness h reduces K_{tn} approximately 12%. For small radii the reduction is approximately linearly proportional to the radius. Thus, for example, for $r/h = 0.35$, the reduction is approximately 6%.

4.3.8 Symmetrically Reinforced Circular Hole in a Biaxially Stressed Wide, Thin Element

Pressure vessels, turbine casings, deep sea vessels, aerospace devices, and other structures subjected to pressure require perforation of the shell by holes for introduction of control mechanisms, windows, access to personnel, and so on. Although these designs involve complicating factors such as vessel curvature and closure details, some guidance can be obtained from the work on flat elements, especially for small openings, including those for leads and rods.

The state of stress in a pressurized thin spherical shell is biaxial, $\sigma_1 = \sigma_2$. For a circular hole in a biaxially stressed thin element with $\sigma_1 = \sigma_2$, from Eqs. (4.17) or (4.18), $K_t = 2$. The stress state in a pressurized cylindrical shell is $\sigma_2 = \sigma_1/2$, where σ_1 is the hoop stress and σ_2 is the longitudinal (axial) stress. For the corresponding flat panel, $K_t = 2.5$ (Eq. 4.18, with $\alpha = \sigma_2/\sigma_1 = 1/2$). By proper reinforcement design, these factors can be reduced to 1, with a resultant large gain in strength. It has long been the practice to reinforce holes, but design information for achieving a specific K value, and in an optimum way, has not been available.

The reinforcement considered here is a ring type of rectangular cross section, symmetrically disposed on both sides of the panel (Chart 4.13). The results are for flat elements and applicable for pressure vessels only when the diameter of the hole is small compared to

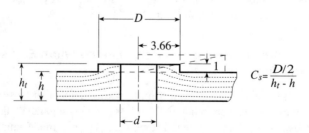

Figure 4.10 Shape factor for a nonsymmetric reinforced circular hole.

the vessel diameter. The data should be useful in optimization over a fairly wide practical range.

A considerable number of theoretical analyses have been made (Gurney 1938; Beskin 1944; Levy et al. 1948; Reissner and Morduchow 1949; Wells 1950; Mansfield 1953; Hicks 1957; Wittrick 1959a; Savin 1961; Houghton and Rothwell 1961; Davies 1967). In most of the analyses it has been assumed that the edge of the hole, in an infinite sheet, is reinforced by a "compact" rim (one whose round or square cross-sectional dimensions are small compared to the diameter of the hole). Some of the analyses (Gurney 1938; Beskin 1944; Davies 1967) do not assume a compact rim. Most analyses are concerned with stresses in the sheet. Where the rim stresses are considered, they are assumed to be uniformly distributed in the thickness direction.

The curves in Chart 4.13 provide the stress concentration factors for circular holes with symmetrical reinforcement. This chart is based on the theoretical (analytical) derivation of Gurney (1938). The maximum stresses occur at the hole edge and at the element to reinforcement junction. Because of the complexity of the stress fields at the junction of the element and the reinforcement, the von Mises stress of Section 1.8 is used as the basis to define the stress concentration factor. Suppose that σ_1 and σ_2 represent the principal stresses in the element remote from the hole and reinforcement. The corresponding von Mises (equivalent) stress is given by (Eq. 1.35)

$$\sigma_{eq} = \sqrt{\sigma_1^2 - \sigma_1\sigma_2 + \sigma_2^2} \qquad (4.28)$$

The stress concentration factors based on σ_{eq} are defined as

$$K_{ted} = \frac{\sigma_{\max d}}{\sigma_{eq}} \qquad (4.29)$$

$$K_{teD} = \frac{\sigma_{\max D}}{\sigma_{eq}} \qquad (4.30)$$

where K_{ted} is the stress concentration factor at the edge of the hole, and K_{teD} is the stress concentration factor at the junction of the element and reinforcement.

The plots of K_{ted} and K_{teD} versus h_r/h for various values of D/d are provided in Chart 4.13. For these curves, $\nu = 0.25$ and $h_r < (D - d)$. The highest equivalent stress occurs at the edge of a hole for the case of low values of h_r/h. For high values of h_r/h, the highest stress is located at the junction of the element and the reinforcement.

If the reinforcement and the element have different Young's moduli, introduce a modulus-weighted h_r/h (Pilkey 1994), that is, multiply h_r/h by E_r/E for use in entering the charts. The quantities E_r and E are the Young's moduli of the reinforcement and the element materials, respectively.

Example 4.3 Reinforced Circular Thin Element with In-Plane Loading A 10-mm thick element has a 150-mm diameter hole. It is reinforced symmetrically about the midplane of the element with two 20-mm-thick circular rings of 300-mm outer diameter and 150-mm inner diameter. The stresses $\sigma_x = 200$ MN/m^2, $\sigma_y = 100$ MN/m^2, and $\tau_{xy} = 74.83$ MN/m^2 are applied on this element as shown in Fig. 4.11. Find the equivalent stress at the edge of the hole and at the junction of the reinforcement and the element.

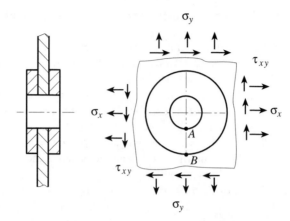

Figure 4.11 Symmetrically reinforced circular hole in an infinite in-plane loaded thin element.

For this element

$$\frac{h_r}{h} = \frac{2 \cdot 20}{10} = 4, \frac{D}{d} = \frac{300}{150} = 2 \tag{1}$$

If there were no hole, the principal stresses would be calculated as

$$\sigma_1 = \frac{1}{2}(200 + 100) + \frac{1}{2}\sqrt{(200 - 100)^2 + 4 \cdot 74.83^2} = 240 \text{ MN/m}^2 \tag{2}$$

$$\sigma_2 = \frac{1}{2}(200 + 100) - \frac{1}{2}\sqrt{(200 - 100)^2 + 4 \cdot 74.83^2} = 60 \text{ MN/m}^2 \tag{3}$$

The ratio of the principal stresses is $\sigma_2/\sigma_1 = 60/240 = 0.25$, and from Eq. (4.28), the corresponding equivalent stress is

$$\sigma_{eq} = \sqrt{240^2 - 240 \cdot 60 + 60^2} = 216.33 \text{ MN/m}^2 \tag{4}$$

The stress concentration factors for this case can not be obtained from the curves in Chart 4.13 directly. First, read the stress concentration factors for $D/d = 2$ and $h_r/h = 4$ in Chart 4.13 to find

	$\sigma_2 = \sigma_1$	$\sigma_2 = \sigma_1/2$	$\sigma_2 = 0$	$\sigma_2 = -\sigma_1/2$	$\sigma_2 = -\sigma_1$
$K_{teB} = K_{teD}$	1.13	1.33	1.63	1.74	1.76
$K_{teA} = K_{ted}$	0.69	1.09	1.20	1.09	0.97

Use the table values and the Lagrangian 5-point interpolation method (Kelly 1967) to find, for $\sigma_2/\sigma_1 = 0.25$,

$$K_{teA} = 1.18 \tag{5}$$

$$K_{teB} = 1.49$$

with the equivalent stresses

$$\sigma_{eqB} = 1.49 \cdot 216.33 = 322.33 \text{ MN/m}^2 \quad (6)$$
$$\sigma_{eqA} = 1.18 \cdot 216.33 = 255.27 \text{ MN/m}^2$$

The results of strain gage tests made at NASA by Kaufman et al. (1962) on in-plane loaded flat elements with noncompact reinforced circular holes can be used for design purposes. The diameter of the holes is eight times the thickness of the element. The connection between the panel and the reinforcement included no fillet. The actual case, using a fillet, would in some instances be more favorable. They found that the degree of agreement with the theoretical results of Beskin (1944) varied considerably with the variation of reinforcement parameters. Since in these strain gage tests the width of the element is 16 times the hole diameter, it can be assumed that for practical purposes an invariant condition corresponding to an infinite element has been attained. Since no correction has been made for the section removal by the hole, $K_{tg} = \sigma_{max}/\sigma_1$ is used.

The Charts 4.14 to 4.17 are based on the strain gage results of Kaufman and developed in a form more suitable for the types of problem encountered in turbine and pressure vessel design. These show stress concentration factors for given D/d and h_t/h. These charts involved interpolation in regions of sparse data. For this reason the charts are labeled as giving approximate stress concentration values. Further interpolation can be used to obtain K_{tg} values between the curves.

In Charts 4.14 to 4.19 the stress concentration factor $K_{tg} = \sigma_{max}/\sigma_1$ has been used instead of $K_{te} = \sigma_{max}/\sigma_{eq}$. The former is perhaps more suitable where the designer wishes to obtain σ_{max} as simply and directly as possible. For $\sigma_1 = \sigma_2$ the two factors are the same. For $\sigma_2 = \sigma_1/2$, $K_{te} = (2/\sqrt{3})K_{tg} = 1.157 K_{tg}$.

In drawing Charts 4.14 to 4.17, it has been assumed that as D/d is increased an invariant condition is approached where $h_t/h = 2/K_{tg}$ for $\sigma_1 = \sigma_2$; $h_t/h = 2.5/K_{tg}$ for $\sigma_2 = \sigma_1/2$. It has also been assumed that for relatively small values of D/d, less than about 1.7, constant values of K_{tg} are reached as h_t/h is increased; that is, the outermost part of the reinforcement in the thickness direction becomes stress free (dead photoelastically) (Fig. 4.12).

Charts 4.14 to 4.17 are plotted in terms of two ratios defining the reinforcement proportions D/d and h_t/h. When these ratios are not much greater than 1.0, the stress in the rim of the reinforcement exceeds the stress in the element. The basis for this conclusion can be observed in the charts. To the left of and below the dashed line $K_{tg} \approx 1$, K_{tg} is greater than 1, so the maximum stress in the rim is higher than in the element. When the ratios are large, the reverse is true. Also note in Charts 4.14 to 4.17 the crossover, or limit, line (dotted line denoted $K_{tg} \approx 1$) dividing the two regions. Beyond the line (toward the upper right) the maximum stress in the reinforcement is approximately equal to the applied nominal stress, $K_{tg} = 1$. In the other direction (toward the lower left) the maximum stress is in the rim, with K_{tg} increasing from approximately 1 at the crossover line to a maximum (2 for $\sigma_1 = \sigma_2$ and 2.5 for $\sigma_2 = \sigma_1/2$) at the origin. It is useful to consider that the left-hand and lower straight line edges of the diagrams (Charts 4.14 to 4.17) also represent the above maximum conditions. Then one can readily interpolate an intermediate curve, as for $K_{tg} = 1.9$ in Charts 4.14 and 4.15 or $K_{tg} = 2.3$ in Charts 4.16 and 4.17.

The reinforcement variables D/d and h_t/h can be used to form two dimensionless ratios:

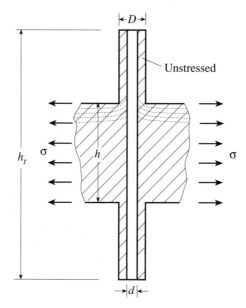

Figure 4.12 Effect of narrow reinforcement.

$A/(hd)$ = cross-sectional area of added reinforcement material/cross-section area of the hole,

$$\frac{A}{hd} = \frac{(D-d)(h_t - h)}{hd} = \left(\frac{D}{d} - 1\right)\left(\frac{h_t}{h} - 1\right) \quad (4.31)$$

V_R/V_H = volume of added reinforcement material/volume of hole,

$$\frac{V_R}{V_H} = \frac{(\pi/4)(D^2 - d^2)(h_t - h)}{(\pi/4)d^2 h} = \left[\left(\frac{D}{d}\right)^2 - 1\right]\left[\frac{h_t}{h} - 1\right] \quad (4.32)$$

The ratio $F = A/(hd)$ is used in pressure vessel design in the form (ASME 1974)

$$A = Fhd \quad (4.33)$$

where $F \geq 1$. Then Eq. (4.33) becomes

$$\frac{A}{dh} \geq 1 \quad (4.34)$$

Although for certain specified conditions (ASME 1974) F may be less than 1, usually $F = 1$. The ratio V_R/V_H is useful in arriving at optimum designs where weight is a consideration (aerospace devices, deep sea vehicles, etc.).

In Charts 4.14 and 4.16 a family of A/hd curves has been drawn, and in Charts 4.15 and 4.17 a family of V_R/V_H curves has been drawn, each pair for $\sigma_1 = \sigma_2$ and $\sigma_2 = \sigma_1/2$ stress states. Note that there are locations of tangency between the $A/(hd)$ or V_R/V_H curves and the K_{tg} curves. These locations represent optimum design conditions, that is, for any given value of K_{tg}, such a location is the minimum cross-sectional area or weight of reinforcement. The dot-dash curves, labeled "locus of minimum," provide the full range optimum conditions. For example, for $K_{tg} = 1.5$ in Chart 4.15, the minimum V_R/V_H occurs at the point where the dashed line ($K_{tg} = 1.5$) and the solid line (V_R/V_H) are tangent. This occurs at $(D/d, h_t/h) = (1.55, 1.38)$. The corresponding value of V_R/V_H is 1/2. Any other point corresponds to larger K_{tg} or V_R/V_H. It is clear that K_{tg} does not depend solely on the reinforcement area A (as assumed in a number of analyses) but also on the shape (rectangular cross-sectional proportions) of the reinforcement.

In Charts 4.18 and 4.19 the K_{tg} values corresponding to the dot-dash locus curves are presented in terms of $A/(hd)$ and V_R/V_H. Note that the largest gains in reducing K_{tg} are made at relatively small reinforcements and that to reduce K_{tg} from, say, 1.2 to 1.0 requires a relatively large volume of material.

The pressure vessel codes (ASME 1974) formula (Eq. 4.34) may be compared with the values of Charts 4.14 and 4.16, which are for symmetrical reinforcements of a circular hole in a flat element. For $\sigma_1 = \sigma_2$ (Chart 4.14) a value of K_{tg} of approximately 1 is attained at $A/(hd) = 1.6$. For $\sigma_2 = \sigma_1/2$ (Chart 4.16) a value of K_{tg} of approximately 1 is attained at A/hd a bit higher than 3.

It must be borne in mind that the tests (Kaufman et al. 1962) were for $d/h = 8$. For pressure vessels d/h may be less than 8, and for aircraft windows d/h is greater than 8. If d/h is greater than 8, the stress distribution would not be expected to change markedly; furthermore the change would be toward a more favorable distribution.

However, for a markedly smaller d/h ratio, the optimal proportions corresponding to $d/h = 8$ are not satisfactory. To illustrate, Fig. 4.13a shows the approximately optimum proportions $h_t/h = 3$, $D/d = 1.8$ from Chart 4.14 where $d/h = 8$. If we now consider a case where $d/h = 4$ (Fig. 4.13b), we see that the previous proportions $h_t/h = 3$, $D/d = 1.8$, are unsatisfactory for spreading the stress in the thickness direction. As an interim procedure, for $\sigma_1 = \sigma_2$ it is suggested that the optimum h_t/h value be found from Chart 4.14 or 4.16 and D/d then be determined in such a way that the same reinforcement shape factor $[(D - d)/2]/[(h_t - h)/2]$ is maintained. For $\sigma_1 = \sigma_2$ the stress pattern is symmetrical, with the principal stresses in radial and tangential (circular) directions.

From Chart 4.14, for $\sigma_1 = \sigma_2$, the optimum proportions for $K_{tg} \approx 1$ are approximately $D/d = 1.8$ and $h_t/h = 3$. The reinforcement shape factor is

$$C_1 = \frac{(D-d)/2}{(h_t-h)/2} = \frac{[(D/d)-1]d}{[(h_t/h)-1]h} \tag{4.35}$$

For $D/d = 1.8$, $h_t/h = 3$, and $d/h = 8$, the shape factor C_1 is equal to 3.2.

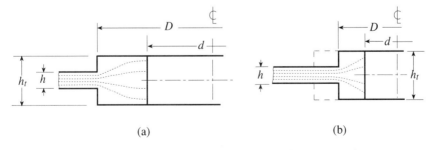

(a)   (b)

Figure 4.13 Effects of different d/h ratios: (a) $d/h = 8$; (b) $d/h = 4$.

For $\sigma_1 = \sigma_2$, suggested tentative reinforcement proportions for d/h values less than 8, which is the basis of Charts 4.14 to 4.17, are found as follows:

$$\frac{h_t}{h} = 3 \tag{4.36}$$

$$\frac{D}{d} = \frac{C_1[(h_t/h) - 1]}{d/h} + 1 \tag{4.37}$$

Substitute $h_t/h = 3$ into Eq. (4.37), retaining the shape factor of $C_1 = 3.2$, to find

$$\frac{D}{d} = \frac{6.4}{d/h} + 1 \tag{4.38}$$

For $d/h = 4$, Eq. (4.38) reduces to $D/d = 2.6$ as shown by the dashed line in Fig. 4.13b.

For $\sigma_2 = \sigma_1/2$ and $d/h < 8$, it is suggested as an interim procedure that the shape factor $C_s = (D/2)/(h_t - h)$ of Eq. (4.27) for $d/h = 8$ be maintained for the smaller values of d/h (see Eq. 4.27, uniaxial tension):

$$C_s = \frac{D/2}{h_t - h} = \frac{D/d}{2[(h_t/h) - 1]}\left(\frac{d}{H}\right) \tag{4.39}$$

For $D/d = 1.75$, $h_t/h = 5$, and $d/h = 8$, $C_s = 1.75$. For d/h less than 8 and $h_t/h = 5$, D/d can be obtained from Eq. (4.39):

$$\frac{D}{d} = \frac{2C_s[(h_t/h) - 1]}{d/h} = \frac{14}{d/h} \tag{4.40}$$

The foregoing formulas are based on $K_{tg} \approx 1$. If a higher value of K_{tg} is used, for example, to obtain a more favorable V_R/V_H ratio (i.e., less weight), the same procedure may be followed to obtain the corresponding shape factors.

Example 4.4 Weight Optimization through Adjustment of K_{tg}   Consider an example of a design trade-off. Suppose for $\sigma_2 = \sigma_1/2$, the rather high reinforcement thickness ratio of $h_t/h = 5$ is reduced to $h_t/h = 4$. We see from Chart 4.16 that the K_{tg} factor increases from about 1.0 to only 1.17. Also from Chart 4.19 the volume of reinforcement material is reduced 33% (V_R/V_H of 8.4 to 5.55).

The general formula for this example, based on Eq. (4.39), for $h_t/h = 4$ and $d/h < 8$ is

$$\frac{D}{d} = \frac{2C_s[(h_t/h) - 1]}{d/h} = \frac{10.5}{d/h} \tag{1}$$

Similarly for $\sigma_1 = \sigma_2$, if we accept $K_{tg} = 1.1$ instead of 1.0, we see from the locus of minimum $A/(hd)$, Chart 4.14, that $h_t/h = 2.2$ and $D/d = 1.78$. From Chart 4.19 the volume of reinforcement material is reduced 41% (V_R/V_H of 4.4 to 2.6).

The general formula for this example, based on Eq. (4.37), for d/h values less than 8 is

$$\frac{h_t}{h} = 2.2 \tag{2}$$

$$\frac{D}{d} = \frac{6.25}{d/h} + 1 \tag{3}$$

The foregoing procedure may add more weight than is necessary for cases where $d/h < 8$, but from a stress standpoint, the procedure would be on the safe side. The same procedure applied to d/h values larger than 8 would go in the direction of lighter, more "compact" reinforcements. However, owing to the planar extent of the stress distribution around the hole, it is not recommended to extend the procedure to relatively thin sheets, $d/h > 50$, such as in an airplane structure. Consult Gurney (1938), Beskin (1944), Levy et al. (1948), Reissner and Morduchow (1949), Wells (1950), Mansfield (1953), Hicks (1957), Wittrick (1959a, b), Savin (1961), Houghton and Rothwell (1961), and Davies (1967).

Where weight is important, some further reinforcements may be worth considering. Due to the nature of stress-flow lines, the outer corner region is unstressed (Fig. 4.14a). An ideal contour would be similar to Fig. 4.14b.

Kaufman et al. (1962) studied a reinforcement of triangular cross section, Fig. 4.14c. The angular edge at A may not be practical, since a lid or other member often is used. A compromise shape may be considered (Fig. 4.14d). Dhir and Brock (1970) present results for a shape like Fig. 4.14d and point out the large savings of weight that is attained.

Studies of a "neutral hole," that is, a hole that does not create stress concentration (Mansfield 1953), and of a variation of sheet thickness that results in uniform hoop stress for a circular hole in a biaxial stressed sheet (Mansfield 1970) are worthy of further consideration for certain design applications (i.e., molded parts).

4.3.9 Circular Hole with Internal Pressure

As illustrated in Example 1.7, the stress concentration factor of an infinite element with a circular hole with internal pressure (Fig. 4.15) may be obtained through superposition of the solutions for the cases of Figs. 1.28b and c. At the edge of the hole, this superposition

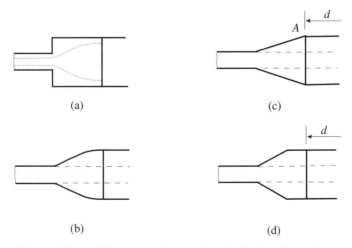

Figure 4.14 Reinforcement shape optimal design based on weight.

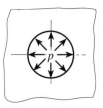

Figure 4.15 An infinite element with a hole with internal pressure.

provides

$$\sigma_r = \sigma_{r1} + \sigma_{r2} = -p \qquad (4.41)$$
$$\sigma_\theta = \sigma_{\theta 1} + \sigma_{\theta 2} = p$$
$$\tau_{r\theta} = \tau_{r\theta 1} + \tau_{r\theta 2} = 0$$

so that the corresponding stress concentration factor is $K_t = \sigma_\theta/p = 1$.

The case of a square plate with a pressurized central circular hole could be useful as a cross section of a construction conduit. The $K_t = \sigma_{max}/p$ factors (Riley et al. 1959; Durelli and Kobayashi 1958) are given in Chart 4.20. Note that for the thinner walls ($a/e > 0.67$), the maximum stress occurs on the outside edge at the thinnest section (point A). For the thicker wall ($a/e < 0.67$), the maximum stress occurs on the hole edge at the diagonal location (point B). As a matter of interest the K_t based on the Lamè solution (Timoshenko and Goodier 1970) is shown, although for $a/e > 0.67$. These are not the maximum values. A check at a/e values of $1/4$ and $1/2$ with theoretical factors (Sekiya 1955) shows good agreement.

A more recent analysis (Davies 1965) covering a wide a/e range is in good agreement with Chart 4.20. By plotting $(K_t-1)(1-a/e)/(a/e)$ versus a/e, linear relations are obtained for small and large a/e values. Extrapolation is made to $(K_t - 1)(1 - a/e)/(a/e) = 2$ at $a/e \to 1$, as indicated by an analysis by Koiter (1957) and to 0 for $a/e \to 0$.

The upper curve (maximum) values of Chart 4.20 are in reasonably good agreement with other recently calculated values (Slot 1972). For a pressurized circular hole near a corner of a large square panel (Durelli and Kobayashi 1958), the maximum K_t values are quite close to the values for the square panel with a central hole.

For the hexagonal panel with a pressurized central circular hole (Slot, 1972), the K_t values are somewhat lower than the corresponding values of the upper curve of Chart 4.20, with $2a$ defined as the width across the sides of the hexagon. For other cases involving a pressurized hole, see Sections 4.3.16 and 4.4.5.

For an eccentrically located hole in a circular panel, see Table 4.2 (Section 4.3.16) and Charts 4.48 and 4.49.

4.3.10 Two Circular Holes of Equal Diameter in a Thin Element in Uniaxial Tension or Biaxial In-plane Stresses

Consider stress concentration factors for the case of two equal holes in an infinite thin element subjected to uniaxial tension σ. The holes lie along a line that is perpendicular to the direction of stress σ, as shown in Fig. 4.16. From the conclusion for a single hole

CIRCULAR HOLES WITH IN-PLANE STRESSES   201

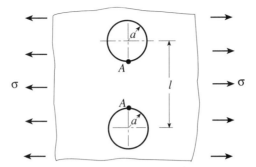

Figure 4.16 Two circular holes of equal diameter, aligned on a line perpendicular to the direction of stress σ.

(Section 4.3.2), the stress concentration at point A will be rather high if the distance between the two holes is relatively small. However, when $l > 6a$, the influence between the two holes will be weak. Then it is reasonable to adopt the results for a single hole with $K_t = 3$. If the holes lie along a line that is parallel to the stress σ, as shown in Fig. 4.17, the situation is different. As discussed in Section 4.3.2, for a single hole,, the maximum stress occurs at point A and decreases very rapidly in the direction parallel to σ (Fig. 4.5). For two holes there is some influence between the two locations A if l is small. The stress distribution for σ_θ tends to become uniform more rapidly than in the case of a single hole. The stress concentration factor is less than 3. However, as l increases, the influence between the two holes decreases, so K_t increases. At the location $l = 10a$, $K_t = 2.98$, which is quite close to the stress concentration factor of the single hole case and is consistent with the distribution of Fig. 4.5. The stress concentration factors for two equal circular holes are presented in Charts 4.21 to 4.25 (Ling 1948a; Haddon 1967).

If it is assumed that section B–B of Chart 4.22 carries a load corresponding to the distance between center lines, we obtain

$$K_{tnB} = \frac{\sigma_{\max B}(1 - d/l)}{\sigma} \qquad (4.42)$$

This corresponds to the light K_{tnB} lines of Charts 4.22 and 4.24. It will be noted that near $l/d = 1$, the factor becomes low in value (less than 1 for the biaxial case). If the same basis is used as for Eq. (4.11) (i.e., actual load carried by minimum section), the heavy K_{tnB}

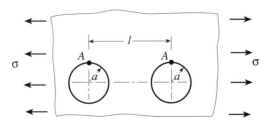

Figure 4.17 Two circular holes of equal diameter, aligned along σ.

202   HOLES

curves of Charts 4.22 and 4.24 are obtained. For this case

$$K_{tnB} = \frac{\sigma_{\max B}(1 - d/l)}{\sigma\sqrt{1 - (d/l)^2}} \tag{4.43}$$

Note that K_{tnB} in Charts 4.22 and 4.24 approaches 1.0 as l/d approaches 1.0, the element tending to become, in effect, a uniformly stressed tension member. A photoelastic test by North (1965) of a panel with two holes having $l/d = 1.055$ and uniaxially stressed transverse to the axis of the holes showed nearly uniform stress in the ligament.

In Chart 4.21, $\sigma_{\max}$ is located at $\theta = 90°$ for $l/d = 0$, $\theta = 84.4°$ for $l/d = 1$, and θ approaches 90° as l/d increases. In Chart 4.23, $\sigma_{\max}$ for $\alpha = 0°$ is same as in Chart 4.21 ($\theta = 84.4°$ for $l/d = 1.055$, $\theta = 89.8°$ for $l/d = 6$); $\sigma_{\max}$ for $\alpha = 45°$ is located at $\theta = 171.8°$ at $l/d = 1.055$ and decreases toward 135° with increasing values of l/d; $\sigma_{\max}$ for $\alpha = 90°$ is located at $\theta = 180°$.

Numerical determination of K_t (Christiansen 1968) for a biaxially stressed plate with two circular holes with $l/d = 2$ is in good agreement with the corresponding values of Ling (1948a) and Haddon (1967). For the more general case of a biaxially stressed plate in which the center line of two holes is inclined 0°, 15°, 30°, 45°, 60°, 75°, 90°, to the stress direction, the stress concentration factors are given in Chart 4.25 (Haddon 1967). These curves represent the relation between K_t and a/l for various values of the principal stress ratio σ_1/σ_2. It is assumed that the σ_1 and σ_2 are uniform in the area far from the holes. If the minimum distance between an element edge and the center of either hole is greater than $4a$, these curves can be used without significant error. There are discontinuities in the slopes of some of the curves in Chart 4.25, which correspond to sudden changes in the positions of the maximum (or minimum) stress.

Example 4.5   Flat Element with Two Equal-sized Holes under Biaxial Stresses   Suppose that a thin flat element with two 0.5-in. radius holes is subjected to uniformly distributed stresses $\sigma_x = 3180$ psi, $\sigma_y = -1020$ psi, $\tau_{xy} = 3637$ psi, along the straight edges far from the holes as shown in Fig. 4.18a. If the distance between the centers of the holes is 1.15 in., find the maximum stresses at the edges of the holes.

For an area far from the holes, resolution of the applied stresses gives the principal stresses

$$\sigma_1 = \frac{1}{2}(\sigma_x + \sigma_y) + \frac{1}{2}\sqrt{(\sigma_x - \sigma_y)^2 + 4\tau_{xy}^2} = 5280 \text{ psi} \tag{1}$$

$$\sigma_2 = \frac{1}{2}(\sigma_x + \sigma_y) - \frac{1}{2}\sqrt{(\sigma_x - \sigma_y)^2 + 4\tau_{xy}^2} = -3120 \text{ psi} \tag{2}$$

The angle θ_1 between σ_x and the principal stress σ_1 is given by (Pilkey 1994)

$$\tan 2\theta_1 = \frac{2\tau_{xy}}{\sigma_x - \sigma_y} = 1.732 \tag{3}$$

or

$$\theta_1 = 30° \tag{4}$$

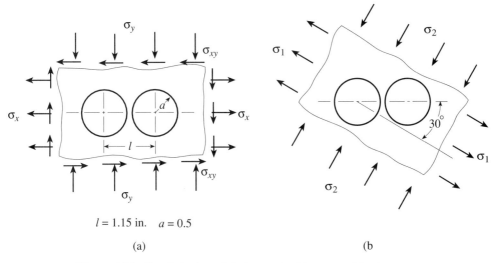

$l = 1.15$ in. $a = 0.5$

(a) (b)

Figure 4.18 Two holes in an infinite panel subject to combined stresses.

This problem can now be considered as a problem of finding the maximum stress of a flat element under biaxial tensile stresses σ_1 and σ_2, where σ_1 forms a 30° angle with the line connecting the hole centers (Fig. 4.18b). Chart 4.25 applies to this case. We need

$$\frac{\sigma_2}{\sigma_1} = \frac{-3120}{5280} = -0.591 \tag{5}$$

$$\frac{a}{l} = \frac{0.5}{1.15} = 0.435 \tag{6}$$

It can be found from Chart 4.25c that when the abscissa value $a/l = 0.435$, $K_t = 4.12$ and -5.18 for $\sigma_2/\sigma_1 = -0.5$, and $K_t = 4.45$ and -6.30 for $\sigma_2/\sigma_1 = -0.75$. The stress concentration factors for $\sigma_2/\sigma_1 = -0.591$ can be obtained through interpolation

$$K_t = 4.24 \quad \text{and} \quad -5.58 \tag{7}$$

The extreme stresses at the edges of the holes are

$$\sigma_{\max} = 4.24 \cdot 5280 = 22{,}390 \text{ psi} \quad \text{(tension)}$$

$$\sigma_{\max} = -5.58 \cdot 5280 = -29{,}500 \text{ psi} \quad \text{(compression)} \tag{8}$$

where $\sigma_1 = 5280$ psi is the nominal stress.

Example 4.6 Two Equal-Sized Holes Lying at an Angle in a Flat Element under Biaxial Stresses Figure 4.19a shows a segment of a flat thin element containing two holes of 10-mm diameter. Find the extreme stresses near the holes.

The principal stresses are (Pilkey 1994)

$$\sigma_{1,2} = \frac{1}{2}(\sigma_x + \sigma_y) \pm \frac{1}{2}\sqrt{(\sigma_x - \sigma_y)^2 + 4\tau_{xy}^2} = -38.2, 28.2 \text{ MPa} \tag{1}$$

204 HOLES

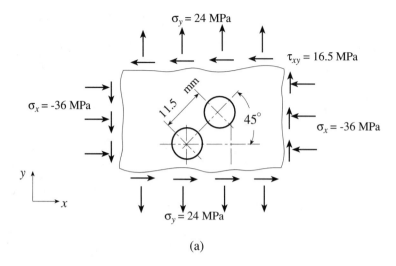

(a)

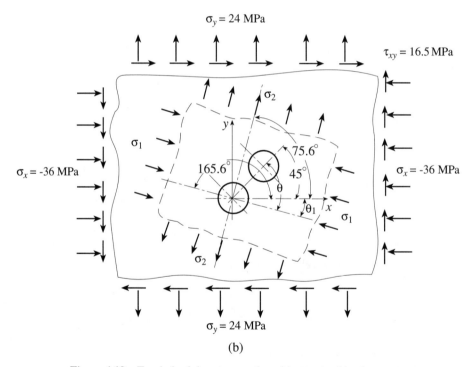

(b)

Figure 4.19 Two holes lying at an angle, subject to combined stresses.

and occur at

$$\theta = \frac{1}{2}\tan^{-1}\frac{2\tau_{xy}}{\sigma_x - \sigma_y} = -14.4° \quad (2)$$

See Fig. 4.19b. Use Chart 4.25 to find the extreme stresses. In this chart

$$\theta = 45° + 14.4° = 59.4° \quad (3)$$

Since Chart 4.25 applies only for angles $\theta = 0, 15°, 30°, 45°, 60°, 75°$, and $90°$, the stress concentration factors for $\theta = 60°$ can be considered to be adequate approximations. Alternatively, use linear interpolation. This leads to, for $a/l = 5/11.5 = 0.435$ and $\sigma_2/\sigma_1 = 56.5/(-80.5) = -0.702$,

$$K_t = 6.9 \quad \text{and} \quad -3.7 \quad (4)$$

Thus

$$\sigma_{max} = \sigma_1 K_t = \begin{cases} -80.5 \times 6.9 = -263.9 \text{ MPa} \\ -80.5 \times (-3.7) = 141.5 \text{ MPa} \end{cases} \quad (5)$$

are the extreme stresses occurring at each hole boundary.

4.3.11 Two Circular Holes of Unequal Diameter in a Thin Element in Uniaxial Tension or Biaxial In-plane Stresses

Stress concentration factors have been developed for two circular holes of unequal diameters in panels in uniaxial and biaxial tension. Values for K_{tg} for uniaxial tension in an infinite element have been obtained by Haddon (1967). His geometrical notation is used in Charts 4.26 and 4.27, since this is convenient in deriving expressions for K_{tn}.

For Chart 4.26, to obtain K_{tn} exactly, one must know the exact loading of the ligament between the holes in tension and bending and the relative magnitudes of these loadings. For two equal holes the loading is tensile, but its relative magnitude is not known. In the absence of this information, two methods were proposed to determine if reasonable K_{tn} values could be obtained.

Procedure A arbitrarily assumes (Chart 4.26) that the unit thickness load carried by s is $\sigma(b + a + s)$. Then

$$\sigma_{net}s = \sigma(b + a + s)$$

$$K_{tn} = \frac{\sigma_{max}}{\sigma_{net}} = \frac{K_{tg} \cdot s}{b + a + s} \quad (4.44)$$

Procedure B assumes, based on Eq. (4.11), that the unit thickness load carried by s is made up of two parts: $\sigma c_1 \sqrt{1 - (b/c_1)^2}$ from the region of the larger hole, carried over distance $c_R = bs/(b + a)$; $\sigma c_2 \sqrt{1 - (b/c_2)^2}$ from the region of the smaller hole, carried over distance $c_a = as/(b + a)$. In the foregoing $c_1 = b + c_b$; $c_2 = a + c_a$. For either the smaller or larger hole,

$$K_{tn} = \frac{K_{tg}}{\left(1 + \frac{(b/a) + 1}{s/a}\right) \sqrt{1 - \left(\frac{(b/a) + 1}{(b/a) + 1 + (s/a)}\right)^2}} \quad (4.45)$$

Referring to Chart 4.26, procedure A is not satisfactory in that K_{tn} for equal holes is less than 1 for values of s/a below 1. As s/a approaches 0, for two equal holes the ligament becomes essentially a tension specimen, so one would expect a condition of uniform stress ($K_{tn} = 1$) to be approached. Procedure B is not satisfactory below $s/a = 1/2$, but it does provide K_{tn} values greater than 1. For s/a greater than $1/2$, this curve has a reasonable shape, assuming that $K_{tn} = 1$ at $s/a = 0$.

In Chart 4.27, σ_{max} denotes the maximum tension stress. For $b/a = 5$, σ_{max} is located at $\theta = 77.8°$ at $s/a = 0.1$ and increases to $87.5°$ at $s/a = 10$. Also σ_{max} for $b/a = 10$ is located at $\theta = 134.7°$ at $s/a = 0.1$, $90.3°$ at $s/a = 1$, $77.8°$ at $s/a = 4$, and $84.7°$ at $s/a = 10$. The σ_{max} locations for $b/a = 1$ are given in the discussion of Chart 4.23. Since $s/a = 2[(l/d) - 1]$ for $b/a = 1$, $\theta = 84.4°$ at $s/a = 0.1$, and $89.8°$ at $s/a = 10$. The highest compression stress occurs at $\theta = 100°$.

For biaxial tension, $\sigma_1 = \sigma_2$, K_{tg} values have been obtained by Salerno and Mahoney (1968) in Chart 4.28. The maximum stress occurs at the ligament side of the larger hole.

Charts 4.29 to 4.31 provide more curves for different b/a values and loadings, which are also based on Haddon (1967). These charts can be useful in considering stress concentration factors of neighboring cavities of different sizes. In these charts, stress concentration factors K_{tgb} (for the larger hole) and K_{tga} (for the smaller hole) are plotted against a/c. In the case of Chart 4.30, there are two sets of curves for the smaller hole, which corresponds to different locations on the boundary of the hole. The stress at point A is negative and the stresses at points B or C are positive, when the element is under uniaxial tension load. Equation (4.45) can be used to evaluate K_{tn}, if necessary.

Example 4.7 A Thin Tension Element with Two Unequal Circular Holes   A uniaxial fluctuating stress of $\sigma_{max} = 24$ MN/m^2 and $\sigma_{min} = -62$ MN/m^2 is applied to a thin element as shown in Fig. 4.20. It is given that $b = 98$ mm, $a = 9.8$ mm, and that the centers

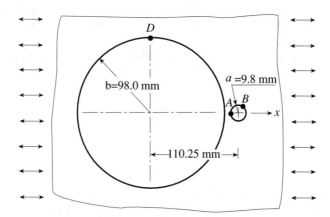

Figure 4.20 A thin element with two circular holes of unequal diameters.

of the two circles are 110.25 mm apart. Find the range of stresses occurring at the edge of the holes.

The geometric ratios are found to be $b/a = 98/9.8 = 10$ and $a/c = 9.8/(110.25 - 98) = 0.8$. From Chart 4.30 the stress concentration factors are found to be $K_{tgb} = 3.0$ at point D, $K_{tga} = 0.6$ at point B, and -4.0 at point A. When $\sigma_{\min} = -62$ MN/m² is applied to the element, the stresses corresponding to points A, B, and D are

$$-4.0 \times (-62) = 248 \text{ MN/m}^2 \qquad \text{at A} \qquad (1)$$

$$0.6 \times (-62) = -37.2 \text{ MN/m}^2 \qquad \text{at B}$$

$$3.0 \times (-62) = -186 \text{ MN/m}^2 \qquad \text{at D}$$

When $\sigma_{\max} = 24$ MN/m² is applied, the stresses corresponding to points A, B, and D are found to be

$$-4.0 \times 24 = -96 \text{ MN/m}^2 \qquad \text{at A} \qquad (2)$$

$$0.6 \times 24 = 14.4 \text{ MN/m}^2 \qquad \text{at B}$$

$$3.0 \times 24 = 72 \text{ MN/m}^2 \qquad \text{at D}$$

The critical stress, which varies between -96 and 248 MN/m², is at point A.

4.3.12 Single Row of Equally Distributed Circular Holes in an Element in Tension

For a single row of holes in an infinite panel, Schulz (1941) developed curves as functions of d/l (Charts 4.32 and 4.33), where l is the distance between the centers of the two adjoining holes, and d is the diameter of the holes. More recently calculated values of Meijers (1967) are in agreement with the Schulz values. Slot (1972) found that when the height of an element is larger than $3d$ (Chart 4.32), the stress distribution agrees well with the case of an element with infinite height.

For the cases of the element stressed parallel to the axis of the holes (Chart 4.33), when $l/d = 1$, the half element is equivalent to having an infinite row of edge notches. This portion of the curve (between $l/d = 0$ and 1) is in agreement with the work of Atsumi (1958) on edge notches.

For a row of holes in the axial direction with $l/d = 3$, and with $d/H = 1/2$, Slot obtained good agreement with the Howland K_t value (Chart 4.1) for the single hole with $a/H = 1/2$. A specific K_t value obtained by Slot for $l/d = 2$ and $d/H = 1/3$ is in good agreement with the Schulz curves (Chart 4.33).

The biaxially stressed case (Chart 4.34), from the work of Hütter (1942), represents an approximation in the midregion of d/l. Hütter's values for the uniaxial case with perpendicular stressing are inaccurate in the mid-region.

For a finite-width panel (strip), Schulz (1942–45) has provided K_t values for the dashed curves of Chart 4.33. The K_t factors for $d/l = 0$ are the Howland (1929–30) values. The K_t factors for the strip are in agreement with the Nisitani (1968) values of Chart 4.57 ($a/b = 1$).

The K_t factors for a single row of holes in an infinite plate in transverse bending are given in Chart 4.86, in shear in Chart 4.93.

4.3.13 Double Row of Circular Holes in a Thin Element in Uniaxial Tension

Consider a double row of staggered circular holes. This configuration is used in riveted and bolted joints. The K_{tg} values of Schultz (1941) are presented in Chart 4.35. Comparable values of Meijers (1967) are in agreement.

In Chart 4.35, as θ increases, the two rows grow farther apart. At $\theta = 90°$ the K_{tg} values are the same as for a single row (Chart 4.32). For $\theta = 0°$, a single row occurs with an intermediate hole in the span l. The curves 0° and 90° are basically the same, except that as a consequence of the nomenclature of Chart 4.35, l/d for $\theta = 0°$ is twice l/d for $\theta = 90°$ for the same K_{tg}. The type of plot used in Chart 4.35 makes it possible to obtain K_{tg} for intermediate values of θ by drawing θ versus l/d curves for various values of K_{tg}. In this way the important case of $\theta = 60°$, shown dashed on Chart 4.35, was obtained. For $\theta < 60°$, σ_{max} occurs at points A, and for $\theta > 60°$, σ_{max} occurs at points B. At $\theta = 60°$, both points A and B are the maximum stress points.

In obtaining K_{tn}, based on a net section, two relations are needed since for a given l/d the area of the net sections A–A and B–B depends on θ (Chart 4.36). For $\theta < 60°$, A–A is the minimum section and the following formula is used:

$$K_{tnA} = \frac{\sigma_{max}}{\sigma}\left[1 - 2\frac{d}{l}\cos\theta\right] \quad (4.46)$$

For $\theta > 60°$, B–B is the minimum section and the formula is based on the net section in the row

$$K_{tnB} = \frac{\sigma_{max}}{\sigma}\left(1 - \frac{d}{l}\right) \quad (4.47)$$

At 60° these formulas give the same result. The K_{tn} values in accordance with Eqs. (4.46) and (4.47) are given in Chart 4.36.

4.3.14 Symmetrical Pattern of Circular Holes in a Thin Element in Uniaxial Tension or Biaxial In-plane Stresses

Symmetrical triangular or square patterns of circular holes are used in heat exchanger and nuclear vessel design (O'Donnell and Langer 1962). The notation used in these fields will be employed here. Several charts here give stress concentration factors versus ligament efficiency. Ligament efficiency is defined as the minimum distance (s) of solid material between two adjacent holes divided by the distance (l) between the centers of the same holes; that is, ligament efficiency $= s/l$. It is assumed here that the hole patterns are repeated throughout the panel.

For the triangular pattern of Chart 4.37, Horvay (1952) obtained a solution for long and slender ligaments, taking account of tension and shear (Chart 4.38). Horvay considers the results as not valid for s/l greater than 0.2. Photoelastic tests (Sampson 1960; Leven 1963, 1964) have been made over the s/l range used in design. Computed values (Meijers 1967; Grigolyuk and Fil'shtinskii 1970; Goldberg and Jabbour 1965) are in good agreement but differ slightly in certain ranges. When this occurs the computed values (Meijers 1967) are used in Charts 4.37 and 4.41. Subsequent computed values (Slot 1972) are in good agreement with the values of Meijers (1967).

A variety of stress concentration factors for several locations on the boundaries of the holes are given in Chart 4.39 (Nishida 1976) for a thin element with a triangular hole pattern.

For the square pattern, Bailey and Hicks (1960), with confirmation by Hulbert and Niedenfuhr (1965) and O'Donnell (1967), have obtained solutions for applied biaxial fields oriented in the square and diagonal directions (Charts 4.40, 4.41). Photoelastic tests by Nuno et al. (1964) are in excellent agreement with mathematical results (Bailey and Hicks 1960) for the square direction of loading but, as pointed out by O'Donnell (1966), are lower than those by Bailey and Hicks (1960) for intermediate values of s/l for the diagonal direction of loading. Check tests by Leven (1967) of the diagonal case resulted in agreement with the previous photoelastic tests (Nuno et al. 1964) and pointed to a recheck of the mathematical solution of this case. This was done by Hulbert under PVRC sponsorship at the instigation of O'Donnell. The corrected results are given in O'Donnell (1967), which is esssentially his paper (O'Donnell 1966) with the Hulbert correction. Later confirmatory results were obtained by Meijers (1967). Subsequently computed values (Slot 1972; Grigolyuk and Fil'shtinskii 1970) are in good agreement with those of Meijers (1967).

The $\sigma_2 = -\sigma_1$ state of stress (O'Donnell 1966 and 1967; Sampson 1960; Bailey and Hicks 1960) shown in Chart 4.41 corresponds to shear stress $\tau = \sigma_1$ at $45°$. For instance, the stress concentration factor of a cylindrical shell with a symmetrical pattern of holes under shear loading can be found from Chart 4.94.

The $\sigma_2 = \sigma_1/2$ state of stress, Chart 4.40, corresponds to the case of a thin cylindrical shell with a square pattern of holes under the loading of inner pressure.

The values of the stress concentration factors K_{tg} are obtained for uniaxial tension and for various states of biaxiality of stress (Chart 4.42) by superposition. Chart 4.42 is approximate. Note that the lines are not straight, but they are so nearly straight that the curved lines drawn should not be significantly in error. For uniaxial tension, Charts 4.43 to 4.45, are for rectangular and diamond patterns (Meijers 1967).

4.3.15 Radially Stressed Circular Element with a Ring of Circular Holes, with or without a Central Circular Hole

For the case of six holes in a circular element loaded by six external radial forces, maximum K_{tg} values are given for four specific cases as shown in Table 4.1. K_{tg} is defined as $R_0 \sigma_{max}/P$ for an element of unit thickness. Good agreement has been obtained for the maximum K_{tg} values of Hulbert (1965) and the corresponding photoelastic and calculated values of Buivol (1960, 1962).

For the case of a circular element with radial edge loading and with a central hole and a ring of four or six holes, the maximum K_t values (Kraus 1963) are shown in Chart 4.46 as a function of a/R_0 for two cases: all holes of equal size ($a = R_i$); central hole 1/4 of outside diameter of the element ($R_i/R_0 = 1/4$). Kraus points out that with appropriate assumptions concerning axial stresses and strains, the results apply to both plane stress and plane strain.

For the case of an annulus flange ($R = 0.9R_0$), the maximum K_t values (Kraus et al. 1966) are shown in Chart 4.47 as a function of hole size and the number of holes. K_t is defined as σ_{max} divided by σ_{nom}, the average tensile stress on the net radial section through a hole. In Kraus's paper K_t factors are given for other values of R_i/R_0 and R/R_0.

TABLE 4.1 Maximum K_{tg} for Circular Holes in Circular Element Loaded Externally with Concentrated Radial Forces

	Pattern	Spacing	Maximum K_{tg}	Location	References
1	(diagram)	$R/R_0 = 0.65$ $a/R_0 = 0.2$	4.745	A	Hulbert 1965 Buivol 1960
		$R/R_0 = 0.7$ $a/R_0 = 0.25$	5.754	B	
2	(diagram)	$R/R_0 = 0.65$ $a/R_0 = 0.2$	9.909	A $\alpha = 50°$	Hulbert 1965 Buivol 1960, 1962
		$R/R_0 = 0.6$ $a/R_0 = 0.2$	7.459	A $\alpha = 50°$	

4.3.16 Thin Element with Circular Holes with Internal Pressure

As stated in Section 4.3.9, the stress concentration factor of an infinite element with a circular hole with internal pressure can be obtained through superposition. This state of stress can be separated into two cases. One case is equal biaxial tension, and the other is equal biaxial compression with the internal pressure p. For the second case, since every point in the infinite element is in a state of equal biaxial compressive stress, $(-p)$, the stress concentration factor is equal to $K_{t2} = \sigma_{max}/p = -p/p = -1$. For the first case when there are multiple holes, the stress concentration factor K_{t1} depends on the number of holes, the geometry of the holes, and the distribution of the holes. Thus, from superposition, the stress concentration factor K_t for elements with holes is $K_t = K_{t1} + K_{t2} = K_{t1} - 1$. That is, the stress concentration factor for an element with circular holes with internal pressure can be obtained by subtracting 1 from the stress concentration factor for the same element with the same holes, but under external equal biaxial tension with stress of magnitude equal to the internal pressure p.

For two holes in an infinite thin element, with internal pressure only, the K_t are found by subtracting 1.0 from the biaxial K_{tg} values of Charts 4.24, 4.25 (with $\sigma_1 = \sigma_2$), and 4.28. For an infinite row of circular holes with internal pressure, the K_t can be obtained by subtracting 1.0 from the K_t of Chart 4.34. For different patterns of holes with internal pressure, the K_t can be obtained the same way from Charts 4.37 (with $\sigma_1 = \sigma_2$), 4.39b, and 4.40 (with $\sigma_1 = \sigma_2$). This method can be used for any pattern of holes in an infinite thin element. That is, as long as the K_t for equal biaxial tension state of stress is known, the K_t for the internal pressure only can be found by subtracting 1.0 from the K_t for the equal biaxial tension case. Maximum K_t values for specific spacings of hole patterns in circular panels are given in Table 4.2. Some other hole patterns in an infinite panel are discussed in Peterson (1974).

For the $a/R_0 = 0.5$ case of the single hole eccentrically located in a circular panel (Hulbert 1965), a sufficient number of eccentricities were calculated to permit Chart 4.48 to be prepared. For a circular ring of three or four holes in a circular panel, Kraus (1962)

TABLE 4.2 Maximum K_t for Circular Holes in Circular Element Loaded with Internal Pressure Only

Pattern	Spacing	Maximum K_t	Location	References
1	$a/R_0 = 0.5$	See Chart 4.48	See Chart 4.48	Hulbert 1965 Timoshenko 1970
2	$R/R_0 = 0.5$ $a/R_0 = 0.2$	See Chart 4.49	See Chart 4.49	Savin 1961 Kraus 1963
3	$R/R_0 = 0.5$ $a/R_0 = 0.2$	See Chart 4.49	See Chart 4.49	Savin 1961 Kraus 1963
	$R/R_0 = 0.5$ $a/R_0 = 0.25$	2.45	A	Hulbert 1965
4	$R/R_0 = 0.6$ $a/R_0 = 0.2$	2.278 Pressure in all holes 1.521 Pressure in center hole only	A B	Hulbert 1965

has obtained K_t for variable hole size (Chart 4.49). With the general finite element codes available, it is relatively straight-forward to compute stress concentration factors for a variety of cases.

4.4 ELLIPTICAL HOLES IN TENSION

Consider an elliptical hole of major axis $2a$ and minor axis $2b$ and introduce the elliptical coordinates (Fig. 4.21a).

$$x = \sqrt{a^2 - b^2} \cosh \alpha \cos \beta$$
$$y = \sqrt{a^2 - b^2} \sinh \alpha \sin \beta \tag{4.48}$$

Let $\tanh \alpha_0 = b/a$ so that

$$\cosh \alpha_0 = \frac{a}{\sqrt{a^2 - b^2}}$$
$$\sinh \alpha_0 = \frac{b}{\sqrt{a^2 - b^2}} \tag{4.49}$$

and Eq. (4.48) becomes

$$x = a \cdot \cos \beta$$
$$y = b \cdot \sin \beta \tag{4.50}$$

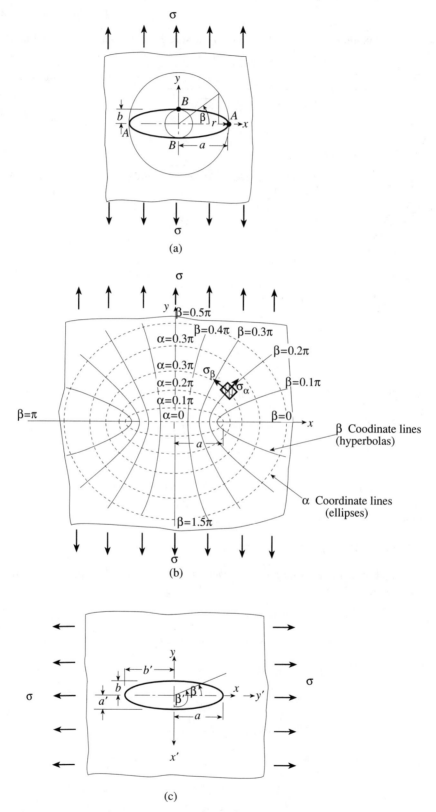

Figure 4.21 Elliptical hole in uniaxial tension: (*a*) Notation; (*b*) elliptical coordinates and stress components; (*c*) stress applied in direction perpendicular to the minor axis of the ellipse.

or

$$\frac{x^2}{a^2} + \frac{y^2}{b^2} = 1$$

This represents all the points on the elliptical hole of major axis $2a$ and minor axis $2b$. As α changes, Eq. (4.48) represents a series of ellipses, which are plotted with dashed lines in Fig. 4.21b. As $\alpha \to 0$, $b \to 0$, and the equation for the ellipse becomes

$$x = a \cdot \cos \beta$$
$$y = 0$$
(4.51)

This corresponds to a crack, (i.e., an ellipse of zero height, $b = 0$) and of length $2a$.

For $\beta = \pi/6$, Eq. (4.48) represents a hyperbola

$$x = \frac{\sqrt{3}}{2}\sqrt{a^2 - b^2} \cosh \alpha$$
$$y = \frac{1}{2}\sqrt{a^2 - b^2} \sinh \alpha$$
(4.52)

or

$$\frac{x^2}{\frac{3}{4}(a^2 - b^2)} - \frac{y^2}{\frac{1}{4}(a^2 - b^2)} = 1$$

As β changes from 0 to 2π, Eq. (4.48) represents a series of hyperbolas, orthogonal to the ellipses as shown in Fig. 4.21b. The elliptical coordinates (α, β) can represent any point in a two-dimensional plane. The coordinate directions are the directions of the tangential lines of the ellipses and of hyperbolas, which pass through that point.

4.4.1 Single Elliptical Hole in Infinite- and Finite-Width Thin Elements in Uniaxial Tension

Define the stress components in elliptic coordinates as σ_α and σ_β as shown in Fig. 4.21b. The elastic stress distribution of the case of an elliptical hole in an infinite-width thin element in uniaxial tension has been determined by Inglis (1913) and Kolosoff (1910). At the edge of the elliptical hole, the sum of the stress components σ_α and σ_β is given by the formula (Inglis 1913)

$$(\sigma_\alpha + \sigma_\beta)_{\alpha_0} = \sigma \frac{\sinh 2\alpha_0 - 1 + e^{2\alpha_0} \cos 2\beta}{\cosh 2\alpha_0 - \cos 2\beta}$$
(4.53)

Since the stress σ_α is equal to zero at the edge of the hole ($\alpha = \alpha_0$), Eq. (4.53) becomes

$$(\sigma_\beta)_{\alpha_0} = \sigma \frac{\sinh 2\alpha_0 - 1 + e^{2\alpha_0} \cos 2\beta}{\cosh 2\alpha_0 - \cos 2\beta}$$
(4.54)

The maximum value of $(\sigma_\beta)_{\alpha_0}$ occurs at $\beta = 0, \pi$, namely at the ends of the major axis of the ellipse (point A, Fig. 4.21a),

$$(\sigma_\beta)_{\alpha_0, \beta=0} = \sigma \frac{\sinh 2\alpha_0 - 1 + e^{2\alpha_0}}{\cosh 2\alpha_0 - 1} = \sigma(1 + 2\coth \alpha_0) = \sigma\left(1 + \frac{2a}{b}\right)$$
(4.55)

214 HOLES

From Eq. (4.54) at point B, Fig. 4.21a,

$$(\sigma_\beta)_{\alpha_0, \beta=\pi/2} = \sigma \frac{\sinh 2\alpha_0 - 1 - e^{2\alpha_0}}{\cosh 2\alpha_0 + 1} = \sigma \frac{-(\cosh 2\alpha_0 + 1)}{\cosh 2\alpha_0 + 1} = -\sigma \quad (4.56)$$

The stress concentration factor for this infinite width case is

$$K_{tg} = \frac{(\sigma_\beta)_{\alpha_0, \beta=0}}{\sigma} = \frac{\sigma[1 + (2a/b)]}{\sigma} = 1 + \frac{2a}{b} \quad (4.57)$$

or

$$K_{tg} = 1 + 2\sqrt{\frac{a}{r}} \quad (4.58)$$

where r is the radius of curvature of the ellipse at point A (Fig. 4.21a).

If $b = a$, then $K_{tg} = 3$, and Eq. (4.57) is consistent with the case of a circular hole. Chart 4.50 is a plot of K_{tg} of Eq. (4.57). Also included in Chart 4.50 are stress concentration curves that represent cases where a hole contains material having different moduli of elasticity that are perfectly bonded to its body material (Donnell 1941).

When the uniaxial tensile stress σ is in the direction perpendicular to the minor axis of an elliptical hole, as shown in Fig. 4.21c, the stress at the edge of the hole can be obtained from a transformation of Eq. (4.54). From Fig. 4.21c it can be seen that this case is equivalent to the configuration of Fig. 4.21a if the coordinate system x', y' (Fig. 4.21c) is introduced. In the new coordinates, the semimajor axis is $a' = b$, the semiminor axis is $b' = a$, and the elliptical coordinate is $\beta' = \beta + (\pi/2)$. In the x', y' coordinates, substitution of Eq. (4.49) into Eq. (4.54) leads to

$$(\sigma_{\beta'})_{\alpha_0'} = \sigma \frac{\dfrac{2a'b'}{a'^2 - b'^2} - 1 + \dfrac{a' + b'}{a' - b'}\cos 2\beta'}{\dfrac{a'^2 + b'^2}{a'^2 - b'^2} - \cos 2\beta'} \quad (4.59)$$

Transformation of Eq. (4.59) into the coordinate system x, y, gives

$$(\sigma_\beta)_{\alpha_0} = \sigma \frac{-\dfrac{2ab}{a^2 - b^2} - 1 - \dfrac{a+b}{a-b}\cos(\pi + 2\beta)}{-\dfrac{a^2 + b^2}{a^2 - b^2} - \cos(\pi + 2\beta)}$$

$$= \sigma \frac{\dfrac{2ab}{a^2 - b^2} + 1 - \dfrac{a+b}{a-b}\cos 2\beta}{\dfrac{a^2 + b^2}{a^2 - b^2} - \cos 2\beta} \quad (4.60)$$

Substitution of Eq. (4.49) into Eq. (4.60) leads to

$$(\sigma_\beta)_{\alpha_0} = \sigma \frac{\sinh 2\alpha_0 + 1 - e^{2\alpha_0}\cos 2\beta}{\cosh 2\alpha_0 - \cos 2\beta} \quad (4.61)$$

For an elliptical hole in a finite-width tension panel, the stress concentration values K_t of Isida (1953, 1955a, b) are presented in Chart 4.51. Stress concentration factors for an elliptical hole near the edge of a finite-width panel are provided in Chart 4.51, while those for a semi-infinite panel (Isida 1955a) are given in Chart 4.52.

4.4.2 Width Correction Factor for a Cracklike Central Slit in a Tension Panel

For the very narrow ellipse approaching a crack (Chart 4.53), a number of "finite-width correction" formulas have been proposed including those by the following: Dixon (1960), Westergaard (1939), Irwin (1958), Brown and Srawley (1966), Fedderson (1965), and Koiter (1965). Correction factors have also been calculated by Isida (1965).

The Brown-Srawley formula for $a/H < 0.3$,

$$\frac{K_{tg}}{K_{t\infty}} = 1 - 0.2 \cdot \frac{a}{H} + \left(\frac{2a}{H}\right)^2 \tag{4.62}$$

$$\frac{K_{tn}}{K_{t\infty}} = \frac{K_{tg}}{K_{t\infty}} \left(1 - \frac{2a}{H}\right)$$

where $K_{t\infty}$ is equal to K_t for an infinitely wide panel.

The Fedderson formula,

$$\frac{K_{tg}}{K_{t\infty}} = \left(\sec \pi \frac{a}{H}\right)^{1/2} \tag{4.63}$$

The Koiter formula,

$$\frac{K_{tg}}{K_{t\infty}} = \left[1 - 0.5 \cdot \frac{2a}{H} + 0.326 \left(\frac{2a}{H}\right)^2\right] \left[1 - \frac{2a}{H}\right]^{-1/2} \tag{4.64}$$

Equations (4.62) to (4.64) represent the ratios of stress-intensity factors. In the small-radius, narrow-slit limit, the ratios are valid for stress concentration (Irwin 1960; Paris and Sih 1965).

Equation (4.64) covers the entire a/H range from 0 to 0.5 (Chart 4.53), with correct end conditions. Equation (4.62) is in good agreement for $a/H < 0.3$. Equation (4.63) is in good agreement (Rooke 1970; generally less than 1% difference; at $a/H = 0.45$, less than 2%). Isida values are within 1% difference (Rooke 1970) for $a/H < 0.4$.

Photoelastic tests (Dixon 1960; Papirno 1962) of tension members with a tranverse slit connecting two small holes are in reasonable agreement with the foregoing, taking into consideration the accuracy limits of the photoelastic test.

Chart 4.53 also provides factors for circular and elliptical holes. Correction factors have been developed (Isida 1966) for an eccentrically located crack in a tension strip.

4.4.3 Single Elliptical Hole in an Infinite, Thin Element Biaxially Stressed

If the element is subjected to biaxial tension σ_1 and σ_2 as shown in Fig. 4.22, the solution can be obtained by superposition of Eqs. (4.54) and (4.61):

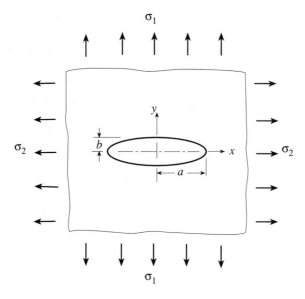

Figure 4.22 Elliptical hole in biaxial tension.

$$(\sigma_\beta)_{\alpha_0} = \frac{(\sigma_1 + \sigma_2)\sinh 2\alpha_0 + (\sigma_2 - \sigma_1)(1 - e^{2\alpha_0}\cos 2\beta)}{\cosh 2\alpha_0 - \cos 2\beta} \qquad (4.65)$$

If the element is subjected to biaxial tension σ_1 and σ_2, while the major axis is inclined an angle θ as shown in Fig. 4.23, the stress distribution at the edge of the elliptic hole, where $\alpha = \alpha_0$, is (Inglis 1913)

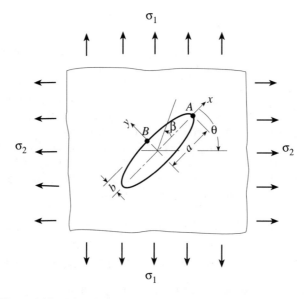

Figure 4.23 Biaxial tension of an obliquely oriented elliptical hole.

$$(\sigma_\beta)_{\alpha_0} = \frac{(\sigma_1 + \sigma_2)\sinh 2\alpha_0 + (\sigma_2 - \sigma_1)[\cos 2\theta - e^{2\alpha_0}\cos 2(\beta - \theta)]}{\cosh 2\alpha_0 - \cos 2\beta} \tag{4.66}$$

Equation (4.66) is a generalized formula for the stress calculation on the edge of an elliptical hole in an infinite element subject to uniaxial, biaxial, and shear stress states. For example, assume the stress state of the infinite element is σ_x, σ_y, and τ_{xy} of Fig. 4.24. The

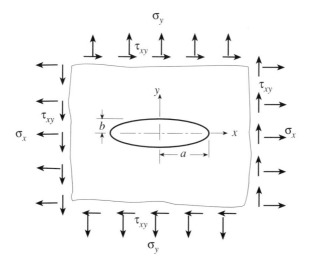

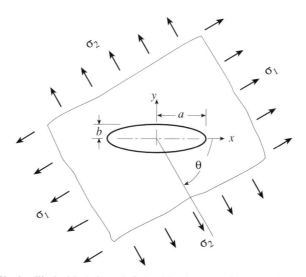

Figure 4.24 Single elliptical hole in an infinite thin element subject to arbitrary stress states.

218 HOLES

two principal stresses σ_1 and σ_2 and the incline angle θ can be found (Pilkey 1994) as

$$\sigma_1 = \frac{\sigma_x + \sigma_y}{2} + \sqrt{\left(\frac{\sigma_x - \sigma_y}{2}\right)^2 + \tau_{xy}^2}$$

$$\sigma_2 = \frac{\sigma_x + \sigma_y}{2} - \sqrt{\left(\frac{\sigma_x - \sigma_y}{2}\right)^2 + \tau_{xy}^2} \quad (4.67)$$

$$\tan 2\theta = \frac{2\tau_{xy}}{\sigma_x - \sigma_y}$$

Substitution of σ_1, σ_2, and θ into Eq. (4.66) leads to the stress distribution along the edge of the elliptical hole. Furthermore the maximum stress along the edge can be found and the stress concentration factor calculated.

Example 4.8 Pure Shear Stress State around an Elliptical Hole Consider an infinite plane element, with an elliptical hole, that is subjected to uniform shear stress τ. The direction of τ is parallel to the major and minor axes of the ellipse as shown in Fig. 4.25a. Find the stress concentration factor.

This two-dimensional element is in a state of stress of pure shear. The principal stresses are $\sigma_1 = \tau$ and $\sigma_2 = -\tau$. The angle between the principal direction and the shear stress τ is $\pi/4$ (Pilkey 1994). This problem then becomes one of calculating the stress concentration factor of an element with an elliptical hole under biaxial tension, with the direction of σ_2 inclined at an angle of $-\pi/4$ to the major axis $2a$ as shown in Fig 4.25b. Substitute $\sigma_1 = \tau$, $\sigma_2 = -\tau$ and $\theta = \pi/4$ into Eq. (4.66):

$$(\sigma_\beta)_{\alpha_0} = \frac{2\tau e^{2\alpha_0} \sin 2\beta}{\cosh 2\alpha_0 - \cos 2\beta} \quad (1)$$

From Eq. (4.49),

$$e^{2\alpha_0} = \frac{a+b}{a-b}, \quad \cosh 2\alpha_0 = \frac{a^2 + b^2}{a^2 - b^2}, \quad \sinh 2\alpha_0 = \frac{2ab}{a^2 - b^2} \quad (2)$$

Substitute (2) into (1)

$$(\sigma_\beta)_{\alpha_0} = \frac{2\tau (a+b)^2 \sin 2\beta}{a^2 + b^2 - (a^2 - b^2)\cos 2\beta} \quad (3)$$

Differentiate (3) with respect to β, and set the result equal to 0. The extreme stresses occur when

$$\cos 2\beta = \frac{a^2 - b^2}{a^2 + b^2} \quad (4)$$

and

$$\sin 2\beta = \pm \frac{2ab}{a^2 + b^2} \quad (5)$$

ELLIPTICAL HOLES IN TENSION   219

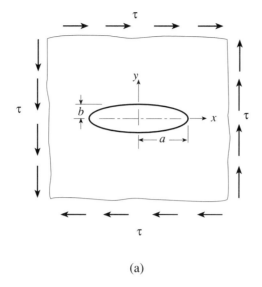

(a)

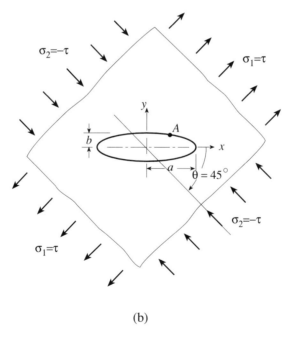

(b)

Figure 4.25 Elliptical hole in pure shear.

The maximum stress occurs at point A, which corresponds to (4) and $\sin 2\beta = 2ab/(a^2 + b^2)$, so

$$\sigma_{\beta\,\max} = \frac{\tau(a+b)^2}{ab} \tag{6}$$

If the stress τ is used as a reference stress, the corresponding stress concentration factor is

$$K_t = \frac{(a+b)^2}{ab} \tag{7}$$

Example 4.9 Biaxial Tension around an Elliptical Hole   Suppose that an element is subjected to tensile stresses σ_1, σ_2 and the direction of σ_2 forms an angle θ with the major axis of the hole, as shown in Fig. 4.23. Find the stress concentration factor at the perimeter of the hole for (1): $\theta = 0$ and (2): $\sigma_1 = 0$, $\theta = \pi/6$.

Equation (4.67) applies to these two cases

$$(\sigma_\beta)_{\alpha_0} = \frac{(\sigma_1 + \sigma_2)\sinh 2\alpha_0 + (\sigma_2 - \sigma_1)[\cos 2\theta - e^{2\alpha_0}\cos 2(\beta - \theta)]}{\cosh 2\alpha_0 - \cos 2\beta} \tag{1}$$

Set the derivative of $(\sigma_\beta)_{\alpha_0}$ with respect to β equal to zero. Then the condition for the maximum stress is

$$(\sigma_2 - \sigma_1)[\sin 2\theta(1 - \cos 2\beta \cdot \cosh 2\alpha_0) - \cos 2\theta \cdot \sin 2\beta \cdot \sinh 2\alpha_0]$$
$$= (\sigma_1 + \sigma_2)e^{-2\alpha_0}\sinh 2\alpha_0 \cdot \sin 2\beta \tag{2}$$

For case 1, $\theta = 0$ and (2) reduces to

$$(\sigma_2 - \sigma_1) \cdot \sin 2\beta = (\sigma_1 + \sigma_2)e^{-2\alpha_0}\sin 2\beta \tag{3}$$

It is evident that only $\beta = 0$, $\pi/2$, which correspond to points A and B of Fig. 4.23, satisfy (3). Thus the extreme values are

$$\sigma_A = \frac{(\sigma_1 + \sigma_2)\sinh 2\alpha_0 + (\sigma_2 - \sigma_1)(1 - e^{2\alpha_0})}{\cosh 2\alpha_0 - 1} \tag{4}$$

$$\sigma_B = \frac{(\sigma_1 + \sigma_2)\sinh 2\alpha_0 + (\sigma_2 - \sigma_1)(1 + e^{2\alpha_0})}{\cosh 2\alpha_0 + 1} \tag{5}$$

Substitute Eq. (4.49) into (4) and (5),

$$\sigma_A = \left(1 + \frac{2a}{b}\right)\sigma_1 - \sigma_2 \tag{6}$$

$$\sigma_B = \left(1 + \frac{2b}{a}\right)\sigma_2 - \sigma_1 \tag{7}$$

With σ_2 as the reference stress, the stress concentration factors are

$$K_{tgA} = \left(1 + \frac{2a}{b}\right)\frac{\sigma_1}{\sigma_2} - 1 \tag{8}$$

$$K_{tgB} = \left(1 + \frac{2b}{a}\right) - \frac{\sigma_1}{\sigma_2} \tag{9}$$

For case 2, using the same reasoning and setting $\sigma_1 = 0$, $\theta = \pi/6$, $n = b/a$, (1) and (2) become

$$(\sigma_\beta)_{\alpha_0} = \sigma_2 \frac{2n + \frac{1}{2}(1 - n^2) - \frac{1}{2}(1 + n)^2(\cos 2\beta - \sqrt{3}\sin 2\beta)}{1 + n^2 - (1 - n^2)\cos 2\beta} \tag{10}$$

$$\frac{\sqrt{3}}{2}(1 - \cos 2\beta) - \frac{\sqrt{3}}{2}n^2(1 + \cos 2\beta) = \frac{n(3 - n)}{1 + n}\sin 2\beta \tag{11}$$

From (11), it can be seen that if $a = b$, the extreme stress points occur at $\beta = -\pi/6, \pi/3$ and the maximum stress point corresponds to $\beta = \pi/3$

$$\sigma_{\beta\,\max} = \frac{2 - \frac{1}{2} \cdot 2^2(-\frac{1}{2} - \frac{3}{2})}{2}\sigma_2 = 3\sigma_2 \tag{12}$$

so that

$$K_t = 3 \tag{13}$$

This is the same result as for a circular hole, with the maximum stress point located at A (Fig. 4.26). For an elliptical hole with $b = a/3$, the maximum stress occurs when

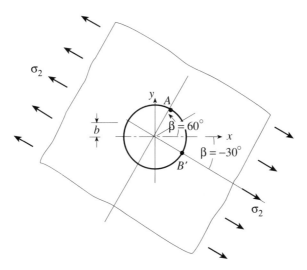

Figure 4.26 Maximum stress location for uniaxial stress.

$\cos 2\beta = 0.8$, that is, $\beta = 161.17°$ or $\beta = 341.17°$.

$$K_t = 3.309 \tag{14}$$

Stress concentration factors corresponding to different b/a values are tabulated in the table below. It can be seen that as the value of b/a decreases, the maximum stress point gradually reaches the tip of the elliptical hole.

b/a	1	0.9	0.8	0.7	0.6	0.5	0.4	0.3	0.1
β	0.33π	0.32π	0.29π	0.27π	0.23π	0.19π	0.14π	0.09π	0.02π
K_t	3.00	2.91	2.83	2.79	2.78	2.76	3.04	3.50	8.16

In the sketch in Chart 4.54, the stress σ_1 is perpendicular to the a dimension of the ellipse, regardless of whether a is larger or smaller than b. The abscissa scale (σ_2/σ_1) goes from -1 to $+1$. In other words, σ_2 is numerically equal to or less than σ_1.

The usual stress concentration factors, based on normal stresses with σ_1 as the reference stress, are taken from Eqs. (6) and (7) of Example 4.9:

$$K_{tA} = \frac{\sigma_A}{\sigma_1} = 1 + \frac{2a}{b} - \frac{\sigma_2}{\sigma_1} \tag{4.68}$$

$$K_{tB} = \frac{\sigma_B}{\sigma_1} = \frac{\sigma_2}{\sigma_1}\left[1 + \frac{2}{a/b}\right] - 1 \tag{4.69}$$

These factors are shown in Chart 4.54.

For $\sigma_1 = \sigma_2$

$$K_{tA} = \frac{2a}{b} \tag{4.70}$$

$$K_{tB} = \frac{2}{a/b} \tag{4.71}$$

Setting Eq. (4.68) equal to Eq. (4.69), we find that the stresses at A and B are equal when

$$\frac{\sigma_2}{\sigma_1} = \frac{a}{b} \tag{4.72}$$

The tangential stress is uniform around the ellipse for the condition of Eq. (4.72). Equation (4.72) is shown by a dot-dash curve on Chart 4.54. This condition occurs only for σ_2/σ_1 between 0 and 1, with the minor axis perpendicular to the major stress σ_1. Equation (4.72) provides a means of design optimization for elliptical openings. For example, for $\sigma_2 = \sigma_1/2$, $\sigma_A = \sigma_B$ for $a/b = 1/2$, with $K_t = 1.5$. Keeping $\sigma_2 = \sigma_1/2$ constant, note that if a/b is *decreased*, K_{tA} becomes less than 1.5 but K_{tB} becomes greater than 1.5. For example, for $a/b = 1/4$, $K_{tA} = 1$, $K_{tB} = 3.5$. If a/b is *increased*, K_{tB} becomes less than 1.5, but K_{tA} becomes greater than 1.5. For example, for $a/b = 1$ (circular opening), $K_{tA} = 2.5$, $K_{tB} = 0.5$.

One usually thinks of a circular hole as having the lowest stress concentration, but this depends on the stress system. We see that for $\sigma_2 = \sigma_1/2$ the maximum stress for a circular hole (Eq. 4.18) greatly exceeds that for the optimum ellipse ($a/b = 1/2$) by a factor of $2.5/1.5 = 1.666$.

An airplane cabin is basically a cylinder with $\sigma_2 = \sigma_1/2$ where $\sigma_1 =$ hoop stress, $\sigma_2 =$ axial stress. This indicates that a favorable shape for a window would be an ellipse of height 2 and width 1. The 2 to 1 factor is for a single hole in an infinite sheet. It should be added that there are other modifying factors, the proximity of adjacent windows, the stiffness of the structures, and so on. A round opening, which is often used, does not seem to be the most favorable design from a stress standpoint, although other considerations may enter.

It is sometimes said that what has a pleasing appearance often turns out to be technically correct. That this is not always so can be illustrated by the following. In the foregoing consideration of airplane windows, a stylist would no doubt wish to orient elliptical windows with the long axis in the horizontal direction to give a "streamline" effect, as was done with the decorative "portholes" in the hood of one of the automobiles of the past. The horizontal arrangement would be most unfavorable from a stress standpoint, where $K_{tA} = 4.5$ as against 1.5 oriented vertically.

The stress concentration factor based on maximum shear stress (Chart 4.54) is defined as

$$K_{ts} = \frac{\sigma_{max}/2}{\tau_{max}}$$

where, from Eq. (1.31)

$$\tau_{max} = \frac{\sigma_1 - \sigma_2}{2} \quad \text{or} \quad \frac{\sigma_1 - \sigma_3}{2} \quad \text{or} \quad \frac{\sigma_2 - \sigma_3}{2}$$

In a sheet, with $\sigma_3 = 0$,

$$\tau_{max} = \frac{\sigma_1 - \sigma_2}{2} \quad \text{or} \quad \frac{\sigma_1}{2} \quad \text{or} \quad \frac{\sigma_2}{2}$$

For $0 \leq (\sigma_2/\sigma_1) \leq 1$,

$$K_{ts} = \frac{\sigma_{max}/2}{\sigma_1/2} = K_t \tag{4.73}$$

For $-1 \leq (\sigma_2/\sigma_1) \leq 0$,

$$K_{ts} = \frac{\sigma_{max}/2}{(\sigma_1 - \sigma_2)/2} = \frac{K_t}{1 - (\sigma_2/\sigma_1)} \tag{4.74}$$

Since σ_2 is negative, the denominator is greater than σ_1, resulting in a lower numerical value of K_{ts} as compared to K_t, as seen in Chart 4.54. For $\sigma_2 = -\sigma_1$, $K_{ts} = K_t/2$.

The stress concentration factor based on equivalent stress is defined as

$$K_{te} = \frac{\sigma_{max}}{\sigma_{eq}}$$

$$\sigma_{eq} = \frac{1}{\sqrt{2}}\sqrt{(\sigma_1 - \sigma_2)^2 + (\sigma_1 - \sigma_3)^2 + (\sigma_2 - \sigma_3)^2}$$

For $\sigma_3 = 0$,

$$\sigma_{eq} = \frac{1}{\sqrt{2}}\sqrt{(\sigma_1 - \sigma_2)^2 + \sigma_1^2 + \sigma_2^2}$$

$$= \sigma_1\sqrt{1 - (\sigma_2/\sigma_1) + (\sigma_2/\sigma_1)^2} \tag{4.75}$$

$$K_{te} = \frac{K_t}{\sqrt{1 - (\sigma_2/\sigma_1) + (\sigma_2/\sigma_1)^2}} \tag{4.76}$$

K_{te} values are shown in Chart 4.55.

For obtaining σ_{max}, the simplest factor K_t is adequate. For mechanics of materials problems, the latter two factors, which are associated with failure theory, are useful.

The condition $\sigma_2/\sigma_1 = -1$ is equivalent to pure shear oriented 45° to the ellipse axes. This case and the case where the shear stresses are parallel to the ellipse axes are discussed in Section 4.7.1, Chart 4.88. Jones and Hozos (1971) provide some values for biaxial stressing of a finite panel with an elliptical hole.

Stresses around an elliptical hole in a cylindrical shell in tension have been studied by Murthy (1969), Murthy and Rao (1970), and Tingleff (1971). Values for an elliptical hole in a pressurized spherical shell are presented in Chart 4.6.

4.4.4 Infinite Row of Elliptical Holes in Infinite- and Finite-Width Thin Elements in Uniaxial Tension

Nisitani (1968) provided the stress concentration factor for an infinite row of elliptical holes in an infinite panel (Chart 4.56). This chart covers a row of holes in the stress direction as well as a row perpendicular to the stress direction. The ordinate values are plotted as K_t/K_{t0}, where $K_{t0} = K_t$ for the single hole (Eq. 4.58). The results are in agreement with Schulz (1941) for circular holes. The effect of finite width is shown in Chart 4.57 (Nisitani 1968). The quantity K_{t0} is the stress concentration factor for a single hole in a finite width element (Chart 4.51). Nisitani concluded that the interference effect of multi-holes K_t/K_{t0}, where K_t is for multi-holes and K_{t0} is for a single hole, is proportional to the square of the major semiaxis of the ellipse over the distance between the centers of the holes, a^2/c.

4.4.5 Elliptical Hole with Internal Pressure

As mentioned in Section 4.3.16 on the thin element with circular holes with internal pressure, the stress concentration factor of an infinite element with circular holes with internal pressure can be found through superposition. This is true for elliptical holes as well. For elliptical holes with internal pressure in an infinite element, as stated in Section 4.3.16, K_t can be found by subtracting 1.0 from the case of Section 4.4.3, Eq. (8), Example 4.9, for $\sigma_1/\sigma_2 = 1$. Thus

$$K_t = \frac{2a}{b} - 1 \qquad (4.77)$$

4.4.6 Elliptical Holes with Bead Reinforcement in an Infinite Thin Element under Uniaxial and Biaxial Stresses

In Chart 4.58 values of K_t for reinforced elliptical holes are plotted against $A_r/[(a+b)h]$ for various values of a/b for uniaxial and biaxial loading conditions (Wittrick 1959; Houghton and Rothwell 1961; ESDU 1981). Here A_r is the cross-sectional area of the bead reinforcement. Care must be taken in attempting to superimpose the maximum equivalent stresses for different loadings. These stresses are not directly additive if the location of the maximum stresses are different for different loading conditions. Stresses in the panel at its junction with the reinforcement are given here. The chart is based on $\nu = 0.33$.

4.5 VARIOUS CONFIGURATIONS WITH IN-PLANE STRESSES

4.5.1 Thin Element with an Ovaloid; Two Holes Connected by a Slit under Tension; Equivalent Ellipse

The "equivalent ellipse" concept (Cox 1953; Sobey 1963; Hirano, 1950) is useful for the ovaloid (slot with semicircular ends, Fig. 4.27a) and other openings such as two holes connected by a slit (Fig. 4.27b). If such a shape is enveloped (Fig. 4.27) by an ellipse (same major axis a and minimum radius r), the K_t values for the shape and the equivalent ellipse may be nearly the same. In the case of the ovaloid, K_t for the ellipse is within 2% of the correct value. The K_t for the ellipse can be calculated using Eq. (4.57).

Another comparison is provided by two tangential circular holes (Fig. 4.27c) of Chart 4.22, where $K_t = 3.869$ for $l/d = 1$. This compares closely with the "equivalent ellipse" value of $K_t = 3.828$ found from Eq. (4.57). The cusps resulting from the enveloping ellipse are, in effect, stress-free ("dead" photoelastically). A similar stress free region occurs for two holes connected by a slit. The round-cornered square hole oriented 45° to the applied uniaxial stress (Isida 1960), not completely enveloped by the ellipse, is approximately represented by the "equivalent ellipse."

Previously published values for a slot with semicircular ends (Frocht and Leven 1951) are low compared with the K_t values for the elliptical hole (Chart 4.51) and for a circular hole (Chart 4.1). It is suggested that the values for the equivalent ellipse be used. It has been shown that although the equivalent ellipse applies for tension, it is not applicable for shear (Cox 1953).

A photoelastic investigation (Durelli et al. 1968) of a slot of constant $a/b = 3.24$ found the optimum elliptical slot end as a function of a/H, where H is the panel width. The optimum shape was an ellipse of a/b about 3 (Chart 4.59), and this resulted in a reduction of K_{tn}, from the value for the semicircular end of about 22% at $a/H = 0.3$ to about 30% for $a/H = 0.1$ with an average reduction of about 26%. The authors state that the results may prove useful in the design of solid propellant grains. Although the numerical conclusions apply only to $a/b = 3.24$, it is clear that the same method of optimization may be useful in other design configurations with the possibility of significant stress reductions.

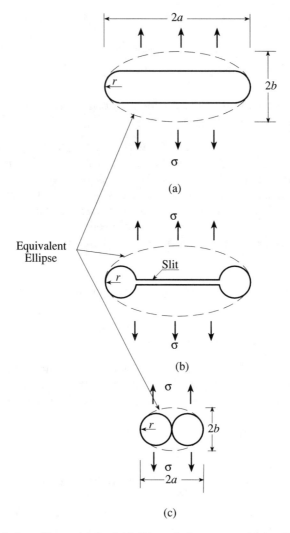

Figure 4.27 Equivalent ellipses: (*a*) Ovaloid; (*b*) two holes connected by a slit; (*c*) two tangential circular holes.

4.5.2 Circular Hole with Opposite Semicircular Lobes in a Thin Element in Tension

Thin tensile elements with circular holes with opposite semicircular lobes have been used for fatigue tests of sheet materials, since the stress concentrator can be readily produced with minimum variation from piece to piece (Gassner and Horstmann 1961; Schultz 1964). Mathematical results (Mitchell 1966) for an infinitely wide panel are shown in Chart 4.60 and are compared with an ellipse of the same overall width and minimum radius (equivalent ellipse).

For a finite-width panel (Chart 4.61) representative of a test piece, the following empirical formula was developed by Mitchell (1966)

VARIOUS CONFIGURATIONS WITH IN-PLANE STRESSES   227

$$K_t = K_{t\infty}\left[1 - \frac{2a}{H} + 4\left(\frac{6}{K_{t\infty}} - 1\right)\left(\frac{a}{H}\right)^2 + 8\left(1 - \frac{4}{K_{t\infty}}\right)\left(\frac{a}{H}\right)^3\right] \quad (4.78)$$

where $K_{t\infty} = K_t$ for infinitely wide panel (see Chart 4.60), a is the half width of hole plus lobes, and H is the width of the panel.

For $H = \infty$, $a/H = 0$, $K_t = K_{t\infty}$.

For $r/d \rightarrow 0$, $K_{t\infty}$ is obtained by multiplying K_t for the hole, 3.0, by K_t for the semicircular notch (Ling 1967), 3.065, resulting in $K_{t\infty} = 9.195$. The Mitchell (1966) value is $3(3.08) = 9.24$.

For r/d greater than about 0.75, the middle hole is entirely swallowed up by the lobes. The resulting geometry, with middle opposite cusps, is the same as in Chart 4.22 ($l/d < 2/3$).

For $r/d \rightarrow \infty$, a circle is obtained, $K_{t\infty} = 3$. Equation (4.78) reduces to the Heywood formula (Heywood 1952).

$$K_t = 2 + \left(1 - \frac{a}{H}\right)^3 \quad (4.79)$$

Photoelastic tests by Cheng (1968) confirm the accuracy of the Mitchell formula.

Miyao (1970) has solved the case for one lobe. The K_t values are lower, varying from 0% at $r/d \rightarrow 0$ to about 10% at $r/d = 0.5$ (ovaloid, see Chart 4.62). Miyao also gives K_t values for biaxial tension.

4.5.3 Infinite Thin Element with a Rectangular Hole with Rounded Corners Subject to Uniaxial or Biaxial Stress

The rectangular opening with rounded corners is often found in structures, such as ship hatch openings and airplane windows. Mathematical results, with specific data obtained by computer, have been published (Sobey 1963; Heller et al. 1958; Heller 1969). For uniaxial tension, K_t is given in Chart 4.62a, where the stress σ_1 is perpendicular to the a dimension. The top dashed curve of Chart 4.62a is for the ovaloid (slot with semicircular ends). In Section 4.5.1 it was noted that for uniaxial tension and for the same a/r, the ovaloid and the equivalent ellipse are the same for all practical purposes (Eq. 4.57). In the published results (Sobey 1963; ESDU 1970) for the rectangular hole, the ovaloid values are close to the elliptical values. The latter are used in Chart 4.62a to give the ovaloid curve a smoother shape. All of Charts 4.62 show clearly the minimum K_t as a function of r/a.

In Charts 4.62 it will be noted that for $a/b > 1$, either the ovaloid represents the minimum K_t (see Chart 4.62c) or the rectangular hole with a particular (optimum) radius (r/a between 0 and 1) represents the minimum K_t (Charts 4.62a, b, and d).

A possible design problem is to select a shape of opening having a minimum K_t within rectangular limits a and b. In Chart 4.63 the following shapes are compared: ellipse, ovaloid, rectangle with rounded corners (for radius giving minimum K_t).

For the uniaxial case (top three dashed curves of Chart 4.63) the ovaloid has a lower K_t than the ellipse when $a/b > 1$ and a higher value when $a/b < 1$. The K_t for the optimum rectangle is lower than (or equal to) the K_t for the ovaloid. It is lower than the K_t for the ellipse when $a/b > 0.85$, higher when $a/b < 0.85$.

One might think that a circular opening in a tension panel would have a lower maximum stress than a round-cornered square opening having a width equal to the circle diameter. From Charts 4.62a and 4.63 it can be seen that *a square opening with corner radii of about a third of the width has a lower maximum stress than a circular opening of the same width.* Photoelastic studies show similar conclusions hold for notches and shoulder fillets. These remarks apply to the uniaxial tension case but not for a biaxial case with $\sigma_1 = \sigma_2$. For $\sigma_2 = \sigma_1/2$, the optimum opening has only a slightly lower K_t.

The full line curves of Chart 4.63, representing $\sigma_2 = \sigma_1/2$, the stress state of a cylindrical shell under pressure, show that the ovaloid and optimum rectangle are fairly comparable and that their K_t values are lower than the ellipse for $a/b > 1$ and $a/b < 0.38$, greater for $a/b > 0.38$ and $a/b < 1$.

Note that for $a/b = 1/2$, K_t reaches the *low value of 1.5* for the ellipse. It is to be noted here that the ellipse is in this case superior to the ovaloid, $K_t = 1.5$ as compared to $K_t = 2.08$.

For the equal biaxial state, found in a pressurized spherical shell, the dot-dash curves of Chart 4.63 show that the ovaloid is the optimum opening in this case and gives a lower K_t than the ellipse (except of course at $a/b = 1$, where both become circles).

For a round-cornered square hole oriented 45° to the applied tension, Hirano (1950) has shown that the "equivalent ellipse" concept (see Section 4.5.1) is applicable.

4.5.4 Finite-Width Tension Thin Element with Round-Cornered Square Hole

In comparing the K_t values of Isida (1960) (for the finite-width strip) with Chart 4.62a, it appears that satisfactory agreement is obtained only for small values of a/H, the half-hole width/element width. As a/H increases, K_t increases in approximately this way: $K_{tg}/K_{t\infty} \approx$ 1.01, 1.03, 1.05, 1.09, 1.13, for $a/H = 0.05, 0.1, 0.15, 0.2$, and 0.25, respectively.

4.5.5 Square Holes with Rounded Corners and Bead Reinforcement in an Infinite Panel under Uniaxial and Biaxial Stresses

The stress concentration factor K_t for reinforced square holes is given as a function of A_r for various values of r/a for the unixial and biaxial loading stress in Chart 4.64 (Sobey 1968; ESDU 1981). A_r is the cross-sectional area of the reinforcement. The curves are based on $\nu = 0.33$.

Care must be taken in attempting to superimpose the maximum equivalent stresses for different loadings. These stresses are not directly additive if the location of the maximum stresses differ for different loading conditions. Stresses in the panel at its junction with the reinforcement are given here.

4.5.6 Round-Cornered Equilateral Triangular Hole in an Infinite Thin Element under Various States of Tension

The triangular hole with rounded corners has been used in some vehicle window designs as well as in certain architectural designs. The stress distribution around a triangular hole with rounded corners has been studied by Savin (1961).

VARIOUS CONFIGURATIONS WITH IN-PLANE STRESSES   229

The K_t values for $\sigma_2 = 0$ (σ_1 only), $\sigma_2 = \sigma_1/2$, and $\sigma_2 = \sigma_1$ in Chart 4.65a were determined by Wittrick (1963) by a complex variable method using a polynomial transformation function for mapping the contour. The corner radius is not constant. The radius r is the minimum radius, positioned symmetrically at the corners of the triangle. For $\sigma_2 = \sigma_1/2$, the equivalent stress concentration factor (von Mises), $K_{te} = (2/\sqrt{3})K_t = 1.157K_t$. For $\sigma_2 = \sigma_1$ and $\sigma_2 = 0$, $K_{te} = K_t$. In Chart 4.65b, the K_t factors of Chart 4.65a are replotted as a function of σ_2/σ_1.

4.5.7 Uniaxially Stressed Tube or Bar of Circular Cross Section with a Transverse Circular Hole

The transverse hole in a tube or bar of circular cross section occurs in engineering practice in lubricant and coolant ducts in shafts, in connectors for control or transmission rods, and in various types of tubular framework. Stress concentration factors K_{tg} and K_{tn} are shown in Chart 4.66. The results of Leven (1955) and of Thum and Kirmser (1943) for the solid shaft are in close agreement. The solid round bar curves of Chart 4.66 represent both sets of data.

The results for the tubes are from British data (Jessop, Snell, and Allison, 1959; ESDU 1965). The factors are defined as follows:

$$K_{tg} = \frac{\sigma_{max}}{\sigma_{gross}} = \frac{\sigma_{max}}{P/A_{tube}} = \frac{\sigma_{max}}{P/[(\pi/4)(D^2 - d_i^2)]} \quad (4.80)$$

$$K_{tn} = \frac{\sigma_{max}}{\sigma_{net}} = \frac{\sigma_{max}}{P/A_{net}} = K_{tg}\frac{A_{net}}{A_{tube}} \quad (4.81)$$

The ratio A_{net}/A_{tube} has been determined mathematically (Peterson 1968). The formulas will not be repeated here, although specific values can be obtained by dividing the Chart 4.66 values of K_{tn} by K_{tg}. If the hole is sufficiently small relative to the shaft diameter, the hole may be considered to be of rectangular cross section. Then

$$\frac{A_{net}}{A_{tube}} = 1 - \frac{4\pi(d/D)[1 - (d_i/D)]}{1 - (d_i/D)^2} \quad (4.82)$$

It can be seen from the bottom curves of Chart 4.66 that the error due to this approximation is small below $d/D = 0.3$.

Thum and Kirmser (1943) found that the maximum stress did not occur on the surface of the shaft but at a small distance inside the hole on the surface of the hole. This was later corroborated by other investigators (Leven 1955; Jessop, Snell, and Allison 1959). The σ_{max} value used in developing Chart 4.66 is the maximum stress inside the hole.

4.5.8 Round Pin Joint in Tension

The case of a pinned joint in an infinite thin element has been solved mathematically by Bickley (1928). The finite-width case has been solved by Knight (1935), where the element width is equal to twice the hole diameter d and by Theocaris (1956) for $d/H = 0.2$ to 0.5. Experimental results (strain gage or photoelastic) have been obtained by Coker and Filon (1931), Schaechterle (1934), Frocht and Hill (1940), Jessop, Snell, and Holister (1958), and Cox and Brown (1964).

Two methods have been used in defining K_{tn}:

Nominal stress based on net section,

$$\sigma_{nd} = \frac{P}{(H-d)h}$$

$$K_{tnd} = \frac{\sigma_{\max}}{\sigma_{nd}} = \sigma_{\max}\frac{(H-d)h}{P} \qquad (4.83)$$

Nominal stress based on bearing area,

$$\sigma_{nb} = \frac{P}{dh}$$

$$K_{tnb} = \frac{\sigma_{\max}}{\sigma_{nb}} = \frac{\sigma_{\max}dh}{P} \qquad (4.84)$$

Note that

$$\frac{K_{tnd}}{K_{tnb}} = \frac{1}{d/H} - 1 \qquad (4.85)$$

In Chart 4.67 the K_{tnb} curve corresponds to the Theocaris (1956) data for $d/H = 0.2$ to 0.5. The values of Frocht and Hill (1940) and Cox and Brown (1964) are in good agreement with Chart 4.67, although slightly lower. From $d/H = 0.5$ to 0.75 the foregoing 0.2–0.5 curve is extended to be consistent with the Frocht and Hill values. The resulting curve is for joints where c/H is 1.0 or greater. For $c/H = 0.5$, the K_{tn} values are somewhat higher.

From Eq. (4.85), $K_{tnd} = K_{tna}$ at $d/H = 1/2$. It would seem more logical to use the lower (full line) branches of the curves in Chart 4.67, since, in practice, d/H is usually less than $1/4$. This means that Eq. (4.84), based on the bearing area, is generally used.

Chart 4.67 is for closely fitting pins. The K_t factors are increased by clearance in the pin fit. For example, at $d/H = 0.15$, K_{tnb} values (Cox and Brown 1964) of approximately 1.1, 1.3, and 1.8 were obtained for clearances of 0.7%, 1.3%, and 2.7%, respectively. (For an in-depth discussion of lug-clevis joint systems, see Chapter 5.) The effect of interference fits is to reduce the stress concentration factor.

A joint having an infinite row of pins has been analyzed (Mori 1972). It is assumed that the element is thin (two-dimensional case), that there are no friction effects, and that the pressure on the hole wall is distributed as a cosine function over half of the hole. The stress concentration factors (Chart 4.68) have been recalculated based on Mori's work to be related to $\sigma_{\text{nom}} = P/d$ rather than to the mean peripheral pressure in order to be defined in the same way as in Chart 4.67. It is seen from Chart 4.68 that decreasing e/d from a value of 1.0 results in a progressively increasing stress concentration factor. Also, as in Chart 4.67, increasing d/l or d/H results in a progressively increasing stress concentration factor.

The end pins in a row carry a relatively greater share of the load. The exact distribution depends on the elastic constants and the joint geometry (Mitchell and Rosenthal 1949).

4.5.9 Inclined Round Hole in an Infinite Panel Subjected to Various States of Tension

The inclined round hole is found in oblique nozzles and control rods in nuclear and other pressure vessels.

The curve for uniaxial stressing and $\nu = 0.5$, second curve from the top of Chart 4.69 (which is for an inclination of 45°), is based on the photoelastic tests of Leven (1970), Daniel (1970), and McKenzie and White (1968) and the strain gage tests of Ellyin (1970). The K_t factors (Leven 1970; McKenzie and White 1968; Ellyin 1970) adjusted to the same K_t definition (to be explained in the next paragraph) for $h/b \sim 1$ are in good agreement (K_t of Daniel 1970 is for $h/b = 4.8$). Theoretical K_t factors (Ellyin 1970) are considerably higher than the experimental factors as the angle of inclination increases. However, the theoretical curves are used in estimating the effect of Poisson's ratio and in estimating the effect of the state of stress. As $h/b \to 0$, the $K_{t\infty}$ values are for the corresponding ellipse(Chart 4.50). For h/b large the $K_{t\infty}$ values at the midsection are for a circular hole ($a/b = 1$ in Chart 4.51). This result is a consequence of the flow lines in the middle region of a thick panel taking a direction perpendicular to the axis of the hole. For uniaxial stress σ_2, the midsection $K_{t\infty}$ is the maximum value. For uniaxial stress σ_1, the surface $K_{t\infty}$ is the maximum value.

For design use it is desirable to start with a factor corresponding to infinite width and then have a method of correcting this to the a/H ratio involved in any particular design (a = semimajor width of surface hole; H = width of panel). This can be done, for design purposes, in the following way: For any inclination the surface ellipse has a corresponding a/b ratio. In Chart 4.53 we obtain K_{tn}, K_{tg}, and $K_{t\infty}$ for the a/H ratio of interest ($K_{t\infty}$ is the value at $a/H \to 0$). Ratios of these values were used to adjust the experimental values to $K_{t\infty}$ in Charts 4.69 and 4.70. In design the same ratio method is used in going from $K_{t\infty}$ to the K_t corresponding to the actual a/H ratio.

In Chart 4.70 the effect of inclination angle θ is given. The $K_{t\infty}$ curve is based on the photoelastic K_{tg} values of McKenzie and White (1968) adjusted to $K_{t\infty}$ as described above. The curve is for $h/b = 1.066$, corresponding to the flat peak region of Chart 4.69. The effect of Poisson's ratio is estimated in Ellyin's work.

For uniaxial stress σ_1 in panels, the maximum stress is located at A, Chart 4.69. An attempt to reduce this stress by rounding the edge of the hole with a contour radius $r = b$ produced the surprising result (Daniel 1970) of increasing the maximum stress (for $h/b = 4.8$, 30% higher for $\theta = 45°$, 50% higher for $\theta = 60°$). The maximum stress was located at the point of tangency of the contour radius with a line perpendicular to the panel surfaces. The stress increase has been explained (Daniel 1970) by the stress concentration due to the egg-shaped cross section in the horizontal plane. For $\theta = 75.5°$ and $h/b = 1.066$, it was found (McKenzie and White 1968) that for $r/b < 2/9$, a small decrease in stress was obtained by rounding the corner, but above $r/b = 2/9$, the stress increased rapidly, which is consistent with the b result (Daniel 1970).

Strain gage tests were made by Ellyin and Izmiroglu (1973) on 45° and 60° oblique holes in 1 in.-thick steel panels subjected to tension. The effects of rounding the corner A (Chart 4.69) and of blunting the corner with a cut perpendicular to the panel surface were evaluated. In most of the tests $h/b \approx 0.8$. For $\theta = 45°$, obtained in the region $r/h < 0.2$. However, $h/b > 0.2$, a small decrease in maximum stress was increased by rounding.

It is difficult to compare the various investigations of panels with an oblique hole having a rounded corner because of large variations in h/b. Also the effect depends on r/h and

h/b. Leven (1970) has obtained a 25% maximum stress reduction in a 45° oblique nozzle in a pressure vessel model by blunting the acute nozzle corner with a cut perpendicular to the vessel axis. From a consideration of flow lines, it appears that the stress lines would not be as concentrated for the vertical cut as for the "equivalent" radius.

4.5.10 Pressure Vessel Nozzle (Reinforced Cylindrical Opening)

A nozzle in pressure vessel and nuclear reactor technology denotes an integral tubular opening in the pressure vessel wall (see Fig. 4.28). Extensive strain gage (Hardenberg and Zamrik 1964) and photoelastic tests (WRCB 1966; Seika et al. 1971) have been made of various geometric reinforcement contours aimed at reducing stress concentration. Figure 4.28 is an example of a resulting "balanced" design (Leven 1966). Stress concentration factors for oblique nozzles (nonperpendicular intersection) are also available (WRCB 1970).

4.5.11 Spherical or Ellipsoidal Cavities

Stress concentration factors for cavities are useful in evaluating the effects of porosity in materials (Peterson 1965). The stress distribution around a cavity having the shape of an

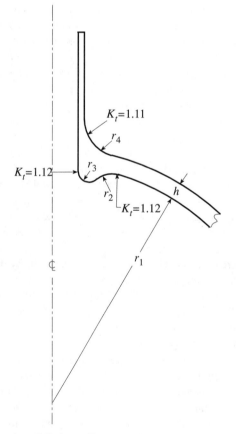

Figure 4.28 Half section of "balanced" design of nozzle in spherical vessel (Leven 1966).

ellipsoid of revolution has been obtained by Neuber (1958) for various types of stressing. The case of tension in the direction of the axis of revolution is shown in Chart 4.71. Note that the effect of Poisson's ratio ν is relatively small. It is seen that high K_t factors are obtained as the ellipsoid becomes thinner and approaches the condition of a disk-shaped transverse crack.

The case of stressing perpendicular to the axis has been solved for an internal cavity having the shape of an elongated ellipsoid of revolution (Sadowsky and Sternberg 1947). From Chart 4.72 it is seen that for a circularly cylindrical hole ($a = \infty$, $b/a \to 0$) the value of $K_t = 3$ is obtained and that this reduces to $K_t = 2.05$ for the spherical cavity ($b/a = 1$). If we now consider an elliptical shape, $a/b = 3$, ($a/r = 9$), from Eq. (4.57) and Chart 4.71, we find that for a cylindrical hole of elliptical cross section, $K_t = 7$. For a circular cavity of elliptical cross section (Chart 4.71), $K_t = 4.6$. And for an ellipsoid of revolution (Chart 4.72), $K_t = 2.69$. The order of the factors quoted above seems reasonable if one considers the course streamlines must take in going around the shapes under consideration.

Sternberg and Sadowsky (1952) studied the "interference" effect of two spherical cavities in an infinite body subjected to hydrostatic tension. With a space of one diameter between the cavities, the factor was increased less than 5%, $K_t = 1.57$, as compared to infinite spacing (single cavity), $K_t = 1.50$. This compares with approximately 20% for the analogous plane problem of circular holes in biaxially stressed panels of Chart 4.24.

In Chart 4.73 stress concentration factors K_{tg} and K_{tn} are given for tension of a circular cylinder with a central spherical cavity (Ling 1959). The value for the infinite body is (Timoshenko and Goodier 1970)

$$K_t = \frac{27 - 15\nu}{14 - 10\nu} \tag{4.86}$$

where ν is Poisson's ratio. For $\nu = 0.3$, $K_t = 2.045$.

For a large spherical cavity in a round tension bar, Ling shows that $K_t = 1$ for $d/D \to 1$. Koiter (1957) obtains the following for $d/D \to 1$:

$$K_t = (6 - 4\nu)\frac{1 + \nu}{5 - 4\nu^2} \tag{4.87}$$

In Chart 4.73 a curve for K_{tg} is given for a biaxially stressed moderately thick element with a central spherical cavity (Ling 1959). For infinite thickness (Timoshenko and Goodier 1970)

$$K_{tg} = \frac{12}{7 - 5\nu} \tag{4.88}$$

This value corresponds to the pole position on the spherical surface perpendicular to the plane of the applied stress.

The curve for the flat element of Chart 4.73 was calculated for $\nu = 1/4$. The value for $d/h = 0$ and $\nu = 0.3$ is also shown.

The effect of spacing for a row of "disk-shaped" ellipsoidal cavities (Nisitani 1968) is shown in Chart 4.74 in terms of K_t/K_{t0}, where $K_{t0} = K_t$ for the single cavity (Chart 4.71). These results are for Poisson's ratio 0.3. Nisitani (1968) concludes that the interference effect is proportional to the cube of the ratio of the major semiwidth of the cavity over the distance between the centers of the cavities. In the case of holes in thin elements (Section 4.4.4), the proportionality was as the square of the ratio.

4.5.12 Spherical or Ellipsoidal Inclusions

The evaluation of the effect of inclusions on the strength of materials, especially in fatigue and brittle fracture, is an important consideration in engineering technology. The stresses around an inclusion have been analyzed by considering that the hole or cavity is filled with a material having a different modulus of elasticity, E', and that adhesion between the two materials is perfect.

Donnell (1941) has obtained relations for cylindrical inclusions of elliptical cross section in a panel for E'/E varying from 0 (hole) to ∞ (rigid inclusion). Donnell found that for Poisson's ratio $\nu = 0.25$ to 0.3, the plane stress and plain strain values were sufficiently close for him to use a formulation giving a value between the two cases (approximation differs from exact values 1.5% or less). Edwards (1951) extended the work of Goodier (1933) and Donnell (1941) to cover the case of the inclusion having the shape of an ellipsoid of revolution.

Curves for E'/E for 1/4, 1/3, and 1/2 are shown in Charts 4.50 and 4.72. These ratios are in the range of interest in considering the effect of silicate inclusions in steel. It is seen that the hole or cavity represents a more critical condition than a corresponding inclusion of the type mentioned.

For a rigid spherical inclusion, $E'/E = \infty$, in an infinite member, Goodier (1933) obtained the following relations for uniaxial tension.

For the maximum adhesion (radial) stress at the axial (pole) position,

$$K_t = \frac{2}{1+\nu} + \frac{1}{4-5\nu} \qquad (4.89)$$

For $\nu = 0.3$, $K_t = 1.94$.

For the tangential stress at the equator (position perpendicular to the applied stress),

$$K_t = \frac{\nu}{1+\nu} - \frac{5\nu}{8-10\nu} \qquad (4.90)$$

For $\nu = 0.3$, $K_t = -0.69$.

For $\nu = 0.2$, $K_t = 0$. For $\nu > 0.2$, K_t is negative—that is, the tangential stress is compressive. The same results have been obtained (Chu and Conway 1970) by using a different method.

The case of a rigid circular cylindrical inclusion may be useful in the design of plastic members and concrete structures reinforced with steel wires or rods. Goodier (1933) has obtained the following plane strain relation for a circular cylindrical inclusion, with $E'/E = \infty$:

$$K_t = \frac{1}{2}\left(3 - 2\nu + \frac{1}{3-4\nu}\right) \qquad (4.91)$$

For $\nu = 0.3$, $K_t = 1.478$.

Studies have been made of the stresses in an infinite body containing a circular cylindrical inclusion of length one and two times the diameter d, with a corner radius r and with the cylinder axis in line with the applied tension (Chu and Conway 1970). The results may provide some guidance for a design condition where a reinforcing rod ends within a concrete member. For a length/diameter ratio of 2 and a corner radius/diameter ratio of 1/4, the following K_{ta} values were obtained ($K_{ta} = \sigma_a/\sigma$ = maximum normal stress/applied stress): $K_{ta} = 2.33$ for $E'/E = \infty$, $K_{ta} = 1.85$ for $E'/E = 8$, $K_{ta} = 1.63$ for $E'/E = 6$. For a length/diameter ratio = 1, K_{ta} does not vary greatly with corner radius/diameter ratio varying from 0.1 to 0.5 (spherical, $K_{ta} = 1.94$). Below $r/d = 0.1$, K_{ta} rises rapidly ($K_{ta} = 2.85$ at $r/d = 0.05$). Defining $K_{tb} = \tau_{max}/\sigma$ for the bond shear stress, the following values were obtained: $K_{tb} = 2.35$ at $r/d = 0.05$, $K_{tb} = 1.3$ at $r/d = 1/4$, $K_{tb} = 1.05$ at $r/d = 1/2$ (spherical).

Donnell (1941) obtained the following relations for a rigid elliptical cylindrical inclusion:

Pole position A, Chart 4.75,

$$K_{tA} = \frac{\sigma_{max A}}{\sigma} = \frac{3}{16}\left(1 - \frac{b}{a}\right) \quad (4.92)$$

Midposition B,

$$K_{tB} = \frac{\sigma_{max B}}{\sigma} = \frac{3}{16}\left(5 + 3\frac{a}{b}\right) \quad (4.93)$$

These stresses are radial (normal to the ellipse), adhesive tension at A and compression at B. The tangential stresses are one-third of the foregoing values.

It would seem that for the elliptical inclusion with its major axis in the tension direction, failure would start at the pole by rupture of the bond, with the crack progressing perpendicular to the applied stress. For the inclusion with its major axis perpendicular to the applied tensile stress, it would seem that for a/b less than about 0.15, the compressive stress at the end of the ellipse would cause plastic deformation but that cracking would eventually occur at position A by rupture of the bond, followed by progressive cracking perpendicular to the applied tensile stress.

Nisitani (1968) has obtained exact values for the plane stress and plane strain radial stresses for the pole position A, Chart 4.75, of the rigid elliptical cylindrical inclusion:

$$K_t = \frac{(\gamma + 1)[(\gamma + 1)(a/b) + (\gamma + 3)]}{8\gamma} \quad (4.94)$$

where $\gamma = 3 - 4\nu$ for plane strain, $\gamma = (3 - \nu)/(1 + \nu)$ for plane stress, a is the ellipse half-width parallel to applied stress, b is the ellipse half-width perpendicular to applied stress, and ν is Poisson's ratio. For plane strain

$$K_t = \frac{(1 - \nu)[2(1 - \nu)(a/b) + 3 - 2\nu]}{3 - 4\nu} \quad (4.95)$$

Equation (4.95) reduces to Eq. (4.89) for the circular cylindrical inclusion. As stated, Eqs. (4.94) and (4.95) are sufficiently close to Eq. (4.92) so that a single curve can be used in Chart 4.75. A related case of a panel having a circular hole with a bonded cylindrical insert ($r_i/r_o = 0.8$) having a modulus of elasticity 11.5 times the modulus of elasticity of the panel has been studied by a combined photoelasticity and Moiré analysis (Durelli and Riley 1965).

The effect of spacing on a row of rigid elliptical inclusions (Nisitani 1968) is shown in Chart 4.76 as a ratio of the K_t for the row and the K_{t0} for the single inclusion (Chart 4.75). Shioya (1971) has obtained the K_t factors for an infinite tension panel with two circular inclusions.

4.5.13 Cylindrical Tunnel

Mindlin (1939) has solved the following cases of an indefinitely long cylindrical tunnel: (1) hydrostatic pressure, $-cw$, at the tunnel location before the tunnel is formed (c = distance from the surface to the center of the tunnel, w = weight per unit volume of material); (2) material restrained from lateral displacement; (3) no lateral stress.

Results for case 1 are shown in Chart 4.77 in dimensionless form, $\sigma_{\max}/2wr$ versus c/r, where r is the radius of the tunnel. It is seen that the minimum value of the peripheral stress $\sigma_{\max}$ is reached at values of $c/r = 1.2, 1.25,$ and 1.35 for $\nu = 0, 1/4,$ and $1/2$, respectively. For smaller values of c/r, the increased stress is due to the thinness of the "arch" over the hole, whereas for larger values of c/r, the increased stress is due to the increased pressure created by the material above.

An arbitrary stress concentration factor may be defined as $K_t = \sigma_{\max}/p = \sigma_{\max}/(-cw)$, where p = hydrostatic pressure, equal to $-cw$. Chart 4.77 may be converted to K_t as shown in Chart 4.78 by dividing $\sigma_{\max}/2wr$ ordinates of Chart 4.77 by $c/2r$, half of the abscissa values of Chart 4.77. It is seen from Chart 4.78 that for large values of c/r, K_t approaches 2, the well-known K_t for a hole in a hydrostatic or biaxial stress field.

For a deep tunnel, c/r large (Mindlin 1939),

$$\sigma_{\max} = -2cw - rw\left[\frac{3-4\nu}{2(1-\nu)}\right] \tag{4.96}$$

By writing (rw) as $(r/c)(cw)$, we can factor out $(-cw)$ to obtain

$$K_t = \frac{\sigma_{\max}}{-cw} = 2 + \frac{1}{c/r}\left[\frac{3-4\nu}{2(1-\nu)}\right] \tag{4.97}$$

The second term arises from the weight of the material removed from the hole. As c/r becomes large, this term becomes negligible and K_t approaches 2, as indicated in Chart 4.78. Solutions for various tunnel shapes (circular, elliptical, rounded square) at depths not influenced by the surface have been obtained with and without a rigid liner (Yu 1952).

4.5.14 Intersecting Cylindrical Holes

The intersecting cylindrical holes (Riley 1964) are in the form of a cross (+), a tee (T), or a round-cornered ell (L) with the plane containing the hole axes perpendicular to the applied uniaxial stress (Fig. 4.29). This case is of interest in tunnel design and in various geometrical arrangements of fluid ducting in machinery.

Three-dimensional photoelastic tests by Riley were made of an axially compressed cylinder with these intersecting cylindrical hole forms located with the hole axes in a midplane perpendicular to the applied uniaxial stress. The cylinder was 8 in. in diameter, and all holes were 1.5 in. in diameter. The maximum nominal stress concentration factor K_{tn} (see Chart 4.66 for a definition of K_{tn}) for the three intersection forms was found to be

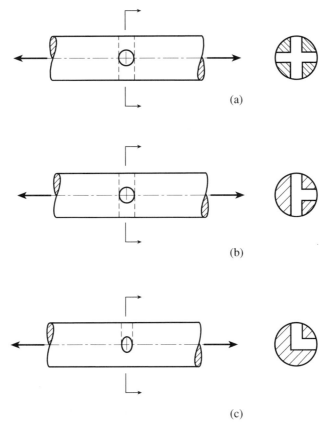

Figure 4.29 Intersecting holes in cylinder: (*a*) Cross hole; (*b*) T hole; (*c*) round-cornered L hole.

3.6, corresponding to the maximum tangential stress at the intersection of the holes at the plane containing the hole axes.

The K_{tn} value of 3.6, based on nominal stress, applies only for the cylinder tested. A more useful value is an estimate of $K_{t\infty}$ in an infinite body. We next attempt to obtain this value.

First, it is observed that K_{tn} for the cylindrical hole away from the intersection is 2.3. The gross (applied) stress concentration factor is $K_{tg} = K_{tn}/(A_{net}/A) = 2.3/(0.665) = 3.46$ for the T intersection (A = cross-sectional area of cylinder, A_{net} = cross-sectional area in plane of hole axes). Referring to Chart 4.1, it is seen that for $d/H = 1 - 0.665 = 0.335$, the same values of $K_{tn} = 2.3$ and $K_{tg} = 3.46$ are obtained and that the $K_{t\infty}$ value for the infinite width, $d/H \to 0$, is 3. The agreement is not as close for the cross and L geometries, about 6% deviation.

Next we start with $K_{tn} = 3.6$ and make the assumption that $K_{t\infty}/K_{tn}$ is the same as in Chart 4.1 for the same d/H. $K_{t\infty} = 3.6(3/2.3) = 4.7$. This estimate is more useful generally than the specific test geometry value $K_{tn} = 3.6$.

Riley (1964) points out that stresses are highly localized at the intersection, decreasing to the value of the cylindrical hole within an axial distance equal to the hole diameter. Also noted is the small value of the axial stresses.

238 HOLES

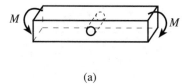

(a)

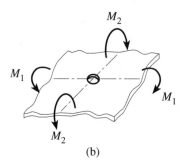

(b)

Figure 4.30 Transverse bending of beam and plate: (*a*) Beam; (*b*) plate.

The experimental determination of maximum stress at the very steep stress gradient at the sharp corner is difficult. It may be that the value just given is too low. For example, $K_{t\infty}$ for the intersection of a small hole into a large one would theoretically[1] be 9.

It would seem that a rounded corner at the intersection (in the plane of the hole axes) would be beneficial in reducing K_t. This would be a practical expedient in the case of a tunnel or a cast metal part, but it does not seem to be practically attainable in the case where the holes have been drilled. An investigation of three-dimensional photoelastic models with the corner radius varied would be of interest. (For pressurized thick-walled cylinders with crossholes and sideholes, see Section 5.19.)

4.5.15 Other Configurations

Photoelastic tests led to stress concentration factors for star-shaped holes in an element under external pressure (Fourney and Parameter 1963). Other photoelastic tests were applied to a tension panel with nuclear reactor hole patterns (Mondina and Falco 1972). These results are treated in Peterson (1974).

4.6 BENDING

Several bending problems for beams and plates are to be considered (Fig. 4.30). For plate bending, two cases are of particular interest: simple bending with $M_1 = M, M_2 = 0$ or in normalized form $M_1 = 1, M_2 = 0$; and cylindrical bending with $M_1 = M, M_2 = \nu M$, or

---

[1] The situation with respect to multiplying of stress concentration factors is somewhat similar to the case discussed in Section 4.5.2 and illustrated in Chart 4.60.

$M_1 = 1$, $M_2 = \nu$. The plate bending moments M_1, M_2, and M are uniformly distributed with dimensions of moment per unit length. The cylindrical bending case removes the anticlastic bending resulting from the Poisson's ratio effect. At the beginning of application of bending, the simple condition occurs. As the deflection increases, the anticlastic effect is not realized, except for a slight curling at the edges. In the region of the hole, it is reasonable to assume that the cylindrical bending condition exists. For design problems the cylindrical bending case is generally more applicable than the simple bending case.

It would seem that for transverse bending, rounding or chamfering of the hole edge would result in reducing the stress concentration factor.

For $M_1 = M_2$, isotropic transverse bending, K_t is independent of d/h, the diameter of a hole over the thickness of a plate. This case corresponds to in-plane biaxial tension of a thin element with a hole.

4.6.1 Bending of a Beam with a Central Hole

An effective method of weight reduction for a beam in bending is to remove material near the neutral axis, often in the form of a circular hole or a row of circular holes. Howland and Stevenson (1933) have obtained mathematically the K_{tg} values for a single hole represented by the curve of Chart 4.79:

$$K_{tg} = \frac{\sigma_{max}}{6M/(H^2 h)} \tag{4.98}$$

For a beam M is the net moment on a cross section. The units of M for a beam are force · length. Symbols are defined in Chart 4.79. The stress concentration factor K_{tg} is the ratio of σ_{max} to σ at the beam edge distant axially from the hole. Photoelastic tests by Ryan and Fischer (1938) and by Frocht and Leven (1951) are in good agreement with Howland and Stevenson's mathematical results.

The factor K_{tn} is based on the section modulus of the net section. The distance from the neutral axis is taken as $d/2$, so that σ_{nom} is at the edge of the hole.

$$K_{tn} = \frac{\sigma_{max}}{6Md/[(H^3 - d^3)h]} \tag{4.99}$$

Another form of K_{tn} has been used where σ_{nom} is at the edge of the beam.

$$K'_{tn} = \frac{\sigma_{max}}{6MH/[(H^3 - d^3)h]} \tag{4.100}$$

The factor K'_{tn} of Eq. (4.100) and Chart 4.79 appears to be a linear function of d/H. Also K'_{tn} is equal to $2d/H$, prompting Heywood (1952) to comment that this configuration has the "curious result that the stress concentration factor is independent of the relative size of the hole, and forms the only known case of a notch showing such independency."

Note from Chart 4.79 that the hole does not weaken the beam for $d/H <\sim 0.45$. For design purposes, $K_{tg} = 1$ for $d/H <\sim 0.45$.

On the outer edge the stress has peaks at A, A. However, this stress is less than at B, except at and to the left of a transition zone in the region of C where $K_t = 1$ is approached. Angle $\alpha = 30°$ was found to be independent of $d/(H - d)$ over the range investigated.

4.6.2 Bending of a Beam with a Circular Hole Displaced from the Center Line

The K_{tg} factor, as defined by Eq. (4.98), has been obtained by Isida (1952) for the case of an eccentrically located hole and is shown in Chart 4.80. At line C–C, $K_{tgB} = K_{tgA}$, corresponding to the maximum stress at B and A, respectively (see the sketch in Chart 4.80). Above C–C, K_{tgB} is the greater of the two stresses. Below C–C, K_{tgA} is the greater, approaching $K_{tg} = 1$ or no effect of the hole.

At $c/e = 1$, the hole is central, with factors as given in the preceding subsection (Chart 4.79). For $a/c \to 0$, K_{tg} is 3 multiplied by the ratio of the distance from the center line to the edge, in terms of c/e:

$$K_{tg} = 3\frac{1 - c/e}{1 + c/e} \qquad (4.101)$$

The calculated values of Isida (1952) are in agreement with the photoelastic results of Nisida (1952).

4.6.3 Bending of a Beam with an Elliptical Hole; Slot with Semicircular Ends (Ovaloid); or Round-Cornered Square Hole

Factors K_{tn} for an ellipse as defined by Eq. (4.100) were obtained by Isida (1953). These factors have been recalculated for K_{tg} of Eq. (4.98), and for K_{tn} of Eq. (4.99), and are presented in Chart 4.81. The photoelastic values of Frocht and Leven (1951) for a slot with semicircular ends are in reasonably good agreement when compared with an ellipse having the same a/r.

Note in Chart 4.81 that the hole does not weaken the beam for a/H values less than at points $C, D,$ and E for $a/r = 4, 2,$ and 1, respectively. For design, use $K_t = 1$ to the left of the intersection points.

On the outer edge, the stress has peaks at A, A. But this stress is less than at B, except at and to the left of a transition zone in the region of $C, D,$ and E, where $K_t = 1$ is approached. In photoelastic tests (Frocht and Leven 1951), angles $\alpha = 35°$, $32.5°$, and $30°$ for $a/r = 4, 2,$ and 1, respectively, were found to be independent of the $a/(H - 2a)$ over the range investigated.

For shapes approximating ovaloids and round-cornered square holes (parallel and at $45°$), K'_{tg} factors have been obtained (Joseph and Brock 1950) for central holes that are small compared to the beam depth

$$K'_{tg} = \frac{\sigma_{max}}{12Ma/(H^3 h)} \qquad (4.102)$$

4.6.4 Bending of an Infinite- and of a Finite-Width Plate with a Single Circular Hole

For simple bending ($M_1 = 1, M_2 = 0$) of an infinite plate with a circular hole, Reissner (1945) obtained K_t as a function of d/h, as shown in Chart 4.82. For $d/h \to 0, K_t = 3$.

For $d/h \to \infty$,

$$K_t = \frac{5 + 3\nu}{3 + \nu} \tag{4.103}$$

giving $K_t = 1.788$ when $\nu = 0.3$.

For cylindrical bending ($M = 1, M_2 = \nu$) of an infinite plate, $K_t = 2.7$ as $d/h \to 0$. For $d/h \to \infty$, Goodier (1936) obtained

$$K_t = (5 - \nu)\frac{1 + \nu}{3 + \nu} \tag{4.104}$$

or $K_t = 1.852$ for $\nu = 0.3$. For design problems, the cylindrical bending case is usually more applicable. For $M_1 = M_2$, isotropic bending, K_t is independent of d/h, and the case corresponds to biaxial tension of a panel with a hole.

For a finite width plate and various d/h values, K_t is given in Chart 4.83, based on Charts 4.1 and 4.82 with the K_{tn} gradient at $d/H = 0$ equal to

$$\frac{\Delta K_{tn}}{\Delta(d/H)} = -K_{tn} \tag{4.105}$$

Photoelastic tests (Goodier and Lee 1941; Drucker 1942) and strain gage measurements (Dumont 1939) are in reasonably good agreement with Chart 4.83.

4.6.5 Bending of an Infinite Plate with a Row of Circular Holes

For an infinite plate with a row of circular holes, Tamate (1957) has obtained K_t values for simple bending ($M_1 = 1, M_2 = 0$) and for cylindrical bending ($M_1 = 1, M_2 = \nu$) with M_1 bending in the x and y directions (Chart 4.84). For design problems the cylindrical bending case is usually more applicable. The K_t value for $d/l \to 0$ corresponds to the single hole (Chart 4.82). The dashed curve is for two holes (Tamate 1958) in a plate subjected to simple bending ($M_1 = 1, M_2 = 0$).

For bending about the x direction nominal stresses are used in Chart 4.84, resulting in K_{tn} curves that decrease as d/l increases. On the other hand, K_{tg} values, $\sigma_{\max}/\sigma$, increase as d/l increases. The two factors are related by $K_{tg} = K_{tn}/(1 - d/l)$.

4.6.6 Bending of an Infinite Plate with a Single Elliptical Hole

Stress concentration factors for the bending of an infinite plate with an elliptical hole (Nisitani 1968; Neuber 1958) are given in Chart 4.85.

For simple bending ($M_1 = 1, M_2 = 0$),

$$K_t = 1 + \frac{2(1 + \nu)(a/b)}{3 + \nu} \tag{4.106}$$

where a is the half width of ellipse perpendicular to M_1 bending direction (Chart 4.85), b is the half width of ellipse perpendicular to half width, a, and ν is Poisson's ratio.

For cylindrical bending (Nisitani 1968; Goodier 1936) ($M_1 = 1, M_2 = \nu$),

$$K_t = \frac{(1 + \nu)[2(a/b) + 3 - \nu]}{3 + \nu} \qquad (4.107)$$

For design problems the cylindrical bending case is usually more applicable. For $M_1 = M_2$, isotropic bending, K_t is independent of a/h, and the case corresponds to in-plane biaxial tension of a thin element with a hole.

4.6.7 Bending of an Infinite Plate with a Row of Elliptical Holes

Chart 4.86 presents the effect of spacing (Nisitani 1968) for a row of elliptical holes in an infinite plate under bending. Stress concentration factor K_t values are given as a ratio of the single hole value (Chart 4.85). The ratios are so close for simple and cylindrical bending that these cases can be represented by a single set of curves (Chart 4.86). For bending about the y axis, a row of edge notches is obtained for $a/c \geq 0.5$. For bending about the x axis, the nominal stress is used, resulting in K_{tn} curves that decrease as a/c increases. Factor K_{tg} values, σ_{max}/σ, increase as a/c increases, where $K_{tg} = K_{tn}/(1 - 2a/c)$.

4.6.8 Tube or Bar of Circular Cross Section with a Transverse Hole

The K_t relations for tubes or bars of circular cross section with a transverse circular hole are presented in Chart 4.87. The curve for the solid shaft is based on blending the data of Thum and Kirmser (1943) and the British data (Jessop, Snell, and Allison 1959; ESDU 1965). There is some uncertainty regarding the exact position of the dashed portion of the curve.

A photoelastic test by Fessler and Roberts (1961) is in good agreement with Chart 4.87. The factors are defined as:

$$K_{tg} = \frac{\sigma_{max}}{\sigma_{gross}} = \frac{\sigma_{max}}{M/Z_{tube}} = \frac{\sigma_{max}}{MD/(2I_{tube})}$$

$$= \frac{\sigma_{max}}{32MD/[\pi(D^4 - d_i^4)]} \qquad (4.108)$$

$$K_{tn} = \frac{\sigma_{max}}{\sigma_{net}} = \frac{\sigma_{max}}{M/Z_{net}} = \frac{\sigma_{max}}{Mc/I_{net}} \qquad (4.109)$$

where $c = \sqrt{(D/2)^2 - (d/2)^2}$ and

$$K_{tn} = K_{tg}\frac{Z_{net}}{Z_{tube}} \qquad (4.110)$$

Quantities Z_{tube} and Z_{net} are the gross and net section moduli ($\sigma = M/Z$). Other symbols are defined in Chart 4.87.

Thum and Kirmser (1943) found that the maximum stress did not occur on the surface of the shaft but at a small distance inside the hole on the surface of the hole. The σ_{max} value used in developing Chart 4.87 is the maximum stress inside the hole. No factors are given for the somewhat lower stress at the shaft surface. If these factors are of interest, Thum and Kirmser's work should be examined.

The ratio $Z_{\text{net}}/Z_{\text{tube}}$ has been determined mathematically (Peterson 1968), although the formulas will not be repeated here. Specific values can be obtained by dividing the Chart 4.87 values of K_{tn} by K_{tg}. If the hole is sufficiently small relative to the shaft diameter, the hole may be considered to be of square cross section with edge length d:

$$\frac{Z_{\text{net}}}{Z_{\text{tube}}} = 1 - \frac{(16/3\pi)(d/D)[1 - (d_i/D)^3]}{1 - (d_i/D)^4} \tag{4.111}$$

It can be seen from the bottom curves of Chart 4.87 that the error due to this approximation is small below $d/D = 0.2$.

4.7 SHEAR, TORSION

4.7.1 Shear Stressing of Infinite Thin Element with Circular or Elliptical Hole, Unreinforced and Reinforced

By superposition of σ_1 and $\sigma_2 = -\sigma_1$ uniaxial stress distributions, the shear case $\tau = \sigma_1$ is obtained. For the circular hole $K_t = \sigma_{\max}/\tau = 4$, as obtained from Eq. (4.66), with $a/b = 1, \sigma_2/\sigma_1 = -1$. This is also found in Chart 4.88, which treats an elliptical hole in an infinite panel subject to shear stress, at $a/b = 1$. Further $K_{ts} = \tau_{\max}/\tau = (\sigma_{\max}/2)/\tau = 2$.

For the elliptical hole Chart 4.88 shows K_t for shear stress orientations in line with the ellipse axes (Godfrey 1959) and at 45° to the axes. The 45° case corresponds to $\sigma_2 = -\sigma_1 = \tau$, as obtained from Eq. (4.66).

The case of shearing forces parallel to the major axis of the elliptical hole, with the shearing force couple counterbalanced by a symmetrical remotely located opposite couple (Fig. 4.31), has been solved by Neuber (1958). Neuber's K_t factors are higher than the parallel shear factors in Chart 4.88. For example, for a circle the "shearing force" K_t factor (Neuber 1958) is 6 as compared to 4 in Chart 4.88.

For symmetrically reinforced elliptical holes, Chart 4.89 provides the stress concentration factors for pure shear stresses. The quantity A_r is the cross-sectional area of the reinforcement.

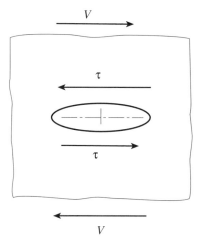

Figure 4.31 Elliptical hole subject to shear force.

4.7.2 Shear Stressing of an Infinite Thin Element with a Round-Cornered Rectangular Hole, Unreinforced and Reinforced

In Chart 4.90, $K_t = \sigma_{max}/\tau$ is given for shear stressing in line with the round-cornered rectangular hole axes (Sobey 1963; ESDU 1970). In Chart 4.64d, $\sigma_2 = -\sigma_1$ is equivalent to shear stress τ at 45° to the hole axes (Heller et al. 1958; Heller 1969).

For symmetrically reinforced square holes, the K_t is shown in Chart 4.91 (Sobey 1968; ESDU 1981). The stress is based on the von Mises stress. The maximum stresses occur at the corner.

4.7.3 Two Circular Holes of Unequal Diameter in a Thin Element in Pure Shear

Stress concentration factors have been developed for two circular holes in panels in pure shear. Values for K_{tg} have been obtained by Haddon (1967). Charts 4.92a and 4.92b show the K_{tg} curves. These charts are useful in considering stress concentration factors of neighboring cavities of different sizes. Two sets of curves are provided for the larger hole and the smaller hole, respectively.

4.7.4 Shear Stressing of an Infinite Thin Element with Two Circular Holes or a Row of Circular Holes

For an infinite thin element with a row of circular holes, Chart 4.93 presents $K_t = \sigma_{max}/\tau$ for shear stressing in line with the hole axis (Meijers 1967; Barrett et al. 1971). The location of σ_{max} varies from $\theta = 0°$ for $l/d \to 1$ to $\theta = 45°$ for $l/d \to \infty$.

4.7.5 Shear Stressing of an Infinite Thin Element with an Infinite Pattern of Circular Holes

In Chart 4.94 for infinite thin elements, $K_t = \sigma_{max}/\tau$ is given for square and equilateral triangular patterns of circular holes for shear stressing in line with the pattern axis (Meijers 1967; Sampson 1960; Bailey and Hicks 1960; O'Donnell 1967). Subsequent computed values (Hooke 1968; Grigolyuk and Fil'shtinskii 1970) are in good agreement with Meijers's results. Note that $\sigma_2 = -\sigma_1$ is equivalent to shear stressing τ at 45° to the pattern axis in Chart 4.41.

In Charts 4.95 and 4.96, $K_t = \sigma_{max}/\tau$ is given for rectangular and diamond (triangular, not limited to equilateral) patterns (Meijers 1967), respectively.

4.7.6 Twisted Infinite Plate with a Circular Hole

In Chart 4.97, $M_x = 1$, $M_y = -1$ corresponds to a twisted plate (Reissner 1945). $K_t = \sigma_{max}/\sigma$, where σ is due to bending moment M_x. For $h/d \to \infty$ ($d/h \to 0$), $K_t = 4$. For $d/h \to \infty$,

$$K_t = 1 + \frac{1 + 3\nu}{3 + \nu} \qquad (4.112)$$

giving $K_t = 1.575$ for $\nu = 0.3$.

4.7.7 Torsion of a Cylindrical Shell with a Circular Hole

Some stress concentration factors for the torsion of a cylindrical shell with a circular hole are given in Chart 4.98. (For a discussion of the parameters, see Section 4.3.3.) The K_t factors (Van Dyke 1965) of Chart 4.98 have been compared with experimental results (Lekkerkerker 1964; Houghton and Rothwell 1962), with reasonably good agreement.

4.7.8 Torsion of a Tube or Bar of Circular Cross Section with a Transverse Circular Hole

For the torsion of a tube or bar of solid circular cross section with a circular hole, the stress concentration factors of Chart 4.99 are based on photoelastic tests (Jessop et al. 1959; ESDU 1965) and strain gage tests (Thum and Kirmser 1943). The factors are defined as follows:

$$K_{tg} = \frac{\sigma_{\max}}{\tau_{\text{gross}}} = \frac{\sigma_{\max}}{TD/(2J_{\text{tube}})} = \frac{\sigma_{\max}}{16TD/[\pi(D^4 - d_i^4)]} \quad (4.113)$$

$$K_{tn} = \frac{\sigma_{\max}}{\tau_{\text{net}}} = \frac{\sigma_{\max}}{TD/(2J_{\text{net}})} = K_{tg} \frac{J_{\text{net}}}{J_{\text{tube}}} \quad (4.114)$$

Other symbols are defined in Chart 4.99. The quantites J_{tube} and J_{net} are the gross and net polar moments of inertia, or in some cases the torsional constants.

Thum and Kirmser (1943) found that the maximum stress did not occur on the surface of the shaft but at a small distance inside the hole on the surface of the hole. This has been corroborated by later investigators (Leven 1955; Jessop et al. 1959). In the chart here, $\sigma_{\max}$ denotes the maximum stress inside the hole. No factors are given for the somewhat lower stress at the shaft surface. If they are of interest, they can be found in Leven and in Jessop et al.

The $J_{\text{net}}/J_{\text{tube}}$ ratios have been determined mathematically (Peterson 1968). Although the formulas and charts will not be repeated here, specific values can be obtained by dividing the Chart 4.99 values of K_{tn} by K_{tg}. If the hole is sufficiently small relative to the shaft diameter, the hole may be considered to be of square cross section with edge length d, giving

$$\frac{J_{\text{net}}}{J_{\text{tube}}} = 1 - \frac{(8/3\pi)(d/D)\{[1 - (d_i/D)^3] + (d/D)^2[1 - (d_i/D)]\}}{1 - (d_i/D)^4} \quad (4.115)$$

The bottom curves of Chart 4.99 show that the error due to the foregoing approximation is small, below $d/D = 0.2$.

The maximum stress $\sigma_{\max}$ is uniaxial and the maximum shear stress $\tau_{\max} = \sigma_{\max}/2$ occurs at 45° from the tangential direction of $\sigma_{\max}$. Maximum shear stress concentration factors can be defined as

$$K_{tsg} = \frac{K_{tg}}{2} \quad (4.116)$$

$$K_{tsn} = \frac{K_{tn}}{2} \quad (4.117)$$

If in Chart 4.99, the ordinate values are divided by 2, maximum shear stress concentration factors will be represented.

246 HOLES

Another stress concentration factor can be defined, based on equivalent stress of the applied system. The applied shear stress τ corresponds to principal stresses σ and $-\sigma$, $45°$ from the shear stress directions. The equivalent stress $\sigma_{eq} = \sqrt{3}\sigma$. The equivalent stress concentration factors are

$$K_{teg} = \frac{\sigma_{max}}{\sigma_{eq}} = \frac{K_{tg}}{\sqrt{3}} \qquad (4.118)$$

$$K_{ten} = \frac{K_{tn}}{\sqrt{3}} \qquad (4.119)$$

Referring to Chart 4.99, the ordinate values divided by $\sqrt{3}$ give the corresponding K_{te} factors.

Factors from Eqs. (4.116) to (4.119) are useful in mechanics of materials problems where one wishes to determine the initial plastic condition. The case of a torsion cylinder with a central spherical cavity has been analyzed by Ling (1952).

REFERENCES

ASME Boiler and Pressure Vessel Code - Pressure Vessels, 1974, Section VIII, ASME, New York, p.30

ASTM, 1994, *Annual Book of ASTM Standards,* Vol. 03.01, ASTM, Philadelphia, PA.

Atsumi, A., 1958, "Stress Concentration in a Strip under Tension and Containing an Infinite Row of Semicircular Notches," *Q. J. Mech. & Appl. Math.,* Vol. 11, p. 478.

Atsumi, A., 1960, "Stresses in a Circular Cylinder Having an Infinite Row of Spherical Cavities under Tension," *Trans. ASME, Appl. Mech. Section,* Vol. 82, p. 87.

Bailey, R., and Hicks, R., 1960, "Behavior of Perforated Plates under Plane Stress," *J. Mech. Eng. Sci.,* Vol. 2, p. 143.

Barrett, R. F., Seth, P. R., and Patel, G. C., 1971, "Effect of Two Circular Holes in a Plate Subjected to Pure Shear Stress," *Trans. ASME, Appl. Mech. Section,* Vol. 93, p. 528.

Belie, R. G., and Appl, F. J., 1972, "Stress Concentration in Tensile Strips with Large Circular Holes," *Exp. Mech.,* Vol. 12, p. 190. Discussion, 1973, Vol. 13, p. 255.

Beskin, L., 1944, "Strengthening of Circular Holes in Plates under Edge Forces," *Trans. ASME, Appl. Mech. Section,* Vol. 66, p. A-140.

Bickley, W. G., 1928, "Distribution of Stress Round a Circular Hole in a Plate," *Phil. Trans. Roy. Soc. (London) A,* Vol. 227, p. 383.

Brown, W. F., and Srawley, S. R., 1966, "Plane Strain Crack Toughness Testing of High Strength Metallic Materials," ASTM, Philadelphia, PA.

Buivol, V. N., 1960, "Experimental Investigations of the Stressed State of Multiple Connected Plates," *Prikladna Mekhanika,* Vol. 3, p. 328 (in Ukranian).

Buivol, V. N., 1962, "Action of Concentrated Forces on a Plate with Holes," *Prikladna Mehkanika,* Vol. 8 (1), p. 42 (in Ukranian).

Cheng, Y. F., 1968, "A Photoelastic Study of Stress Concentration Factors at a Doubly-Symmetric Hole in Finite Strips under Tension," *Trans. ASME, Appl. Mech. Section,* Vol. 90, p. 188.

Christiansen, S., 1968, "Numerical Determination of Stresses in a Finite or Infinite Plate with Several Holes of Arbitrary Form," *Z. angew. Math. u. Mech.* Vol. 48, p. T131.

Chu, W. L., and Conway, D. H., 1970, "A Numerical Method for Computing the Stresses around an Axisymmetrical Inclusion," *Intern. J. Mech. Sci.,* Vol. 12, p. 575.

Coker, E. G., and Filon, L. N. G., 1931, *A Treatise on Photoelasticity,* Cambridge University Press, Cambridge, England, p.486.

Cox, H. L., 1953, "Four Studies in the Theory of Stress Concentration," *Aero Research Council (London),* Rpt. 2704.

Cox, H. L., and Brown, A. F. C., 1964, "Stresses Round Pins in Holes," *Aeronaut. Q.,* Vol. 15, p. 357.

Daniel, I. M., 1970, "Photoelastic Analysis of Stresses Around Oblique Holes," *Exp. Mech.,* Vol. 10, p. 467.

Davies, G. A. O., 1963, "Stresses Around a Reinforced Circular Hole Near a Reinforced Straight Edge," *Aeronaut. Q.,* Vol. XIV, p. 374–386.

Davies, G. A. O., 1965, "Stresses in a Square Plate Having a Central Circular Hole," *J. Royal Aero. Soc.,* Vol. 69, p. 410.

Davies, G. A. O., 1967, "Plate-Reinforced Holes," *Aeronaut. Q.,* Vol. 18, p.43.

Dhir, S. K., and Brock, J. S., 1970, "A New Method of Reinforcing a Hole Effecting Large Weight Savings," *Intern. J. Solids and Structures,* Vol. 6, p. 259.

Dixon, J. R., 1960, "Stress Distribution around a Central Crack in a Plate Loaded in Tension; Effect on Finite Width of Plate," *J. Royal Aero. Soc.,* Vol. 64, p. 141.

Donnell, L. H., 1941, "Stress Concentrations due to Elliptical Discontinuities in Plates under Edge Forces," *Von Karman Anniversary Volume,* California Inst. of Tech., Pasadena, p. 293.

Drucker, D. C., 1942, "The Photoelastic Analysis of Transverse Bending of Plates in the Standard Transmission Polariscope," *Trans. ASME, Appl. Mech. Section,* Vol. 64, p. A-161.

Dumont, C., 1939, "Stress Concentration Around an Open Circular Hole in a Plate Subjected to Bending Normal to the Plane of the Plate," *NACA Tech. Note 740.*

Durelli, A. J., and Kobayashi, A. S., 1958, "Stress Distribution around Hydrostatically Loaded Circular Holes in the Neighborhood of Corners," *Trans. ASME, Appl. Mech. Section,* Vol. 80, p. 178.

Durelli, A. J., and Riley, W. F., 1965, *Introduction to Photomechanics,* Prentice-Hall, Englewood Cliffs, N. J., p. 233.

Durelli, A. J., del Rio, C. J., Parks, V. J.,and Feng, H., 1967, "Stresses in a Pressurized Cylinder with a Hole," *Proc. ASCE,* Vol. 93, p.383.

Durelli, A. J., Parks, V. J., and Uribe, S., 1968, "Optimization of a Slot End Configuration in a Finite Plate Subjected to Uniformly Distributed Load," *Trans. ASME, Appl. Mech. Section,* Vol. 90, p. 403.

Durelli, A. J., Parks, V. J., Chandrashekhara, K., and Norgard, J. S., 1971, "Stresses in a Pressurized Ribbed Cylindrical Shell with a Reinforced Circular Hole Interrupting a Rib," *Trans. ASME, J. Eng. for Industry,* Vol. 93, p. 897.

Edwards, R. H., 1951, "Stress Concentrations Around Spheroidal Inclusions and Cavities," *Trans. ASME, Appl. Mech. Section,* Vol. 75, p. 19.

Ellyin, F., 1970, "Experimental Study of Oblique Circular Cylindrical Apertures in Plates," *Exp. Mech.,* Vol. 10, p. 195.

Ellyin, F., 1970, "Elastic Stresses Near a Skewed Hole in a Flat Plate and Applications to Oblique Nozzle Attachments in Shells," *Welding Research Council Bulletin* 153, p. 32.

Ellyin, F. and Izmiroglu, U. M., 1973, "Effect of Corner Shape on Elastic Stress and Strain Concentration in Plates with an Oblique Hole," *Trans. ASME, J. of Eng. for Industry,* Vol. 95, p. 151.

Eringen, A. C., Naghdi, A. K., and Thiel, C. C., 1965, "State of Stress in a Circular Cylindrical Shell with a Circular Hole," *Welding Research Council Bulletin 102.*

ESDU (Engineering Science Data Unit), 1965, *Stress Concentrations,* London.

ESDU (Engineering Science Data Unit), 1970, *Stress Concentrations,* London.

ESDU (Engineering Science Data Unit), 1981, *Stress Concentrations,* London.

Feddersen, C., 1967, Discussion of "Plain Strain Crack Toughness Testing," *ASTM Special Tech. Publ. 410,* p. 77.

Fessler, H., and Roberts, E. A., 1961, "Bending Stresses in a Shaft with a Tranverse Hole," *Selected Papers on Stress Analysis,*Stress Analysis Conference, Delft, 1959, Reinhold Publ. Co., New York, p. 45.

Fourney, M. E., and Parmerter, R. R., 1961, "Stress Concentrations for Internally Perforated Star Grains," Bur. Naval Weapons, NAVWEPS Rep. 7758.

Fourney, M. E., and Parmerter, R. R., 1963, "Photoelastic Design Data for Pressure Stresses in Slotted Rocket Grains," *J. AIAA,* Vol. 1, p. 697.

Fourney, M. E., and Parmerter, R. R., 1966, "Parametric Study of Rocket Grain Configurations by Photoelastic Analysis," AFSC Rep. AFRPL-TR-66-52.

Frocht, M. M., and Hill, H. N., 1940, "Stress Concentration Factors Around a Central Circular Hole in a Plate Loaded Through a Pin in the Hole," *Trans. ASME, Appl. Mech. Section,* Vol. 62, p. A-5.

Frocht, M. M., and Leven, M. M., 1951, "Factors of Stress Concentration for Slotted Bars in Tension and Bending," *Trans. ASME, Appl. Mech. Section,* Vol. 73, p. 107.

Gassner, E., and Horstmann, K. F., 1961, "The Effect of Ground to Air to Ground Cycle on the Life of Transport Aircraft Wings which are Subject to Gust Loads," *RAE Translation* 933

Godfrey, D. E. R., 1959, *Theoretical Elasticity and Plasticity for Engineers*, Thames and Hudson, London, p. 109.

Goldberg, J.E., and Jabbour, K.N., 1965, "Stresses and Displacements in Perforated Plates," *Nuc. Structural Eng.*, Vol. 2, p. 360.

Goodier, J. N., 1933, "Concentration of Stress Around Spherical and Cylindrical Inclusions and Flaws," *Trans. ASME, Appl. Mech. Section,* Vol. 55, p. A-39.

Goodier, J. N., 1936, "Influence of Circular and Elliptical Holes on Transverse Flexure of Elastic Plates," *Phil. Mag.*, Vol. 22, p. 69.

Goodier, J. N., and Lee, G. H., 1941, "An Extension of the Photoelastic Method of Stress Measurement to Plates in Transverse Bending," *Trans ASME, Appl. Mech. Section*, Vol. 63, p. A-27.

Grigolyuk, E. I., and Fil'shtinskii, L. A., 1970, *Perforirorannye Plastiny i Obolocki* (Perforated Plates and Shells), Nauka Publishing House, Moscow.

Gurney, C., 1938, "An Analysis of the Stresses in a Flat Plate with Reinforced Circular Hole under Edge Forces," *ARC*, R&M 1834, London.

Haddon, R. A. W., 1967, "Stresses in an Infinite Plate with Two Unequal Circular Holes," *Q. J. Mech. Appl. Math.*, Vol. 20, pp. 277-291.

Hamada, M., Yokoya, K., Hamamoto, M., and Masuda, T., 1972, "Stress Concentration of a Cylindrical Shell With One or Two Circular Holes," *Bull. Japan Soc. Mech. Eng.*, Vol. 15, p. 907.

Hanzawa, H., Kishida, M., Murai, M., and Takashina, K., 1972, "Stresses in a Circular Cylindrical Shell having Two Circular Cutouts," *Bull. Japan Soc. Mech. Eng.*, Vol. 15, p. 787.

Hardenberg, D. E., and Zamrik, S. Y., 1964, "Effects of External Loadings on Large Outlets in a Cylindrical Pressure Vessel," *WRCB* 96.

Heller, S. R., Brock, J. S., and Bart, R., 1958, "The Stresses Around a Rectangular Opening with Rounded Corners in a Uniformly Loaded Plate," *Proc. 3rd U. S. Nat. Congr. Appl. Mech.*, ASME, p.357.

Heller, S. R., 1969, "Stress Concentration Factors for a Rectangular Opening with Rounded Corners in a Biaxially Loaded Plate," *J. Ship Research*, Vol. 13, p. 178.

Hennig, A., 1933, "Polarisationsoptische Spannungsuntersuchungen am gelochten Zugstab und am Nietloch," *Forsch. Gebiete Ingenieur.*, *VDI*, Vol. 4, p. 53.

Heywood, R. B., 1952, *Designing by Photoelasticity*, Chapman and Hall, London.

Hicks, R., 1957, "Reinforced Elliptical Holes in Stressed Plates," *J. Royal Aero. Soc.*, Vol. 61, p. 688.

Hirano, F., 1950, "Study of Shape Coefficients of Two-Dimensional Elastic Bodies," *Trans. Japan Soc. Mech. Eng.*, Vol. 16, No. 55, p. 52 (in Japanese).

Hooke, C. J., 1968, "Numerical Solution of Plane Elastostatic Problems by Point Matching," *J. Strain Anal.*, Vol. 3, p. 109.

Horvay, G., 1952, "The Plane-Stress Problem of Perforated Plates," *Trans. ASME, Appl. Mech. Section*, Vol. 74, p.355.

Houghton, D. S., and Rothwell, A., 1961, "The Analysis of Reinforced Circular and Elliptical Cutouts under Various Loading Conditions," College of Aeronautics, Report 151,Cranfield, England.

Houghton, D. S., and Rothwell A., 1962, "The Effect of Curvature on the Stress Concentration around Holes in Shells," College of Aeronautics, Rpt. 156, Cranfield, England.

Howland, R. C. J., 1929-30, "On the Stresses in the Neighborhood of a Circular Hole in a Strip under Tension," *Phil. Trans. Roy. Soc. (London) A*, Vol. 229, p. 67.

Howland, R. C. J., and Stevenson, A. C., 1933, "Biharmonic Analysis in a Perforated Strip," *Phil. Trans. Royal Soc. A*, Vol. 232, p. 155.

Howland, R. C. J., 1935, "Stresses in a Plate Containing an Infinite Row of Holes," *Proc. Roy. Soc. (London) A*, Vol. 148, p. 471.

Hulbert, L. E., and Niedenfuhr, F. W., 1965, "Accurate Calculation of Stress Distributions in Multi-holed Plates," *Trans. ASME, Industry Section*, Vol. 87, p. 331.

Hulbert, L. E., 1965, *The Numerical Solution of Two-Dimensional Problems of the Theory of Elasticity*, Ohio State Univ., *Eng. Exp. Sta. Bull. 198*, Columbus, Ohio.

Hütter, A., 1942, "Die Spannungsspitzen in gelochten Blechscheiben und Streifen," *Z. angew. Math. Mech.*, Vol. 22, p. 322.

Inglis, C.E., 1913, "Stresses in a Plate Due to the Presence of Cracks and Sharp Corners," *Engrg. (London)*, Vol. 95, p. 415.

Irwin, G. R., 1958, "Fracture," In Vol. 6 Encyclopedia of Physics, Springer-Verlag, Berlin.

Irwin, G. R., 1960, "Fracture Mechanics," in *Structural Mechanics*, Pergamon, New York.

Isida, M., 1952, "On the Bending of an Infinite Strip with an Eccentric Circular Hole," *Proc. 2nd Japan Congr. Appl. Mech.*, p. 57.

Isida, M., 1953, "Form Factors of a Strip with an Elliptic Hole in Tension and Bending," *Scientific Papers of Faculty of Engrg., Tokushima University*, Vol. 4, p. 70.

Isida, M., 1955a, "On the Tension of a Semi-Infinite Plate with an Elliptic Hole," *Scientific Papers of Faculty of Engrg., Tokushima University*, Vol. 5, p. 75. (in Japanese)

Isida, M., 1955b, "On the Tension of a Strip with a Central Elliptic Hole," *Trans. Japan Soc. Mech. Eng.*, Vol. 21, p. 514.

Isida, M., 1960, "On the Tension of an Infinite Strip Containing a Square Hole with Rounded Corners," *Bull. Japan Soc. Mech. Eng.*, Vol. 3, p. 254.

Isida, M., 1965, "Crack Tip Intensity Factors for the Tension of an Eccentrically Cracked Strip," Lehigh University, Dept. Mechanics Rpt.

Isida, M., 1966, "Stress Intensity Factors for the Tension of an Eccentrically Cracked Strip," *Trans. ASME, Appl. Mech. Section*, Vol. 88, p. 674.

Jenkins, W. C., 1968, "Comparison of Pressure and Temperature Stress Concentration Factors for Solid Propellant Grains," *Exp. Mech.*, Vol. 8, p. 94.

Jessop, H. T., Snell, C., and Holister, G. S., 1958, "Photoelastic Investigation of Plates with Single Interference-Fit Pins with Load Applied (a) to Pin Only and (b) to Pin and Plate Simultaneously," *Aeronaut., Q.*, Vol. 9, p. 147.

Jessop, H. T., Snell, C., and Allison, I. M., 1959, "The Stress Concentration Factors in Cylindrical Tubes with Transverse Cylindrical Holes," *Aeronaut. Q.*, Vol. 10, p. 326.

Jones, N., and Hozos, D., 1971, "A Study of the Stress Around Elliptical Holes in Flat Plates," *Trans. ASME, J. Eng. for Industry*, Vol. 93, p.688.

Joseph, J. A., and Brock, J. S., 1950, "The Stresses Around a Small Opening in a Beam Subjected to Pure Bending," *Trans. ASME, Appl. Mech. Section*, Vol. 72, p. 353.

Kaufman, A., Bizon, P. T., and Morgan, W. C., 1962, "Investigation of Circular Reinforcements of Rectangular Cross-Section Around Central Holes in Flat Sheets under Biaxial Loads in the Elastic Range," *NASA TN D-1195*.

Kelly, L.G., 1967, *Handbook of Numerical Methods and Applications*, Addison-Wesley, MA.

Knight, R. C., 1935, "Action of a Rivet in a Plate of Finite Breadth," *Phil. Mag.*, Series 7, Vol. 19, p. 517.

Koiter, W. T., 1957, "An Elementary Solution of Two Stress Concentration Problems in the Neighborhood of a Hole," *Q. Appl. Math.*, Vol. 15, p. 303.

Koiter, W. T., 1965, "Note on the Stress Intensity Factors for Sheet Strips with Crack under Tensile Loads," Rpt. of Laboratory of Engr. Mechanics, Technological University, Delft, Holland.

Kolosoff, G., 1910, Disertation, St. Petersburg.

Kraus, H., 1962, "Pressure in Multibore Bodies," *Intern. J. Mech. Sci.*, Vol. 4, p. 187.

Kraus, H., 1963, "Stress Concentration Factors for Perforated Annular Bodies Loaded in Their Plane," Unpublished Report, Pratt and Whitney Co., E. Hartford, Conn.

Kraus, H., Rotondo, P., Haddon, W. D., 1966, "Analysis of Radially Deformed Perforated Flanges," *Trans. ASME, Appl. Mech. Section*, Vol. 88, p. 172.

Leckie, F. A., Paine, D. J., and Penny, R. K., 1967, "Elliptical Discontinuities in Spherical Shells," *J. Strain Anal.*, Vol. 2, p. 34.

Lekkerkerker, J. G., 1964, "Stress Concentration Around Circular Holes in Cylindrical Shells," *Proc. 11th Intern. Congr. Appl. Mech.*, Springer, Berlin, p. 283.

Leven, M. M., 1955, "Quantitative Three-Dimensional Photoelasticity," *Proc. SESA*, Vol. 12, p. 157.

Leven, M. M., 1963, "Effective Elastic Constants in Plane Stressed Perforated Plates of Low Ligament Efficiency," Westinghouse Research Report 63–917-520-R1.

Leven, M. M., 1964, "Stress Distribution in a Perforated Plate of 10% Ligament Efficiency Subjected to Equal Biaxial Stress," Westinghouse Research Report 64-917-520-R1.

Leven, M. M., 1966, "Photoelastic Determination of the Stresses in Reinforced Openings in Pressure Vessels," *Welding Research Council Bulletin* 113, p. 25

Leven, M. M., 1967, "Photoelastic Analysis of a Plate Perforated with a Square Pattern of Penetrations," Westinghouse Research Memo 67-1D7-TAADS-MI.

Leven, M. M., 1970, "Photoelastic Determination of the Stresses at Oblique Openings in Plates and Shells," *Welding Research Council Bulletin* 153, p. 52

Levy, S., McPherson, A. E., and Smith, F. C., 1948, "Reinforcement of a Small Circular Hole in a Plane Sheet under Tension," *Trans. ASME, Appl. Mech. Section*, Vol. 70, p. 160.

Lind, N. C., 1968, "Stress Concentration of Holes in Pressurized Cylindrical Shells," *AIAA J.*, Vol. 6, p. 1397

Ling, Chi-Bing, 1948a, "On the Stresses in a Plate Containing Two Circular Holes," *J. Appl. Physics*, Vol. 19, p. 77.

Ling, Chi-Bing, 1948b, "The Stresses in a Plate Containing an Overlapped Circular Hole," *J. Appl. Physics*, Vol. 19, p. 405.

Ling, Chi-Bing, 1952, "Torsion of a Circular Cylinder Having a Spherical Cavity," *Q. Appl. Math.*, Vol. 10, p. 149.

Ling, Chi-Bing, 1959, "Stresses in a Stretched Slab Having a Spherical Cavity," *Trans. ASME, Appl. Mech. Section*, Vol. 81, p. 235.

Ling, Chi-Bing, 1956, "Stresses in a Circular Cylinder Having a Spherical Cavity under Tension," *Q. Appl. Math*, Vol. 13, p. 381.

Ling, Chi-Bing, 1967, "On Stress Concentration at Semicircular Notch," *Trans. ASME, Appl. Mech. Section*, Vol. 89, p. 522.

Lingaiah, K., North, W. P. T., and Mantle, J. B., 1966, "Photoelastic Analysis of an Asymmetrically Reinforced Circular Cutout in a Flat Plate Subjected to Uniform Unidirectional Stress," *Proc. SESA*, Vol. 23, No. 2, p. 617.

Mansfield, E. H., 1953, "Neutral Holes in Plane Sheet-Reinforced Holes which are Elastically Equivalent to the Uncut Sheet," *Q. J. Mech. Appl. Math.*, Vol. 6, p. 370.

Mansfield, E. H., 1955, *Stress Considerations in the Design of Pressurized Shells*, ARC CP 217.

Mansfield, E. H., 1970, "Analysis of a Class of Variable Thickness Reinforcement Around a Circular Hole in a Flat Sheet," *Aeronaut. Q.*, Vol. 21, p. 303.

McKenzie, H. W., and White, D. J., 1968, "Stress Concentration Caused by an Oblique Round Hole in a Flat Plate Under Uniaxial Tension," *J. Strain Anal.*, Vol. 3, p. 98.

Meijers, P., 1967, "Doubly-Periodic Stress Distributions in Perforated Plates," Dissertation, Tech. Hochschule Delft, Netherlands.

Mindlin, R. D., 1939, "Stress Distribution Around a Tunnel," *Proc. ASCE*, Vol. 65, p. 619.

Mindlin, R. D., 1948, "Stress Distribution Around a Hole near the Edge of a Plate in Tension," *Proc. SESA*, Vol. 5, pp. 56–68.

Mitchell, L. H., 1966, "Stress Concentration Factors at a Doubly-Symmetric Hole," *Aeronaut. Q.*, Vol. 17, p. 177.

Mitchell, W. D., and Rosenthal, D., 1949, "Influence of Elastic Constants on the Partition of Load between Rivets," *Proc. SESA*, Vol. 7, No. 2, p. 17.

Miyamoto, H., 1957, "On the Problem of the Theory of Elasticity for a Region Containing more than Two Spherical Cavities," *Trans. Japan Soc. Mech. Eng.*, Vol. 23, p. 437.

Miyao, K., 1970, "Stresses in a Plate Containing a Circular Hole with a Notch," *Bull. Japan Soc. Mech. Eng.*, Vol. 13, p. 483.

Mondina, A., and Falco, H., 1972, "Experimental Stress Analysis on Models of PWR Internals," *Disegno di Macchine, Palermo*, Vol. 3, p. 77.

Mori, A., 1972, "Stress Distributions in a Semi-Infinite Plate with a Row of Circular Holes," *Bull. Japan Soc. Mech. Eng.*, Vol. 15, p. 899.

Murthy, M. V. V., 1969, "Stresses around an Elliptic Hole in a Cylindrical Shell," *Trans. ASME, Appl. Mech. Section*, Vol. 91, p. 39.

Murthy, M. V. V., and Bapu Rao, M. N., 1970, "Stresses in a Cylindrical Shell Weakened by an Elliptic Hole with Major Perpendicular to Shell Axis," *Trans. ASME, Appl. Mech. Section*, Vol. 92, p. 539.

Neuber, H., 1958, *Kerbspannungslehre*, 2nd ed., Springer, Berlin. Translation, *Theory of Notch Stresses*, Office of Technical Services, Dept. of Commerce, Wash. D.C., 1961.

Nishida, Masataka, 1976, *Stress Concentration*, Morikita Shuppan, Tokyo. (in Japanese).

Nisida, M., 1952, *Rep. Sci. Res. Inst. Japan*, Vol. 28, p. 30.

Nisitani, H., 1968, "Method of Approximate Calculation for Interference of Notch Effect and its Application," *Bull. Japan Soc. Mech. Eng.*, Vol. 11, p. 725.

North, W. E., 1965, "A Photoelastic Study of the Interaction Effect of Two Neighboring Holes in a Plate Under Tension," M.S. Thesis, Univ. of Pittsburgh.

Nuno, H., Fujie, T., and Ohkuma, K., 1964, "Experimental Study on the Elastic Properties of Perforated Plates with Circular Holes in Square Patterns," MAPI Laboratory Research Report 74, Mitsubishi Atomic Power Industries Laboratory.

O'Donnell, W. J., and Langer, B. F., 1962, "Design of Perforated Plates," *Trans. ASME, Industry Section*, Vol. 84, p. 307.

O'Donnell, W. J., 1966, "A Study of Preforated Plates with Square Penetration Patterns," WAPD-T-1957, Bettis Atomic Power Laboratory.

O'Donnell, W. J., 1967, "A Study of Perforated Plates with Square Penetration Patterns," *Welding Research Council Bull.* 124.

Papirno, R., 1962, "Stress Concentrations in Tensile Strips with Central Notches of Varying End Radii," *J. Royal Aero. Soc.*, Vol. 66, p. 323.

Paris, P. C., and Sih, G. C., 1965, "Stress Analysis of Cracks," *ASTM Special Tech. Publ.*, p. 34.

Peterson, R. E., 1965, "The Interaction Effect of Neighboring Holes or Cavities, with Particular Reference to Pressure Vessels and Rocket Cases," *Trans. ASME, Basic Engrg. Section*, Vol. 87, p. 879.

Peterson, R. E., 1968, "Stress Concentration Factors for a Round Bar with a Transverse Hole," Report 68-1D7-TAEUGR8, Westinghouse Research Labs., Pittsburgh, PA.

Peterson, R. E., 1974, *Stress Concentration Factors*, John Wiley, New York

Pierce, D. N., and Chou, S. I., 1973, "Stress State Around an Elliptic Hole in a Circular Cylindrical Shell Subjected to Axial Loads," Presented at SESA Meeting, Los Angeles, CA.

Pilkey, W. D., 1994, *Formulas For Stress, Strain, and Structural Matrices*, John Wiley, New York.

Reissner, E., 1945, "The Effect of Transverse Shear Deformation on the Bending of Elastic Plates," *Trans. ASME, Appl. Mech. Section*, Vol. 67, p. A-69.

Reissner, H., and Morduchow, M., 1949, "Reinforced Circular Cutouts in Plane Sheets," *NACA TN 1852*.

Riley, W. F., Durelli, A. J., and Theocaris, P. S., 1959, "Further Stress Studies on a Square Plate with a Pressurized Central Circular Hole," *Proc. 4th Ann. Conf. on Solid Mech.*, Univ. of Texas, Austin.

Riley, W. F., 1964, "Stresses at Tunnel Intersections," *Proc. ASCE, J. Eng. Mech. Div.*, Vol. 90, p. 167.

Rooke, D. P., 1970, "Width Corrections in Fracture Mechanics," *Engr. Fracture Mech.*, Vol. 1, pp. 727–728.

Ryan, J. J., and Fischer, L. J., 1938, "Photoelastic Analysis of Stress Concentration for Beams in Pure Bending with Central Hole," *J. Franklin Inst.*, Vol. 225, p. 513.

Sadowsky, M. A., and Sternberg, E., 1947, "Stress Concentration Around an Ellipsoidal Cavity in an Infinite Body under Arbitrary Plane Stress Perpendicular to the Axis of Revolution of Cavity," *Trans. ASME, Appl. Mech. Section*, Vol. 69, p. A-191.

Salerno, V. L., and Mahoney, J. B., 1968, "Stress Solution for an Infinite Plate Containing Two Arbitrary Circular Holes under Equal Biaxial Stresses," *Trans. ASME, Industry Section*, Vol. 90, p. 656.

Savin, G. N., 1961, *Stress Concentration Around Holes*, Pergamon Press, New York, p. 234.

Sampson, R. C., 1960, "Photoelastic Analysis of Stresses in Perforated Material Subject to Tension or Bending," *Bettis Technical Review*, WAP-BT-18.

Schaechterle, K., 1934, "On the Fatique Strength of Riveted and Welded Joints and the Design of Dynamically Stressed Structural Members based on Conclusions Drawn from Fatigue Tests," *Intern. Assn. Bridge Structural Eng.*, Vol. 2, p. 312.

Schulz, K. J., 1941, "Over den Spannungstoestand in doorborde Platen," (On the State of Stress in Perforated Plates), Doctoral Thesis, Techn. Hochschule, Delft. (in Dutch).

Schulz, K. J., 1942-1945, "On the State of Stress in Perforated Strips and Plates," *Proc. Koninklÿke Nederlandsche Akademie van Wetenschappen (Netherlands Royal Academy of Science)*, Amsterdam, Vol. 45 (1942), p. 233, 341, 457, 524; Vol. 46–48 (1943–1945), p. 282, 292.
The foregoing six papers were summarized by C. B. Biezeno, "Survey of Papers on Elasticity Published in Holland 1940–1946," *Advances in Appl. Mech.*, Vol. 1, Academic Press, New York (1948), p. 105. Biezeno comments, "The calculations are performed only for a strip with one row of holes, though and extensive program of further investigation was projected, which unfortunately could not be executed; the Jewish author, who lived in Holland, was arrested and marched off to Germany or Poland, which meant his destruction."

Schutz, W., 1964, "Fatigue Test Specimens," Tech. Note TM 10/64, Laboratorium für Betriebsfestigkeit, Darmstadt, Germany.

Seika, M., and Ishii, M., 1964, "Photoelastic Investigation of the Maximum Stress in a Plate with a Reinforced Circular Hole under Uniaxial Tension," *Trans. ASME, Appl. Mech. Section*, Vol. 86, p. 701.

Seika, M., and Amano, A., 1967, "The Maximum Stress in a Wide Plate with a Reinforced Circular Hole under Uniaxial Tension-Effects of a Boss with Fillet," *Trans. ASME, Appl. Mech. Section*, Vol. 89, p. 232.

Seika, M., Isogimi, K., and Inoue, K., 1971, "Photoelastic Study of Stresses in a Cylindrical Pressure Vessel with a Nozzle," *Bull. Japan Soc. Mech. Eng.*, Vol. 14, p. 1036.

Sekiya, T., 1955, "An Approximate Solution in the Problems of Elastic Plates with an Arbitrary External Form and a Circular Hole," *Proc. 5th Japan Nat. Congr. Appl. Mech.*, p. 95.

Shioya, S., 1963, "The Effect of an Infinite Row of Semi-Circular Notches on the Transverse Flexure of a Semi-Infinite Plate," *Ingenieur-Archiv*, Vol. 32, p. 143.

Shioya, S., 1971, "On the Tension of an Infinite Thin Plate Containing a Pair of Circular Inclusions," *Bull. Japan Soc. Mech. Eng.*, Vol. 14, p. 117.

Sjöström, S., 1950, "On the Stresses at the Edge of an Eccentically Located Circular Hole in a Strip under Tension," *Aeronaut. Research Inst. Sweden Report No. 36*, Stockholm.

Slot, T., 1972, *Stress Analysis of Thick Perforated Plates*, Technomic Publ. Co., Westport, Conn.

Sobey, A. J., 1963, "Stress Concentration Factors for Rounded Rectangular Holes in Infinite Sheets," *ARC R&M 3407*. Her Majesties Stationery Office, London.

Sobey, A. J., 1968, "Stress Concentration Factors for Reinforced Rounded-Square Holes in Sheets," *ARC R&M 3546*.

Sternberg, E., and Sadowsky, M. A., 1949, "Three-Dimensional Solution for the Stress Concentration Around a Circular Hole in a Plate of Arbitrary Thickness," *Trans. ASME, Appl. Mech. Section*, Vol. 71, p. 27.

Sternberg, E., and Sadowsky, M. A., 1952, "On the Axisymmetric Problem of the Theory of Elasticity for an Infinite Region Containing Two Spherical Cavities," *Trans. ASME, Appl. Mech. Section*, Vol. 76, p. 19.

Tamate, O., 1957, "Einfluss einer unendichen Reihe gleicher Kreislöcher auf die Durchbiegung einer dünnen Platte," *Z. angew. Math. u. Mech.*, Vol. 37, p. 431.

Tamate, O., 1958, "Transverse Flexure of a Thin Plate Containing Two Holes," *Trans. ASME, Appl. Mech. Section*, Vol. 80, p. 1.

Templin, R. L., 1954, "Fatigue of Aluminum," *Proc. ASTM*, Vol. 54, p. 641.

Theocaris, P. S., 1956, "The Stress Distribution in a Strip Loaded in Tension by Means of a Central Pin," *Trans. ASME, Appl. Mech. Section*, Vol. 78, p. 482.

Thum, A., and Kirmser, W., 1943, "Überlagerte Wechselbeanspruchungen, ihre Erzeugung und ihr Einfluss auf die Dauerbarkeit und Spannungsausbildung quergebohrter Wellen," *VDI-Forschungsheft 419*, Vol. 14(b), p. 1.

Timoshenko, S., 1924, "On Stresses in a Plate with a Circular Hole," *J. Franklin Inst.*, Vol. 197, p. 505.

Timoshenko, S., 1956, *Strength of Materials, Part II*, 3rd ed., Van Nostrand, Princeton, N. J. p. 208.

Timoshenko, S., and Goodier, J. N., 1970, *Theory of Elasticity*, 3rd ed., McGraw Hill, New York, p. 398.

Tingleff, O., 1971, "Stress Concentration in a Cylindrical Shell with an Elliptical Cutout," *AIAA J.*, Vol. 9, p. 2289.

Udoguti, T., 1947, "Solutions of Some Plane Elasticity Problems by Using Dipole Coordinates- Part I," *Trans. Japan Soc. Mech. Eng.*, Vol. 13, p. 17. (in Japanese)

Van Dyke, P., 1965, "Stresses about a Circular Hole in a Cylindrical Shell," *AIAA J.*, Vol. 3, p. 1733.

Wahl, A. W., and Beeuwkes, R., 1934, "Stress Concentration Produced by Holes and Notches," *Trans. ASME, Appl. Mech. Section*, Vol. 56, p. 617.

Wells, A. A., 1950, "On the Plane Stress Distribution in an Infinite Plate with Rim-Stiffened Elliptical Opening," *Q. J. Mech. Appl. Math.*, Vol. 2, p. 23.

Westergaard, H. M., 1939, "Bearing Pressure and Cracks," *Trans. ASME, Appl. Mech. Section*, Vol. 61, p. A-49.

Wilson, H. B., 1961, "Stresses Owing to Internal Pressure in Solid Propellant Rocket Grains," *J. ARS*, Vol. 31, p. 309.

Wittrick, W. H., 1959a, "The Stress Around Reinforced Elliptical Holes in Plane Sheet," *Aero. Res. Lab (Australia) Report*, ARL SM267.

Wittrick, W. H., 1959b, "Stress around Reinforced Elliptical Holes, with Applications to Pressure Cabin Windows," *Aeronaut. Q.*, Vol. 10, p. 373–400.

Wittrick, W. H., 1963, "Stress Concentrations for Uniformly Reinforced Equilateral Triangular Holes with Rounded Corners," *Aeronaut. Q.*, Vol. 14, p. 254.

WRCB, *Welding Research Council Bulletin* 113, 1966.

WRCB, *Welding Research Council Bulletin* 113, 1970.

Youngdahl, C. K., and Sternberg, E., 1966, "Three-Dimensional Stress Concentration Around a Cylindrical Hole in a Semi-Infinite Elastic Body," *Trans. ASME, Appl. Mech. Section*, Vol. 88, p. 855.

Yu, Y. Y., 1952, "Gravitational Stresses on Deep Tunnels," *Trans. ASME, Appl. Mech. Section*, Vol. 74, p. 537.

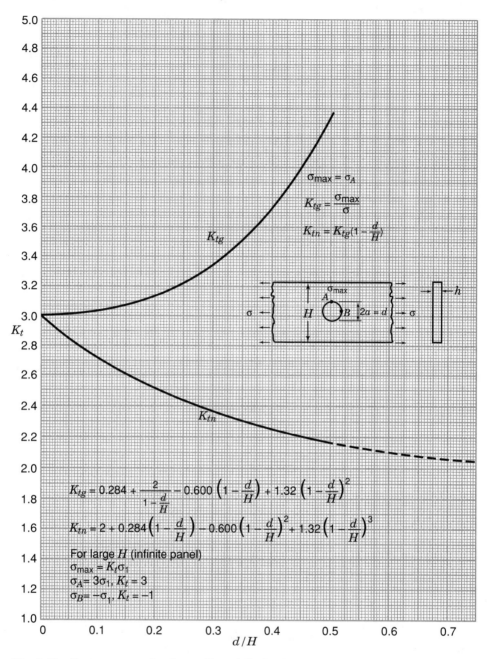

Chart 4.1 Stress concentration factors K_{tg} and K_{tn} for the tension of a finite-width thin element with a circular hole (Howland 1929–30).

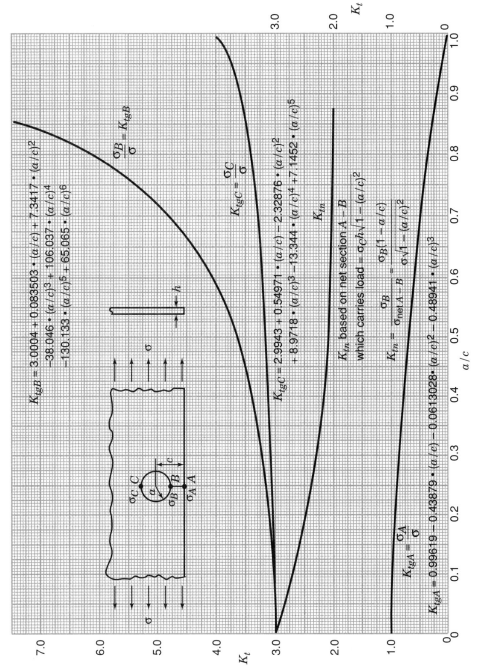

Chart 4.2 Stress concentration factors for the tension of a thin semi-infinite element with a circular hole near the edge (Mindlin 1948; Udoguti 1947; Isida 1955a).

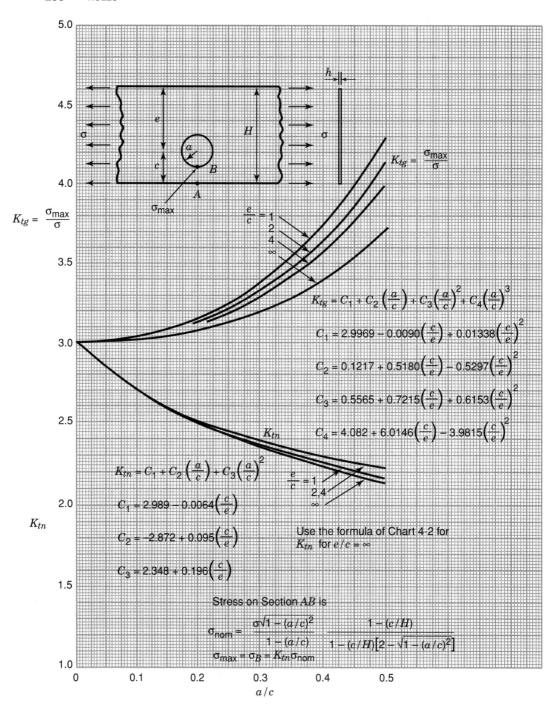

Chart 4.3 Stress concentration factors for the tension of a finite-width element having an eccentrically located circular hole (based on mathematical analysis of Sjöström 1950). $e/c = \infty$ corresponds to Chart 4.2.

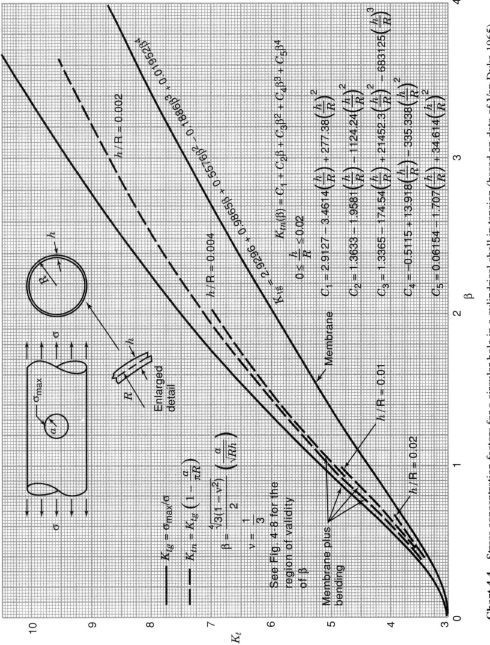

Chart 4.4 Stress concentration factors for a circular hole in a cylindrical shell in tension (based on data of Van Dyke 1965).

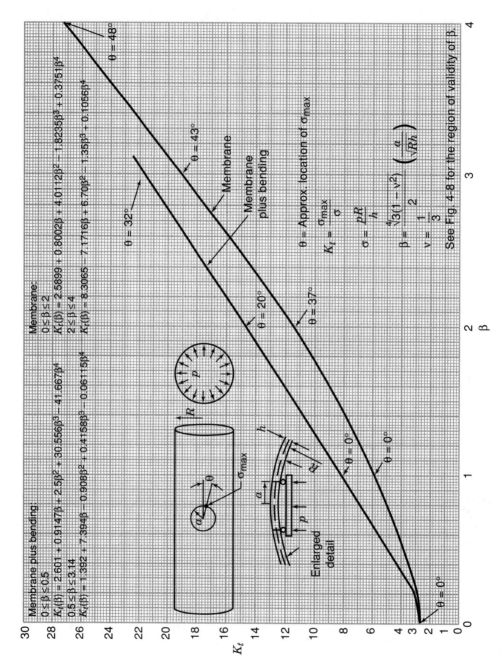

Chart 4.5 Stress concentration factors for a circular hole in a cylindrical shell with internal pressure (based on data of Van Dyke 1965).

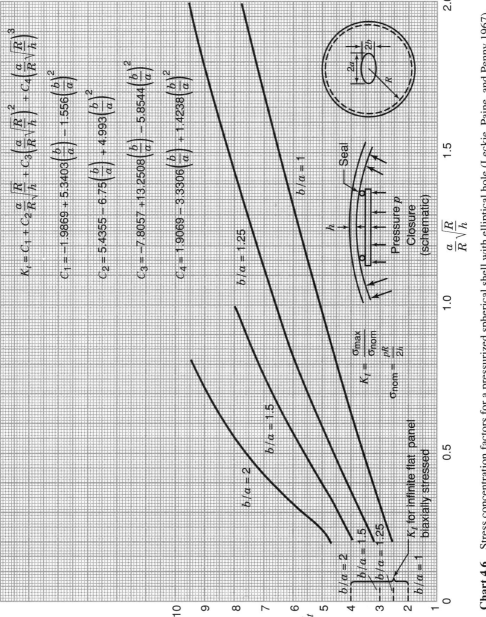

Chart 4.6 Stress concentration factors for a pressurized spherical shell with elliptical hole (Leckie, Paine, and Penny 1967).

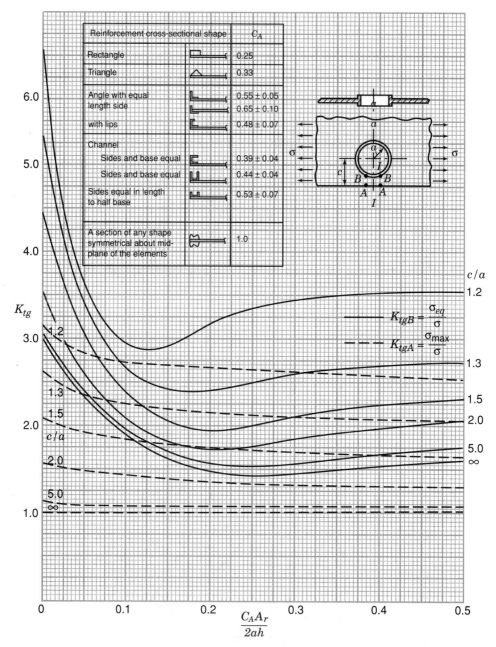

Chart 4.7 Stress concentration factors for various shaped reinforcements of a circular hole near the edge of a semi-infinite element in tension (data from Davies 1963; Wittrick 1959; Mansfield 1955; Mindlin 1948; ESDU 1981).

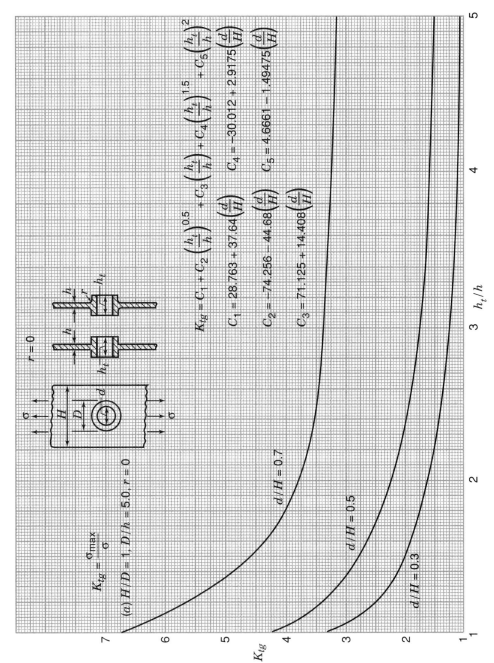

Chart 4.8a Stress concentration factors K_{tg} for a reinforced circular hole in a thin element in tension (Seika and Amano 1967). (a) $H/D = 1$, $D/h = 5.0$, $r = 0$.

(b) $H/D = 4.0$, $D/h = 5.0$

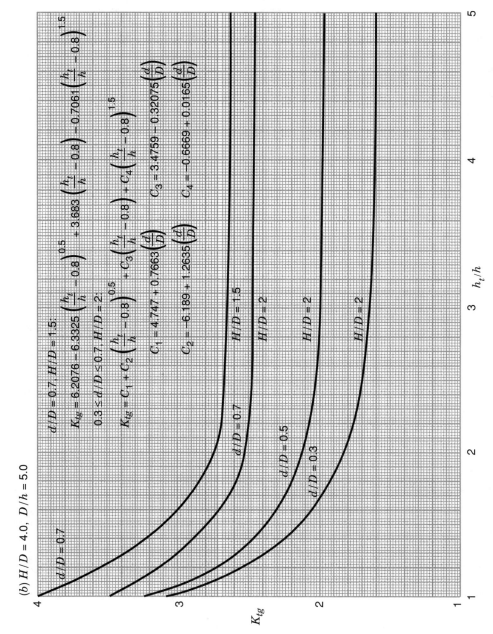

Chart 4.8b Stress concentration factors K_{tg} for a reinforced circular hole in a thin element in tension (Seika and Amano 1967). (b) $H/D = 4.0$, $D/h = 5.0$.

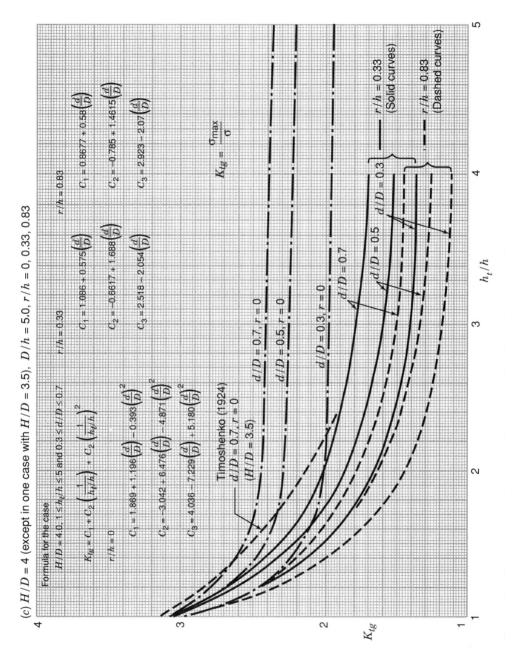

Chart 4.8c Stress concentration factors K_{tg} for a reinforced circular hole in a thin element in tension (Seika and Amano 1967). (c) $H/D = 4$ (except in one case with $H/D = 3.5$), $D/h = 5.0$, $r/h = 0, 0.33, 0.83$.

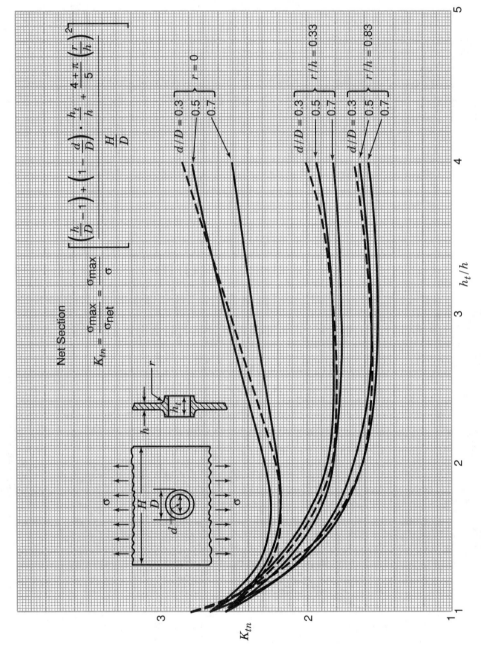

Chart 4.9 Stress concentration factors K_{tn} for a reinforced circular hole in a thin element in tension, $H/D = 4.0$, $D/h = 5.0$ (Seika and Ishii 1964; Seika and Amane 1967).

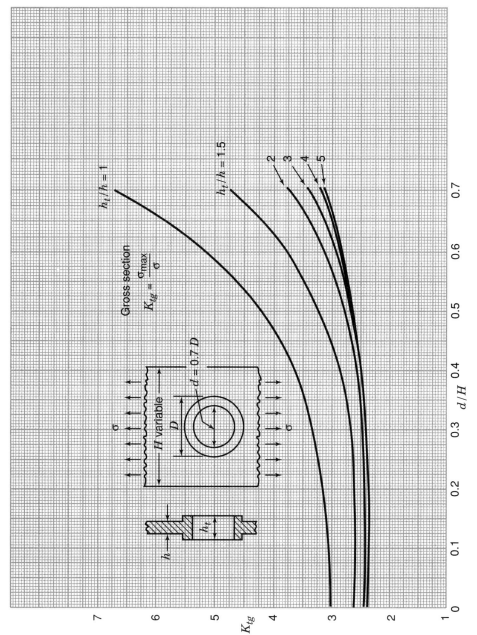

Chart 4.10 Stress concentration factors K_{tg} for a reinforced circular hole in a thin element in tension, $d/D = 0.7$, $D/h = 5.0$ (Seika and Ishii 1964).

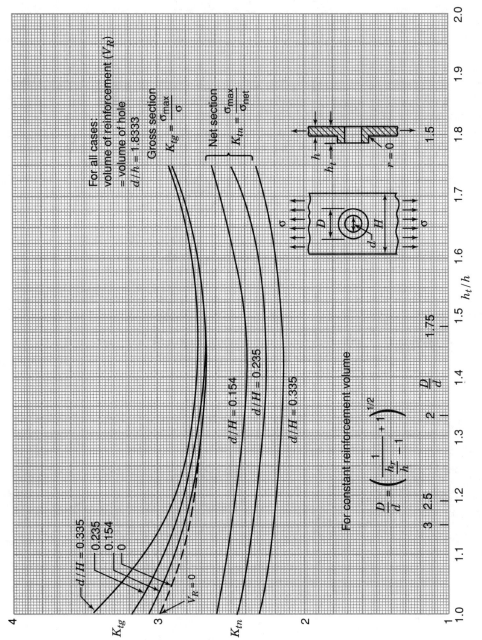

Chart 4.11 Stress concentration factors for a uniaxially stressed thin element with a reinforced circular hole on one side (from photoelastic tests of Lingaiah et al. 1966).

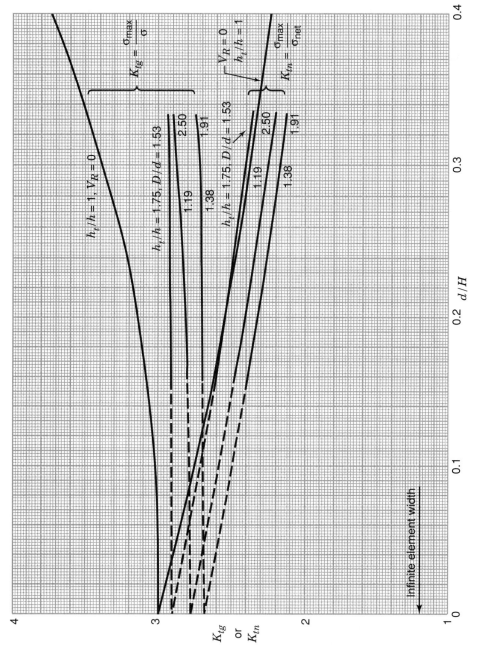

Chart 4.12 Extrapolation of K_{tg} and K_{tn} values of Chart 4.11 to an element of infinite width (from photoelastic tests of Lingaiah et al. 1966).

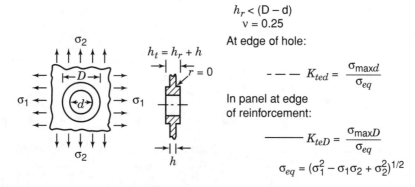

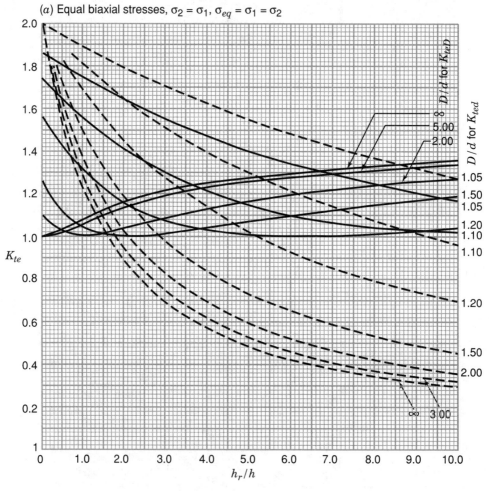

Chart 4.13a Analytical stress concentration factors for a symmetrically reinforced circular hole in a thin element with in-plane biaxial normal stresses, σ_1 and σ_2 (Gurney 1938; ESDU 1981). (a) Equal biaxial stresses, $\sigma_2 = \sigma_1$, $\sigma_{eq} = \sigma_1 = \sigma_2$.

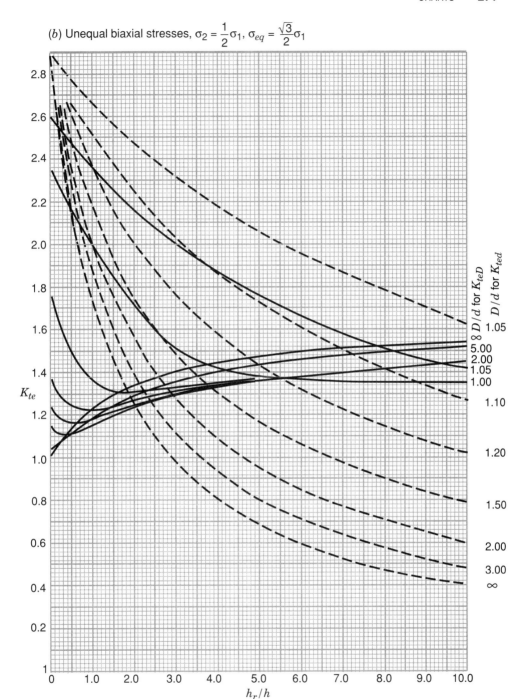

Chart 4.13b Analytical stress concentration factors for a symmetrically reinforced circular hole in a thin element with in-plane biaxial normal stresses, σ_1 and σ_2 (Gurney 1938; ESDU 1981).
(*b*) Unequal biaxial stresses, $\sigma_2 = \frac{1}{2}\sigma_1$, $\sigma_{eq} = \frac{\sqrt{3}}{2}\sigma_1$.

272 HOLES

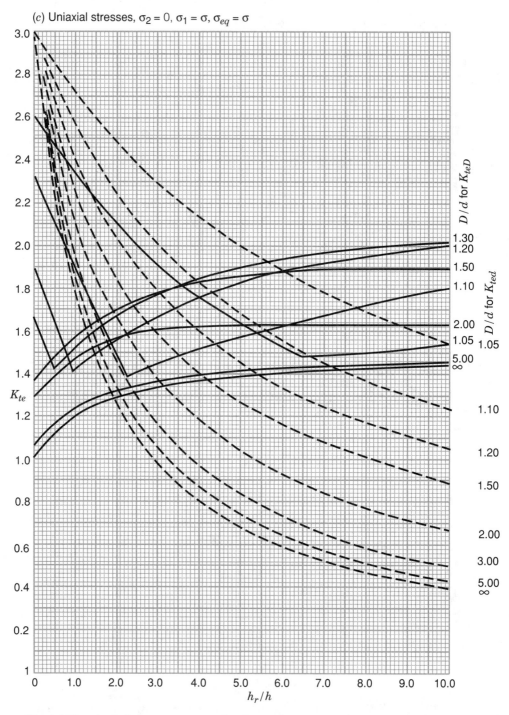

Chart 4.13c Analytical stress concentration factors for a symmetrically reinforced circular hole in a thin element with in-plane biaxial normal stresses, σ_1 and σ_2 (Gurney 1938; ESDU 1981). (c) Uniaxial stress, $\sigma_2 = 0$, $\sigma_1 = \sigma$, $\sigma_{eq} = \sigma$.

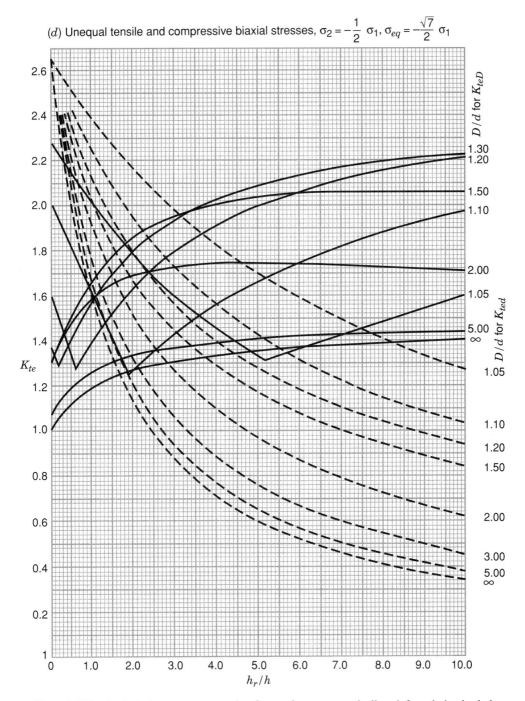

Chart 4.13d Analytical stress concentration factors for a symmetrically reinforced circular hole in a thin element with in-plane biaxial normal stresses, σ_1 and σ_2 (Gurney 1938; ESDU 1981). (d) Unequal tensile and compressive biaxial stresses, $\sigma_2 = -\frac{1}{2}\sigma_1$, $\sigma_{eq} = -\frac{\sqrt{7}}{2}\sigma_1$.

274 HOLES

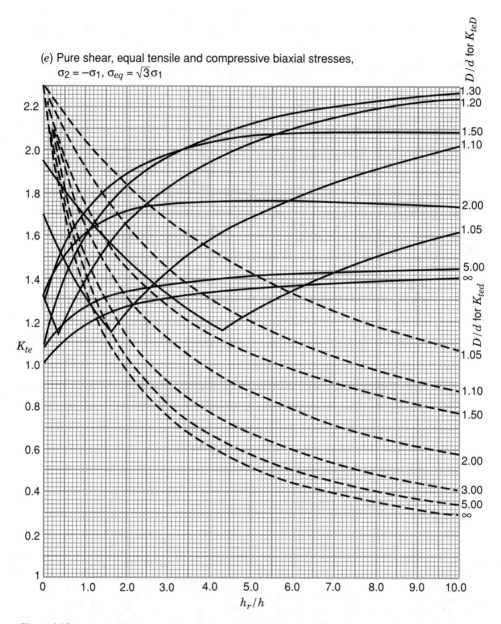

Chart 4.13e Analytical stress concentration factors for a symmetrically reinforced circular hole in a thin element with in-plane biaxial normal stresses, σ_1 and σ_2 (Gurney 1938; ESDU 1981). (*e*) Pure shear, equal tensile and compressive biaxial stresses, $\sigma_2 = -\sigma_1$, $\sigma_{eq} = \sqrt{3}\sigma_1$.

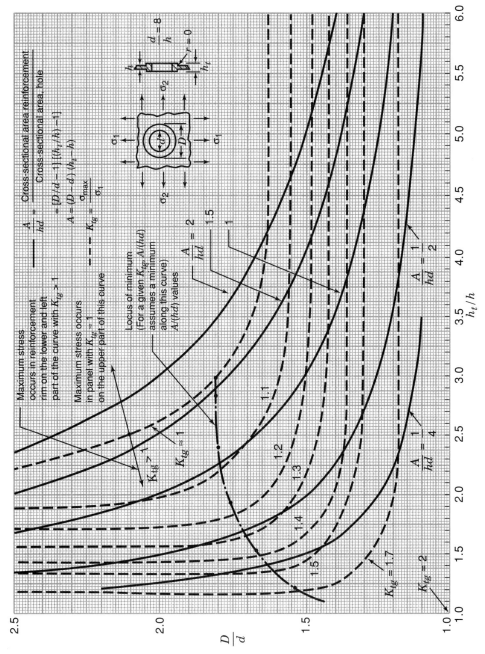

Chart 4.14 Area ratios and experimental stress concentration factors K_{tg} for a symmetrically reinforced circular hole in a panel with equal biaxial normal stresses, $\sigma_1 = \sigma_2$ (approximate results based on strain gage tests by Kaufman et al. 1962).

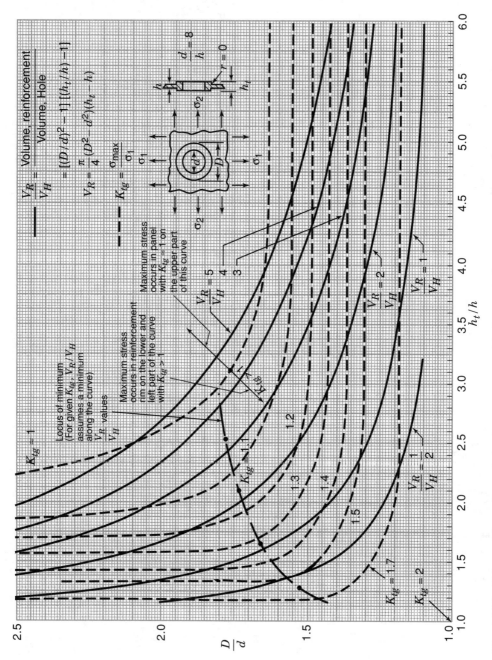

Chart 4.15 Volume ratios and experimental stress concentration factors K_{tg} for a symmetrically reinforced circular hole in a panel with equal biaxial normal stresses, $\sigma_1 = \sigma_2$ (approximate results based on strain gage tests by Kaufman, et al. 1962).

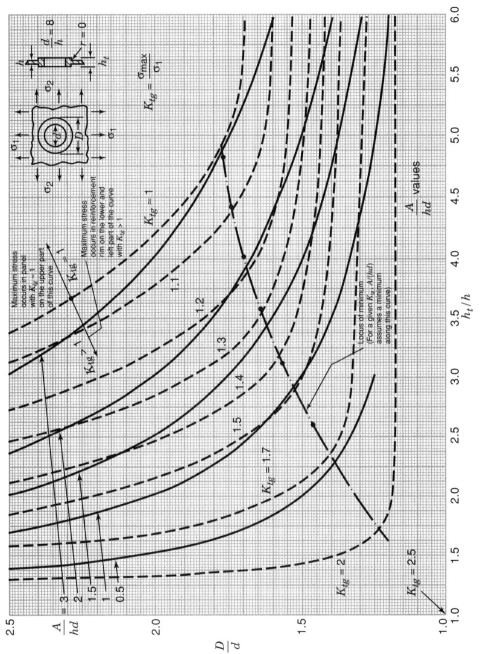

Chart 4.16 Area ratios and experimental stress concentration factors K_{tg} for a symmetrically reinforced circular hole in a panel with unequal biaxial normal stresses, $\sigma_2 = \frac{\sigma_1}{2}$ (approximate results based on strain gage tests by Kaufman, et al. 1962).

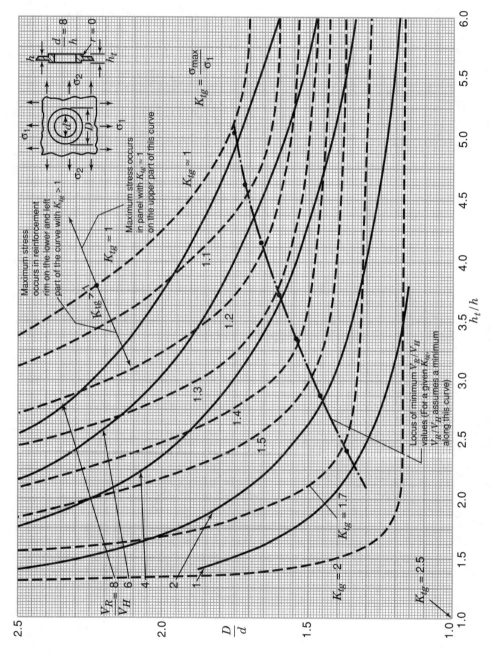

Chart 4.17 Volume ratios and experimental stress concentration factors K_{tg} for a symmetrically reinforced circular hole in a panel with unequal biaxial normal stresses, $\sigma_2 = \sigma_1/2$ (approximate results based on strain gage tests by Kaufman et al., 1962).

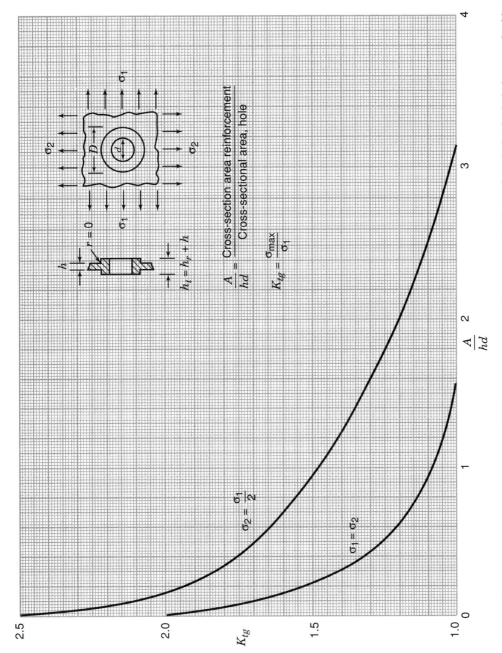

Chart 4.18 Approximate minimum values of K_{tg} versus area ratios for a symmetrically reinforced circular hole in a panel with biaxial normal stresses (based on strain gage tests by Kaufman et al. 1962).

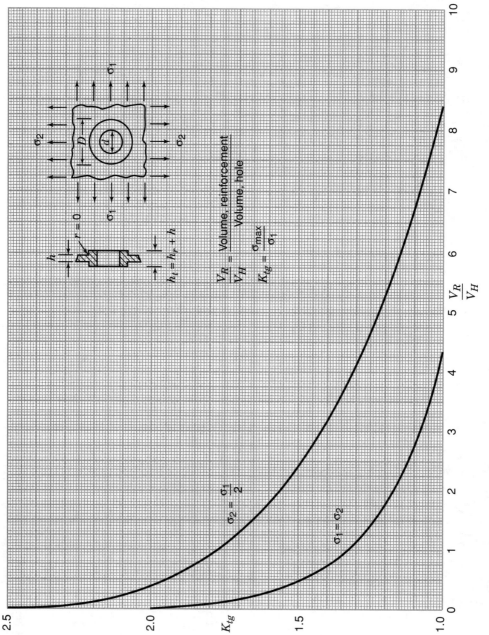

Chart 4.19 Approximate minimum values of K_{tg} versus volume ratios for a symmetrically reinforced circular hole in a panel with biaxial normal stresses (based on strain gage tests by Kaufman et al., 1962).

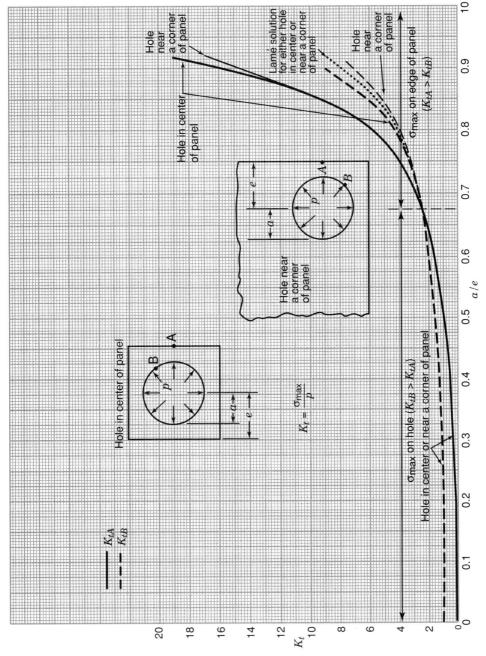

Chart 4.20 Stress concentration factors K_t for a square panel with a pressurized circular hole (Riley et al 1959; Durelli and Kobayashi 1958).

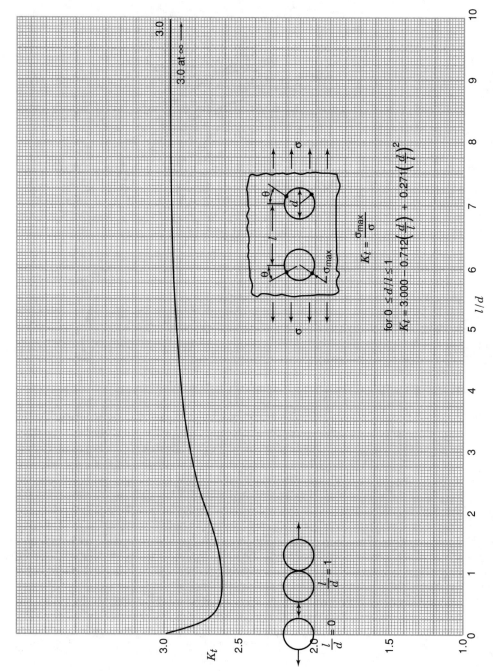

Chart 4.21 Stress concentration factors K_t for uniaxial tension case of an infinite panel with two circular holes (based on mathematical analysis of Ling 1948 and Haddon 1967). Tension parallel to line of holes.

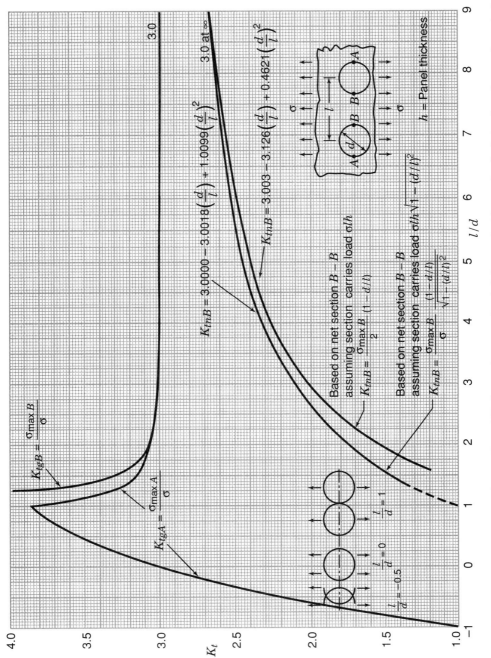

Chart 4.22 Stress concentration factors K_{tg} and K_{tn} for tension case of an infinite panel with two circular holes (based on mathematical analyses of Ling, 1948 and Haddon 1967). Tension perpendicular to line of holes.

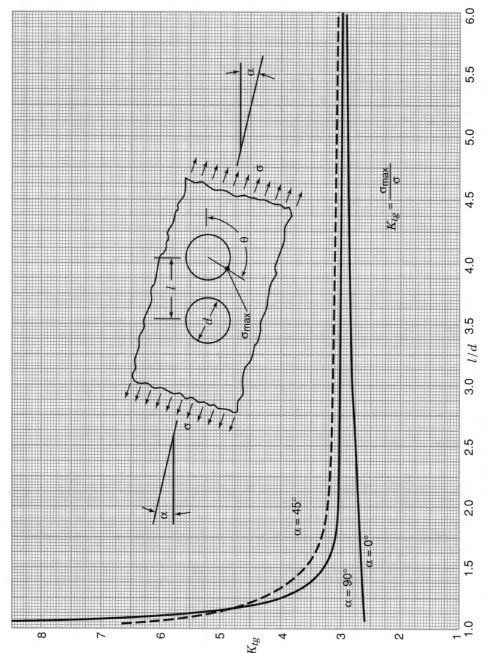

Chart 4.23 Stress concentration factors K_{tg} for tension case of an infinite panel with two circular holes (from mathematical analysis of Haddon 1967). Tension at various angles.

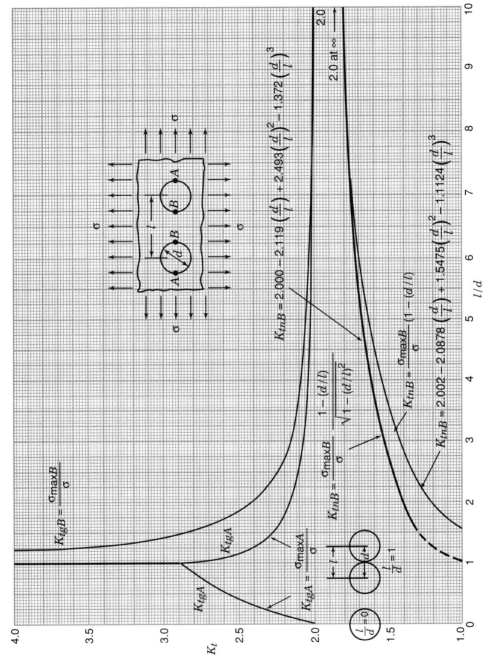

Chart 4.24 Stress concentration factors K_{tg} and K_{tn} for equal biaxial tension case of an infinite panel with two circular holes (based on mathematical analyses of Ling 1948 and Haddon 1967).

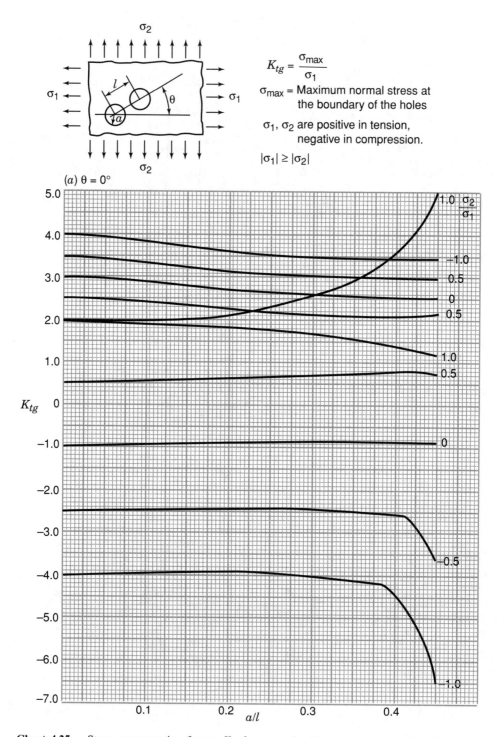

Chart 4.25a Stress concentration factors K_{tg} for a panel with two holes under biaxial stresses (Haddon 1967; ESDU 1981). (a) $\theta = 0°$.

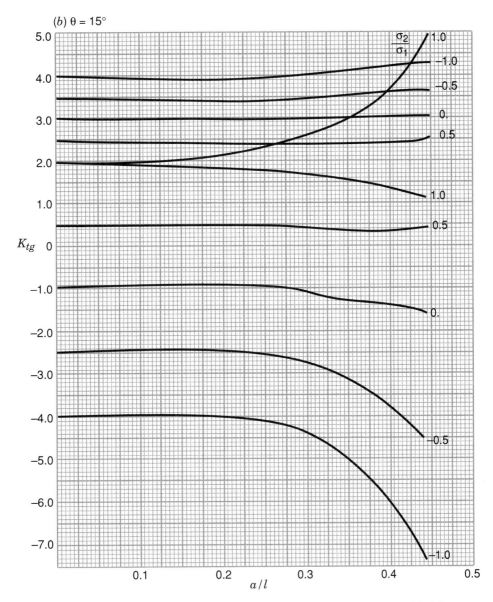

Chart 4.25b Stress concentration factors K_{tg} for a panel with two holes under biaxial stresses (Haddon 1967; ESDU 1981). (b) $\theta = 15°$.

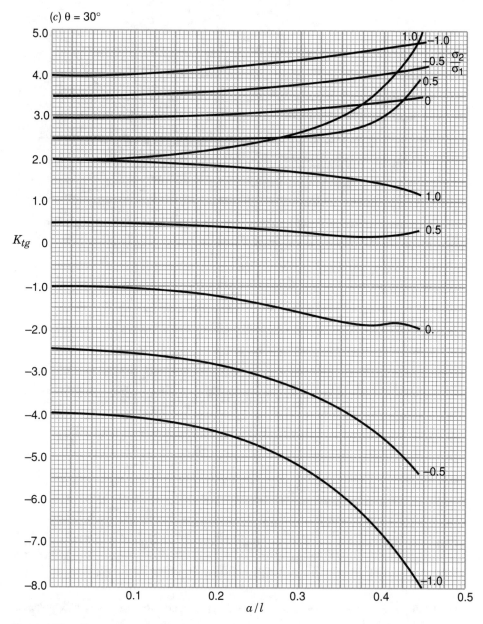

Chart 4.25c Stress concentration factors K_{tg} for a panel with two holes under biaxial stresses (Haddon 1967; ESDU 1981). (c) $\theta = 30°$.

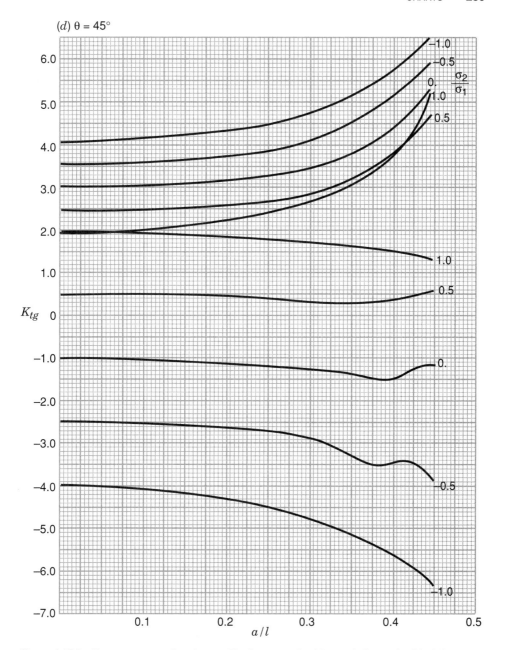

Chart 4.25d Stress concentration factors K_{tg} for a panel with two holes under biaxial stresses (Haddon 1967; ESDU 1981). (d) $\theta = 45°$.

290 HOLES

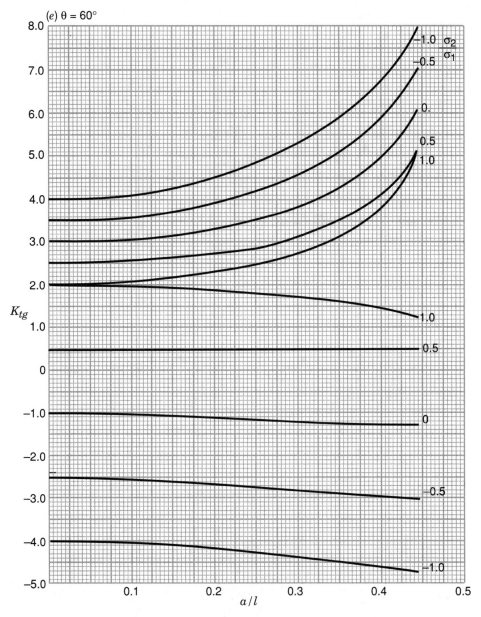

Chart 4.25e Stress concentration factors K_{tg} for a panel with two holes under biaxial stresses (Haddon 1967; ESDU 1981). (e) $\theta = 60°$.

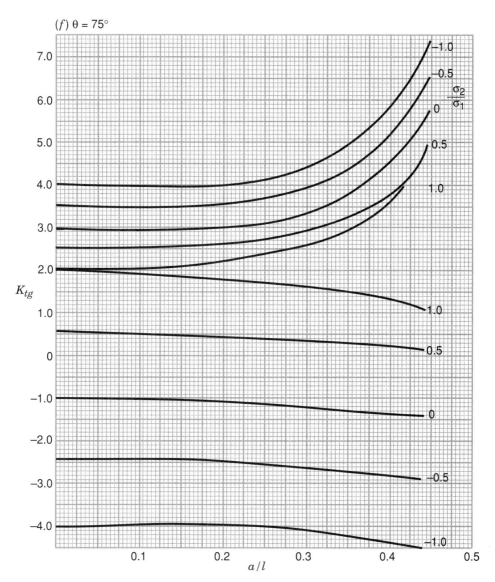

Chart 4.25f Stress concentration factors K_{tg} for a panel with two holes under biaxial stresses (Haddon 1967; ESDU 1981). (f) $\theta = 75°$.

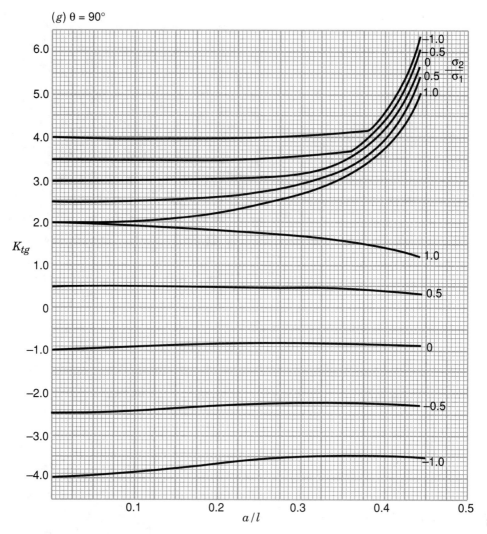

Chart 4.25g Stress concentration factors K_{tg} for a panel with two holes under biaxial stresses (Haddon 1967; ESDU 1981). (g) $\theta = 90°$.

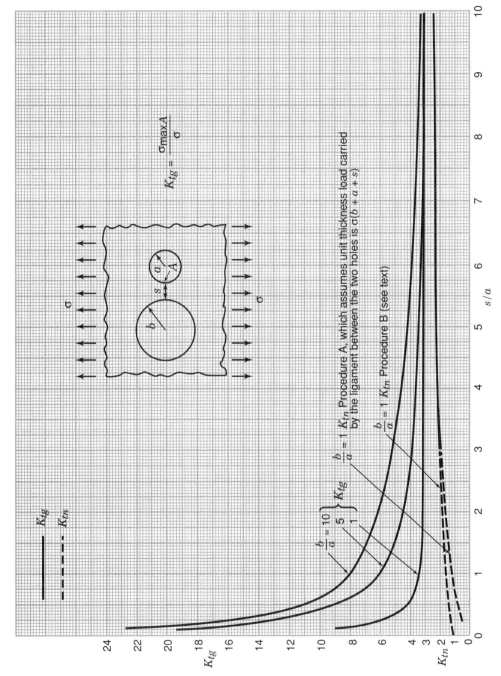

Chart 4.26 Stress concentration factors K_{tg} and K_{tn} for tension in an infinite thin element with two circular holes of unequal diameter (from mathematical analysis of Haddon 1967). Tension perpendicular to the line of holes.

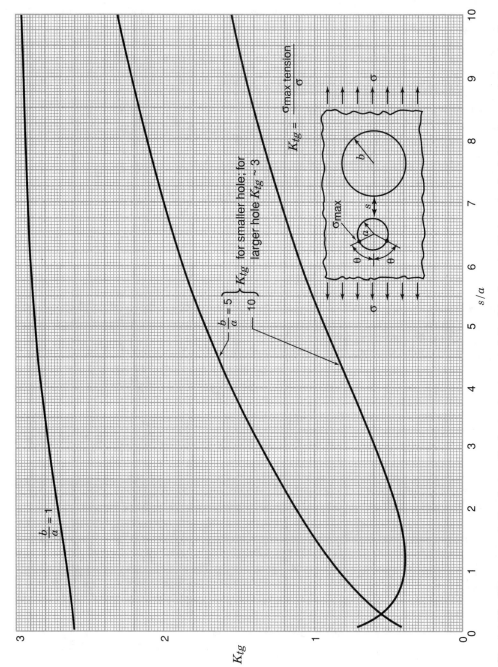

Chart 4.27 Stress concentration factors K_{tg} for tension in an infinite thin element with two circular holes of unequal diameter (from mathematical analysis of Haddon 1967). Tension parallel to the line of holes.

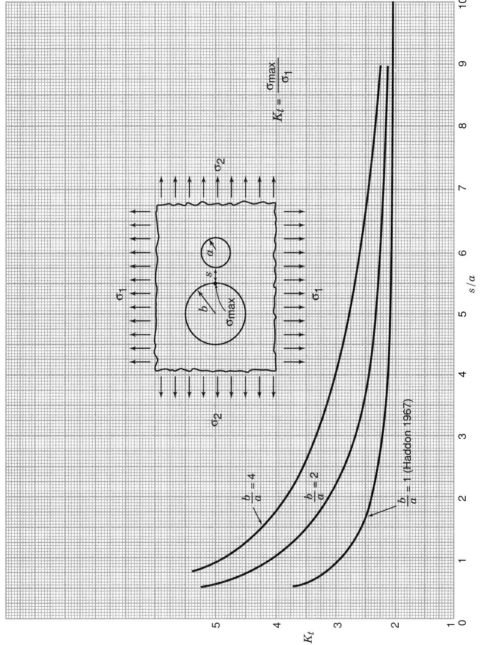

Chart 4.28 Stress concentration factors K_t for biaxial tension in infinite thin element with two circular holes of unequal diameter, $\sigma_1 = \sigma_2$ (Salerno and Mahoney 1968; Haddon 1967).

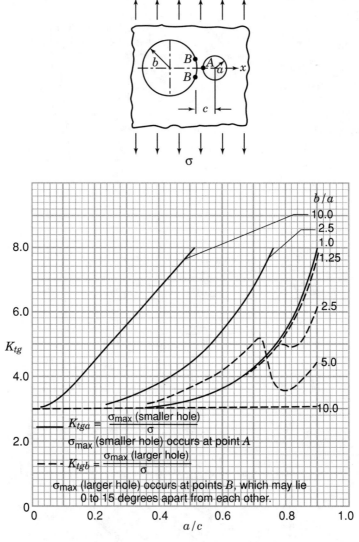

Chart 4.29 Stress concentration factors K_{tg} for tension in infinite thin element with two circular holes of unequal diameter (from mathematical analysis of Haddon 1967; ESDU 1981). Tension perpendicular to line of holes.

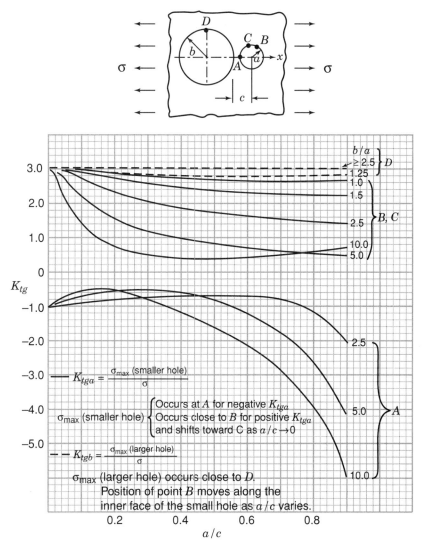

Chart 4.30 Stress concentration factors K_{tg} for tension in infinite thin element with two circular holes of unequal diameter (from mathematical analysis of Haddon 1967; ESDU 1981). Tension parallel to line of holes.

298 HOLES

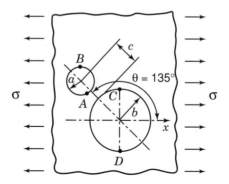

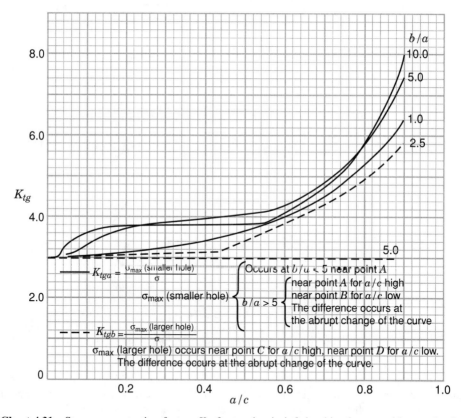

Chart 4.31 Stress concentration factors K_{tg} for tension in infinite thin element with two circular holes of unequal diameter (from mathematical analysis of Haddon 1967; ESDU 1981). Holes aligned diagonal to the loading.

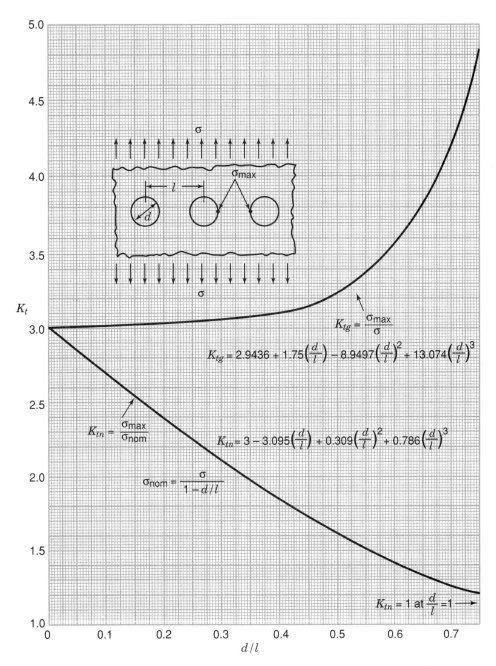

Chart 4.32 Stress concentration factors K_{tg} and K_{tn} for uniaxial tension of an infinite thin element with an infinite row of circular holes (Schulz 1941). Stress perpendicular to the axis of the holes.

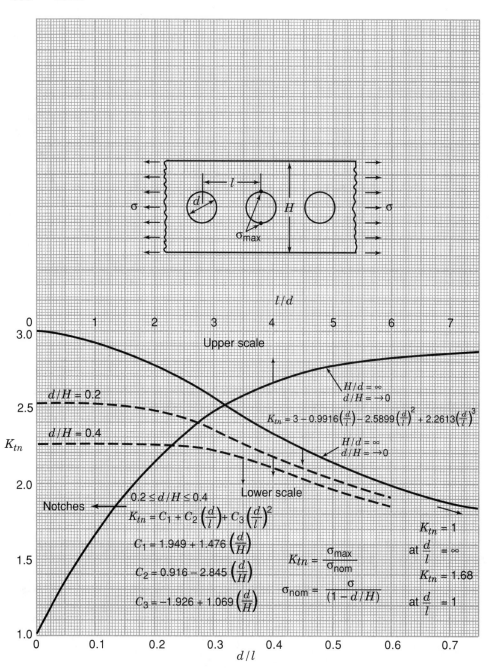

Chart 4.33 Stress concentration factors K_{tn} for uniaxial tension of a finite width thin element with an infinite row of circular holes (Schulz 1941). Stress parallel to the axis of holes.

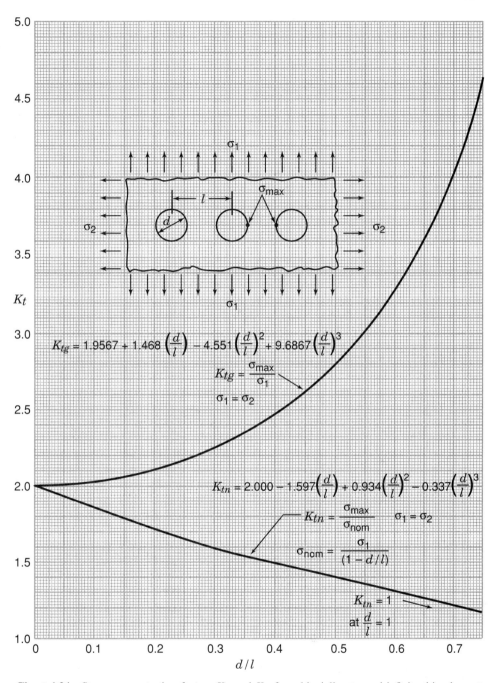

Chart 4.34 Stress concentration factors K_{tg} and K_{tn} for a biaxially stressed infinite thin element with an infinite row of circular holes, $\sigma_1 = \sigma_2$ (Hütter 1942).

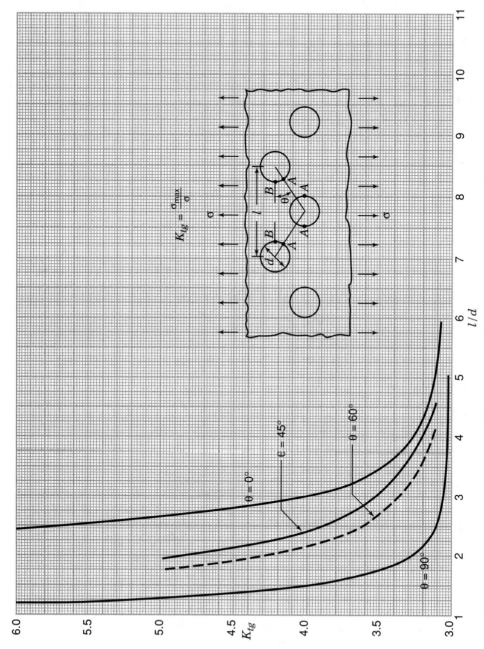

Chart 4.35 Stress concentration factors K_{tg} for a double row of holes in a thin element in uniaxial tension (Schulz 1941). Applied stress perpendicular to the axis of the holes.

302

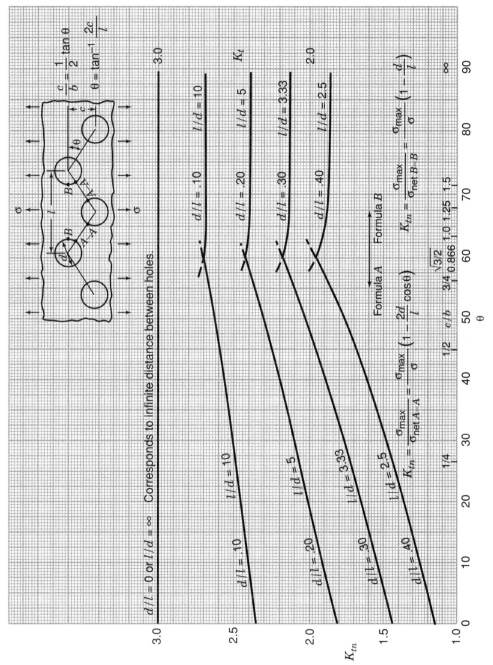

Chart 4.36 Stress concentration factors K_{tn} for a double row of holes in a thin element in uniaxial tension (based on mathematical analysis of Schulz 1941). Applied stress perpendicular to the axes of the holes.

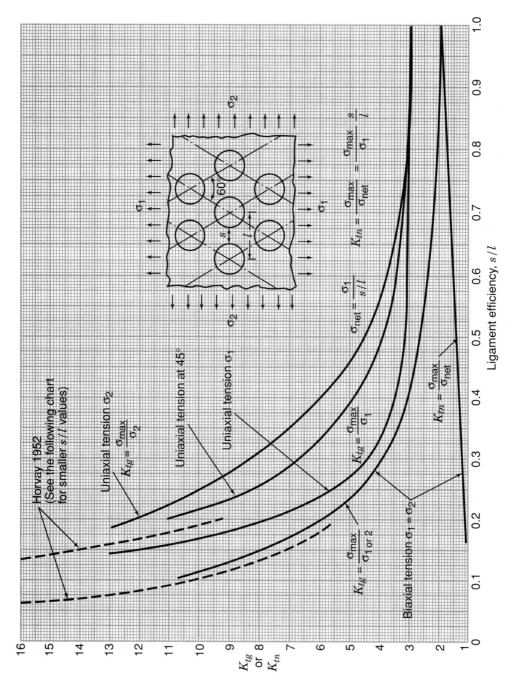

Chart 4.37 Stress concentration factors K_{tg} and K_{tn} for a triangular pattern of holes in a thin element subject to uniaxial and biaxial stresses (Sampson 1960; Meijers 1967). The pattern is repeated throughout the element.

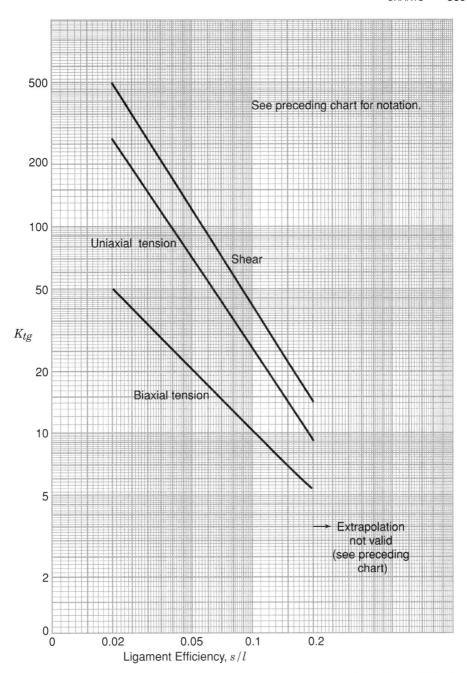

Chart 4.38 Stress concentration factors K_{tg} for the triangular pattern of holes of Chart 4.37 for low values of ligament efficiency (Horvay 1952).

306   HOLES

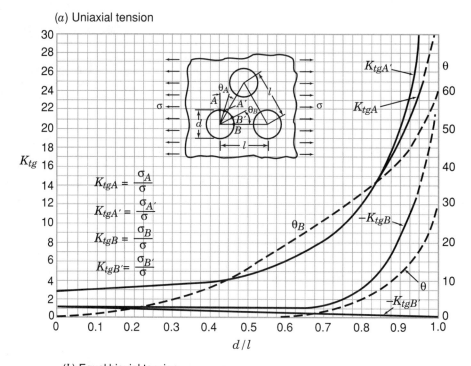

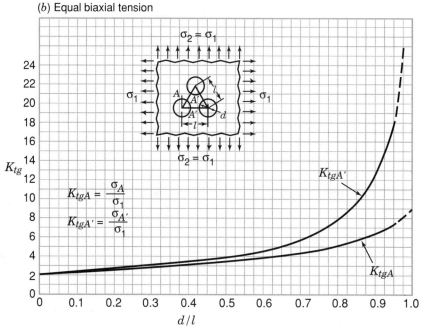

Chart 4.39a,b   Stress concentration factors K_{tg} for particular locations on the holes, for a triangular pattern of holes in a thin element subject to uniaxial and biaxial stresses (Nishida 1976). The pattern is repeated throughout the element. (*a*) Uniaxial tension; (*b*) equal biaxial tension.

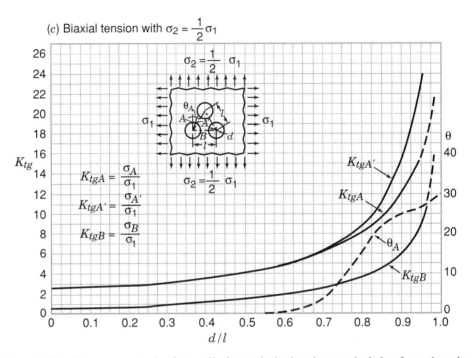

Chart 4.39c Stress concentration factors K_{tg} for particular locations on the holes, for a triangular pattern of holes in a thin element subject to uniaxial and biaxial stresses (Nishida 1976). (c) biaxial tension with $\sigma_2 = \sigma_1/2$.

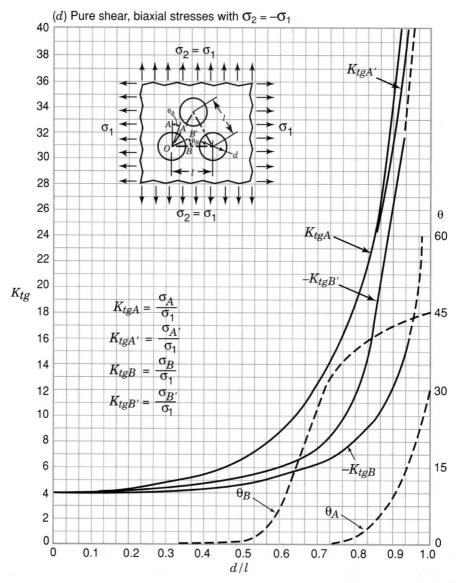

Chart 4.39d Stress concentration factors K_{tg} for particular locations on the holes, for a triangular pattern of holes in a thin element subject to uniaxial and biaxial stresses (Nishida 1976). (d) Pure shear, biaxial stresses with $\sigma_2 = -\sigma_1$.

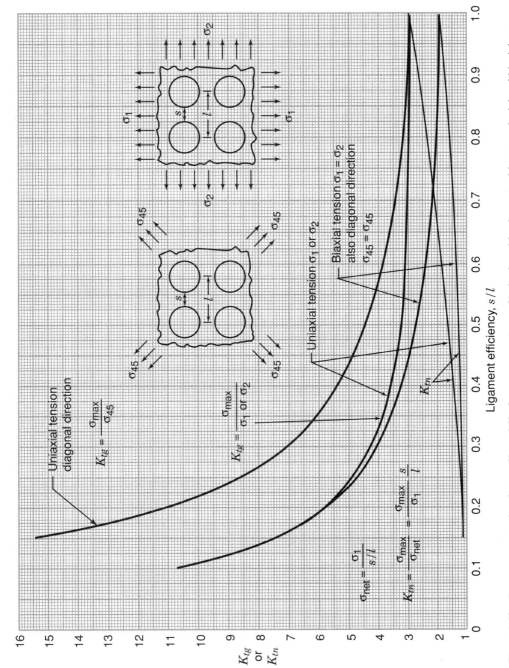

Chart 4.40 Stress concentration factors K_{tg} and K_{tn} for a square pattern of holes in a thin element subject to uniaxial and biaxial stresses (Bailey and Hicks 1960; Hulbert 1965; Meijers 1967). The pattern is repeated throughout the element.

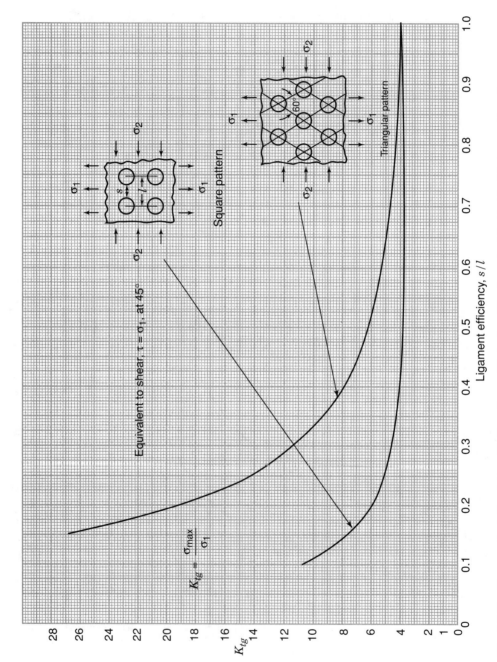

Chart 4.41 Stress concentration factors K_{tg} for patterns of holes in a thin element subject to biaxial stresses. Pure shear $\sigma_2 = -\sigma_1$ (Sampson 1960; Bailey and Hicks 1960; Hulbert and Niedenfuhr 1965; Meijers 1967). The pattern is repeated throughout the element.

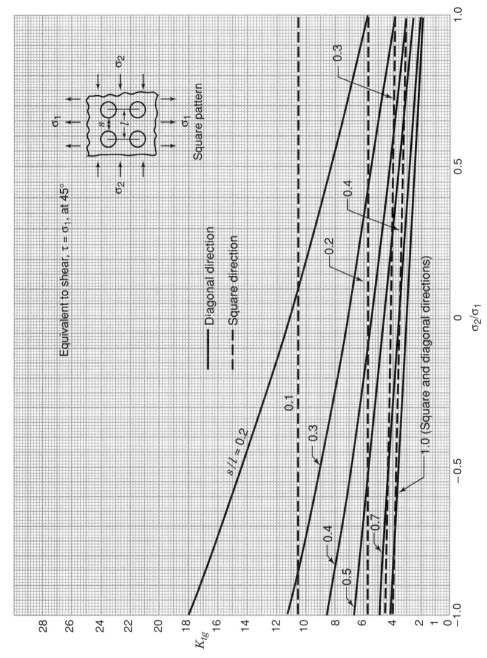

Chart 4.42 Stress concentration factors K_{tg} versus σ_2/σ_1 for a square pattern of holes in a thin element subject to biaxial stresses (Sampson 1960; Bailey and Hicks 1960; Hulbert 1965; Meijers 1967). The pattern is repeated throughout the element.

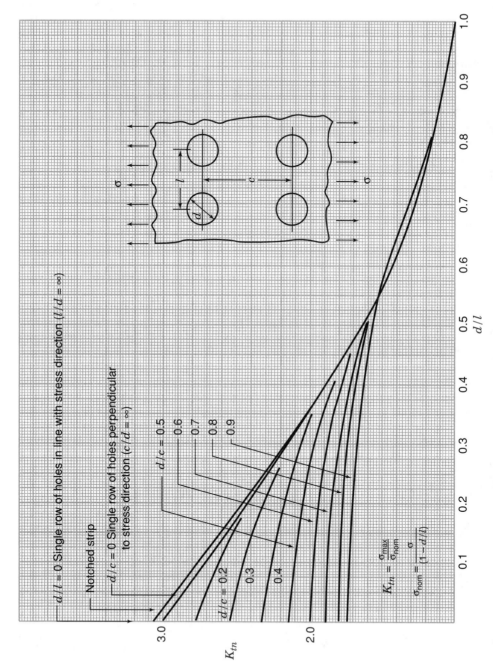

Chart 4.43 Stress concentration factors K_{tn} for a rectangular pattern of holes in a thin element subject to uniaxial stresses (Meijers 1967). The pattern is repeated throughout the element.

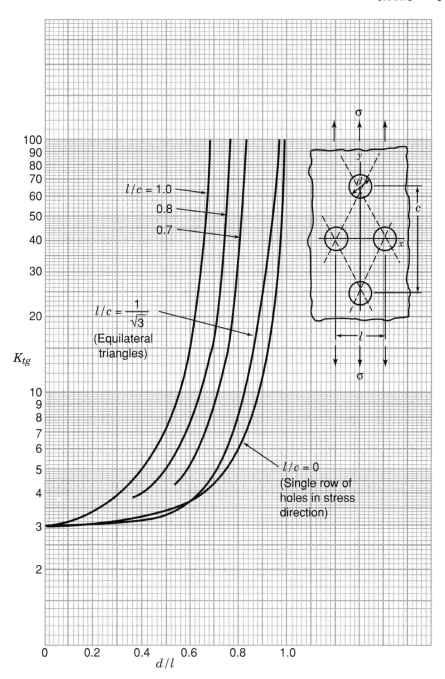

Chart 4.44 Stress concentration factors K_{tg} for a diamond pattern of holes in a thin element subject to uniaxial stresses in y direction (Meijers 1967). The pattern is repeated throughout the element.

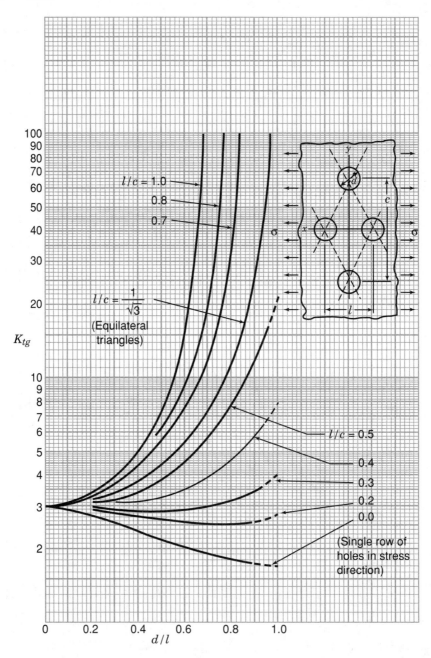

Chart 4.45 Stress concentration factors K_{tg} for a diamond pattern of holes in a thin element subject to uniaxial stresses in x direction (Meijers, 1967). The pattern is repeated throughout the element.

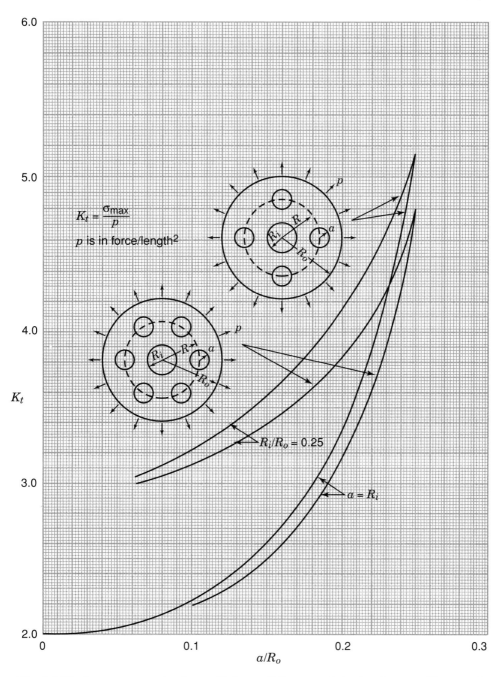

Chart 4.46 Stress concentration factors K_t for a radially stressed circular element, with a central circular hole and a ring of four or six noncentral circular holes, $R/R_0 = 0.625$ (Kraus 1963).

316 HOLES

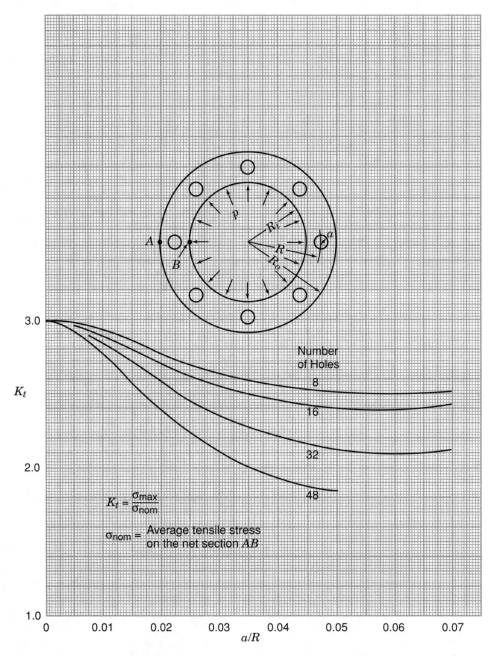

Chart 4.47 Stress concentration factors K_t for a perforated flange with internal pressure, $R_i/R_0 = 0.8$, $R/R_0 = 0.9$ (Kraus et al. 1966).

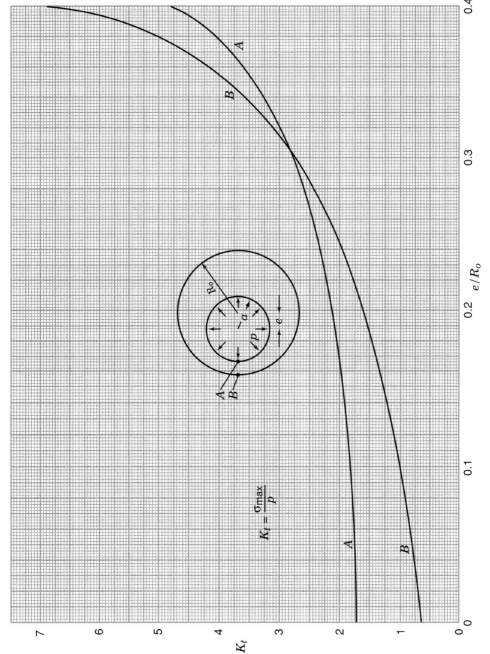

Chart 4.48 Stress concentration factors K_t for a circular thin element with an eccentric circular hole with internal pressure, $a/R_0 = 0.5$ (Savin 1961; Hulbert 1965).

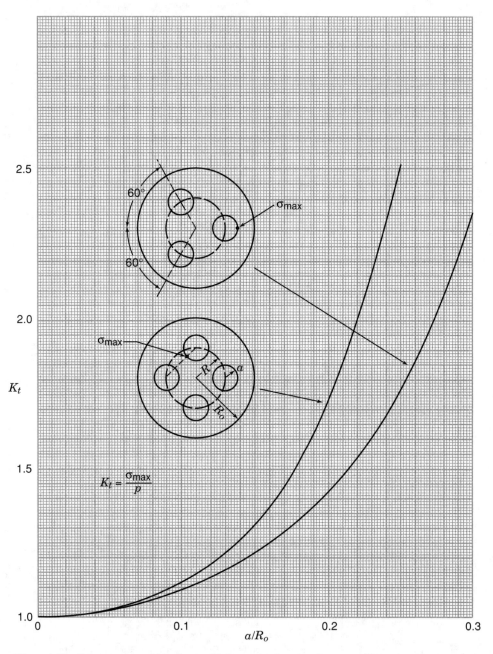

Chart 4.49 Stress concentration factors K_t for a circular thin element with a circular pattern of three or four holes with internal pressure in each hole, $R/R_0 = 0.5$ (Kraus 1962).

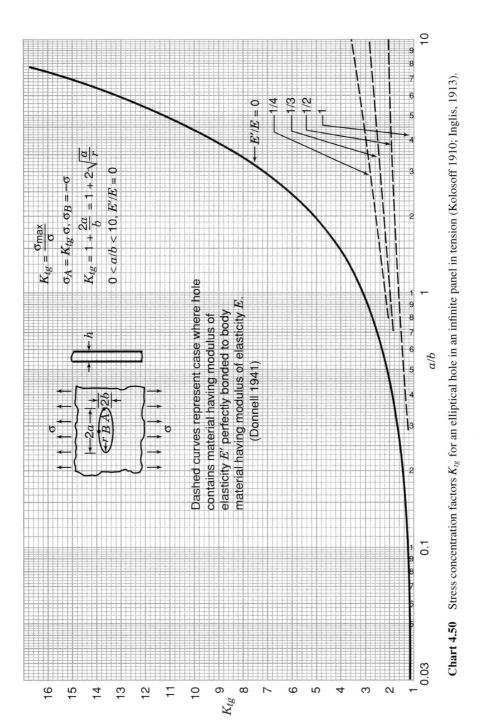

Chart 4.50 Stress concentration factors K_{tg} for an elliptical hole in an infinite panel in tension (Kolosoff 1910; Inglis, 1913).

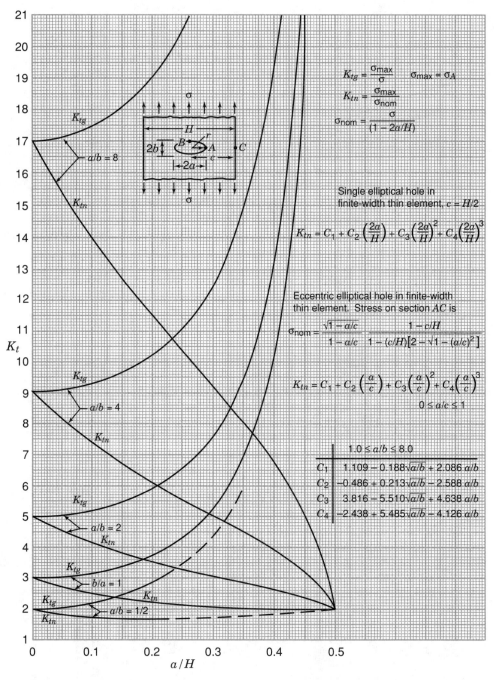

Chart 4.51 Stress concentration factors K_{tg} and K_{tn} of an elliptical hole in a finite-width thin element in uniaxial tension (Isida 1953, 1955b).

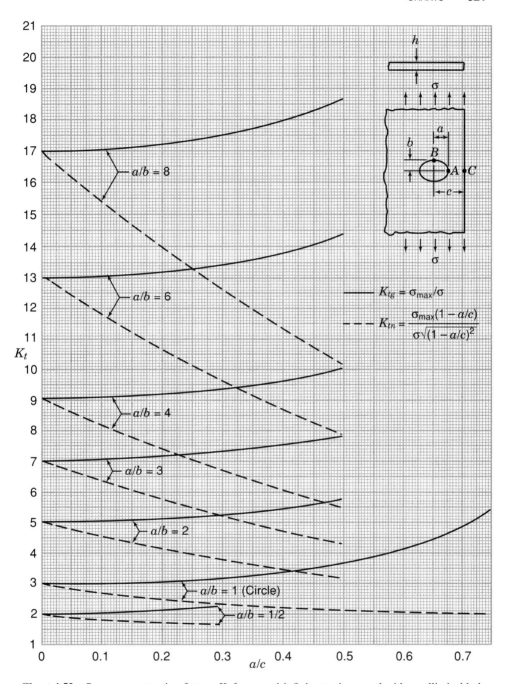

Chart 4.52 Stress concentration factors K_t for a semi-infinite tension panel with an elliptical hole near the edge (Isida 1955a).

322 HOLES

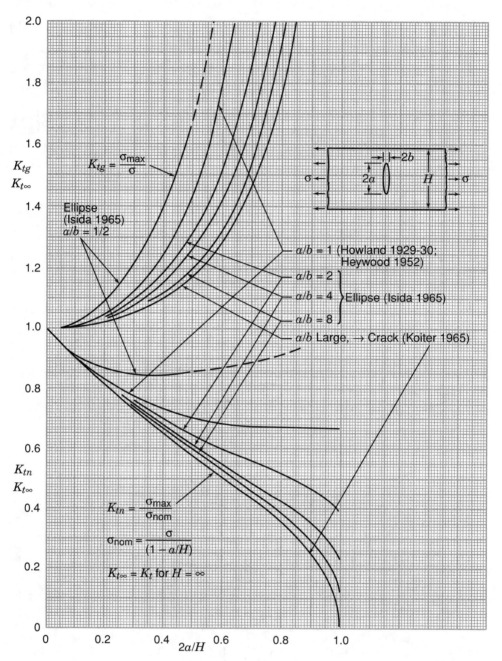

Chart 4.53 Finite-width correction factor $K_t/K_{t\infty}$ for a tension strip with a central opening.

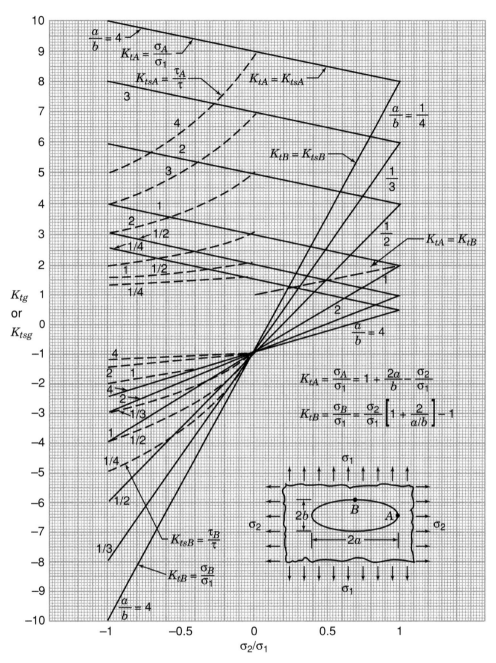

Chart 4.54 Stress concentration factors K_t and K_{ts} for an elliptical hole in a biaxially stressed panel.

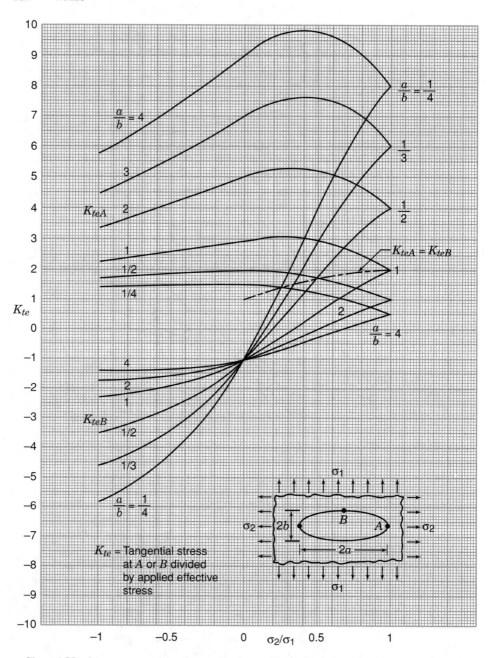

Chart 4.55 Stress concentration factors K_{te} for an elliptical hole in a biaxially stressed panel.

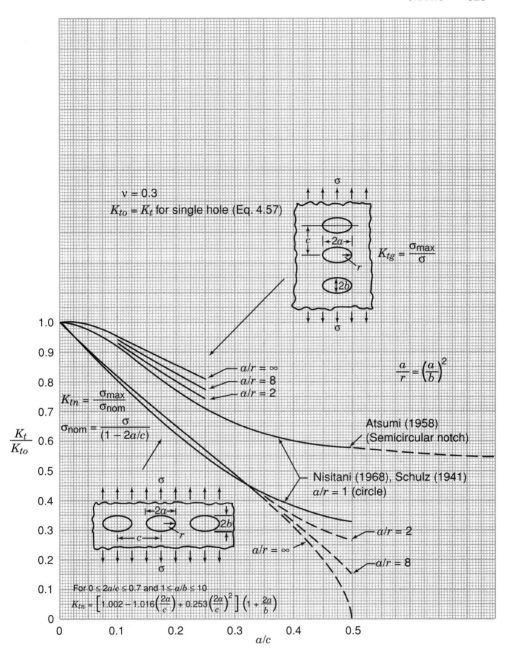

Chart 4.56 Effect of spacing on the stress concentration factor of an infinite row of elliptical holes in an infinite tension member (Nisitani 1968; Schulz 1941).

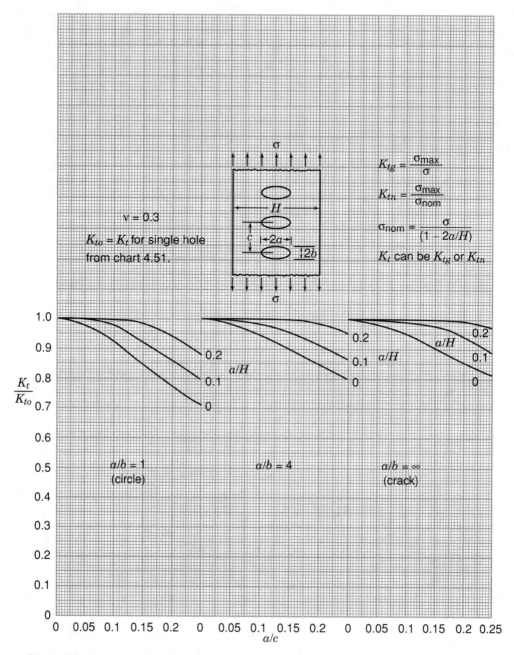

Chart 4.57 Effect of spacing on the stress concentration factor of an infinite row of elliptical holes in a finite-width thin element in tension (Nisitani 1968).

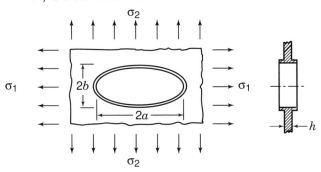

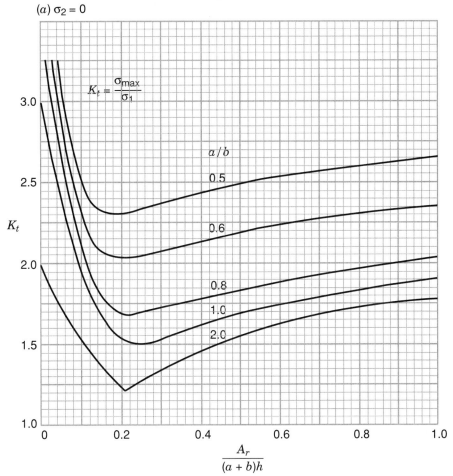

Chart 4.58a Stress concentration factors K_t of elliptical holes with bead reinforcement in an infinite panel under uniaxial and biaxial stresses (Wittrick 1959a, b; Houghton and Rothwell 1961; ESDU 1981). (*a*) $\sigma_2 = 0$.

328 HOLES

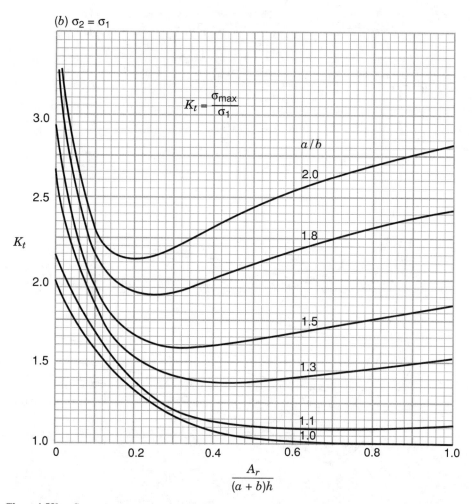

Chart 4.58b Stress concentration factors K_t of elliptical holes with bead reinforcement in an infinite panel under uniaxial and biaxial stresses (Wittrick 1959a, b; Houghton and Rothwell 1961; ESDU 1981). (b) $\sigma_2 = \sigma_1$.

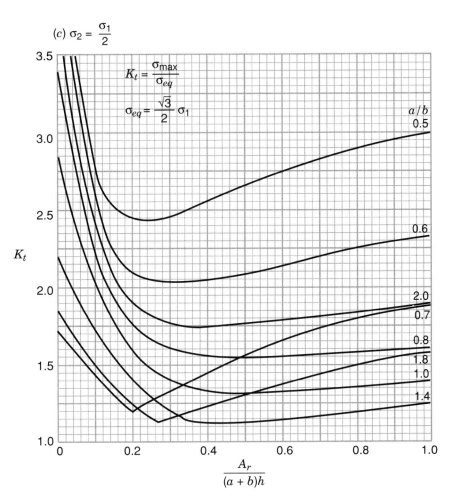

Chart 4.58c Stress concentration factors K_t of elliptical holes with bead reinforcement in an infinite panel under uniaxial and biaxial stresses (Wittrick 1959a, b; Houghton and Rothwell 1961; ESDU 1981). (c) $\sigma_2 = \frac{\sigma_1}{2}$.

330 HOLES

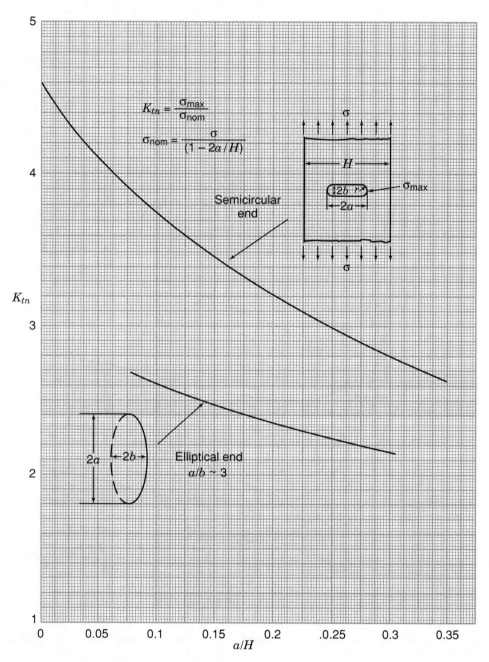

Chart 4.59 Optimization of slot end, $a/b = 3.24$ (Durelli et al. 1968).

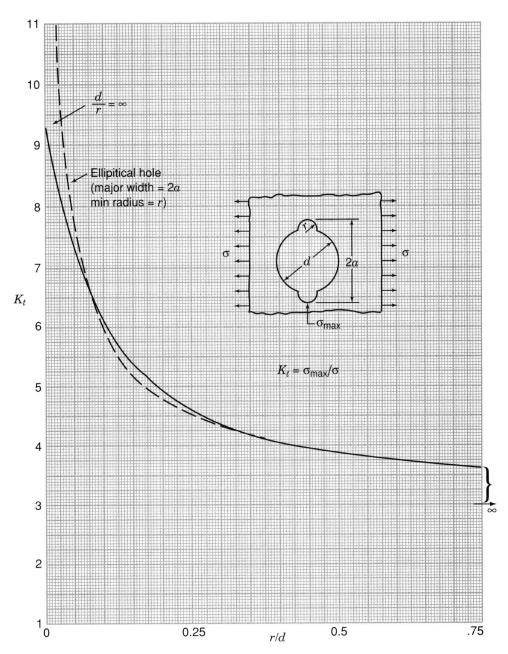

Chart 4.60 Stress concentration factor K_t for an infinitely wide tension element with a circular hole with opposite semicircular lobes (from data of Mitchell 1966).

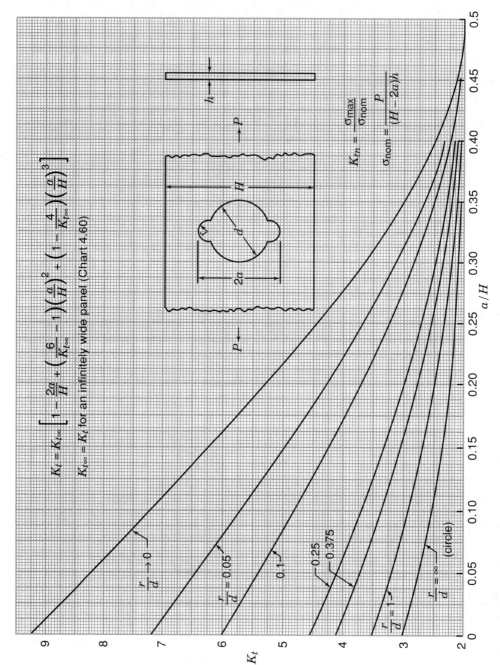

Chart 4.61 Stress concentration factors K_t for a finite-width tension thin element with a circular hole with opposite semicircular lobes (Mitchell 1966 formula).

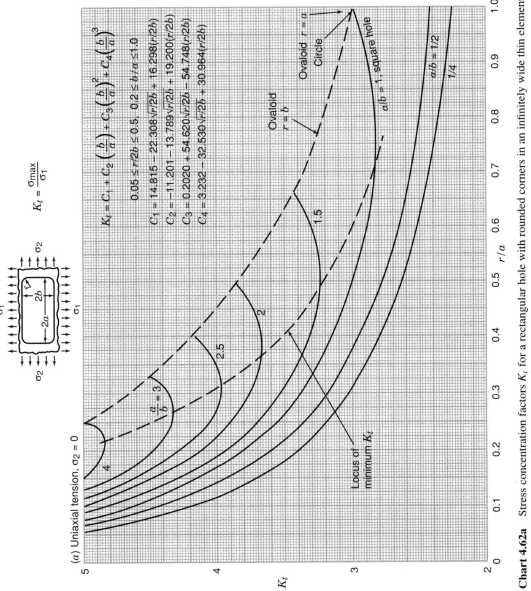

Chart 4.62a Stress concentration factors K_t for a rectangular hole with rounded corners in an infinitely wide thin element (Sobey 1963; ESDU 1970). (*a*) Uniaxial tension, $\sigma_2 = 0$.

(b) Unequal biaxial tension, $\sigma_2 = \sigma_1/2$

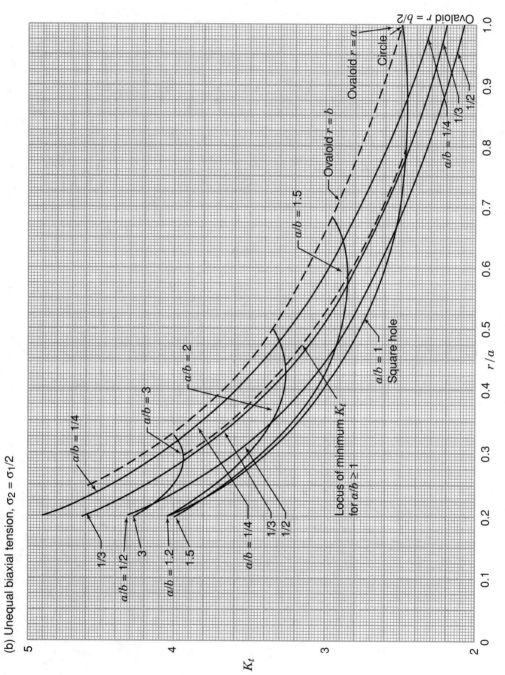

Chart 4.62b Stress concentration factors K_t for a rectangular hole with rounded corners in an infinitely wide thin element (Sobey 1963; ESDU 1970). (b) unequal biaxial tension, $\sigma_2 = \sigma_1/2$.

334

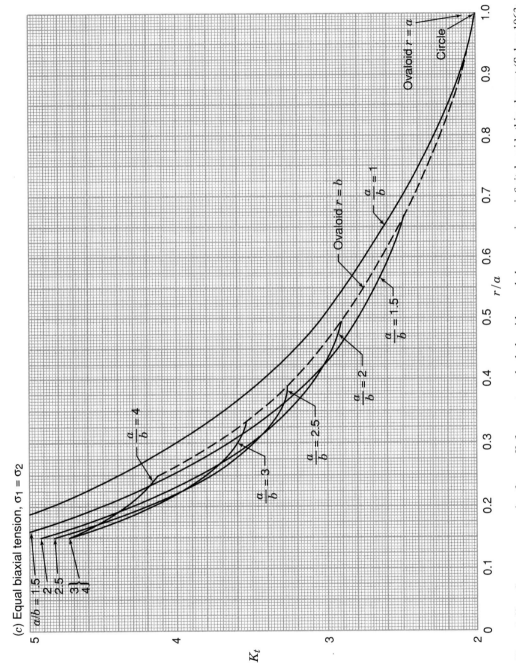

Chart 4.62c Stress concentration factors K_t for a rectangular hole with rounded corners in an infinitely wide thin element (Sobey 1963; ESDU 1970). (c) equal biaxial tension, $\sigma_1 = \sigma_2$.

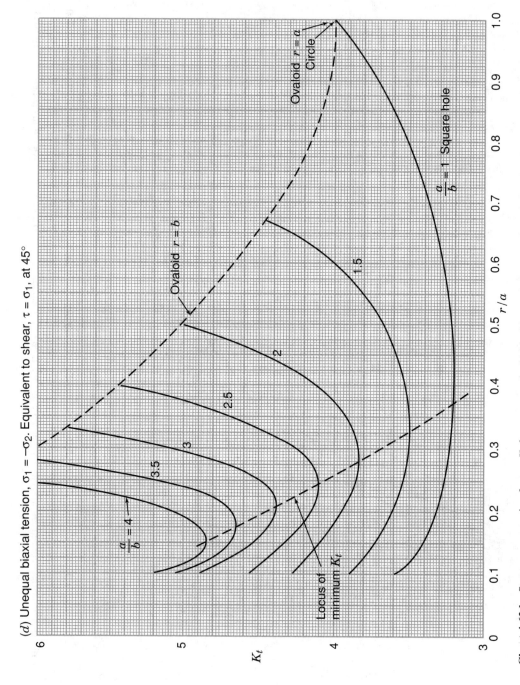

(d) Unequal biaxial tension, $\sigma_1 = -\sigma_2$. Equivalent to shear, $\tau = \sigma_1$, at 45°

Chart 4.62d Stress concentration factors K_t for a rectangular hole with rounded corners in an infinitely wide thin element (Sobey 1963; ESDU 1970). (d) unequal biaxial tension, $\sigma_1 = -\sigma_2$. Equivalent to shear, $\tau = \sigma_1$, at 45°.

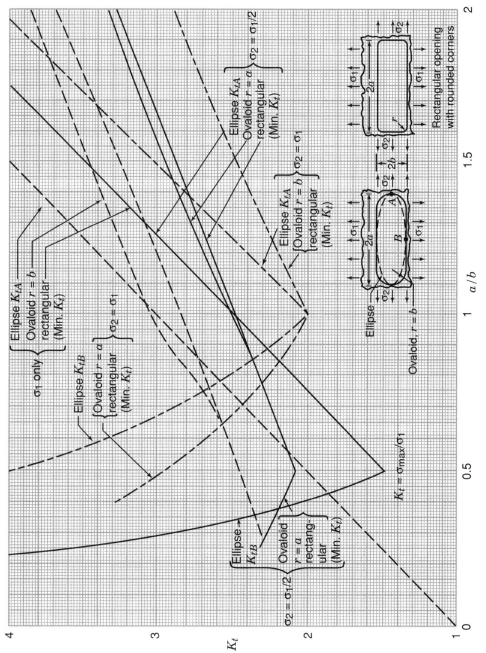

Chart 4.63 Comparison of stress concentration factors of various shaped holes.

338 HOLES

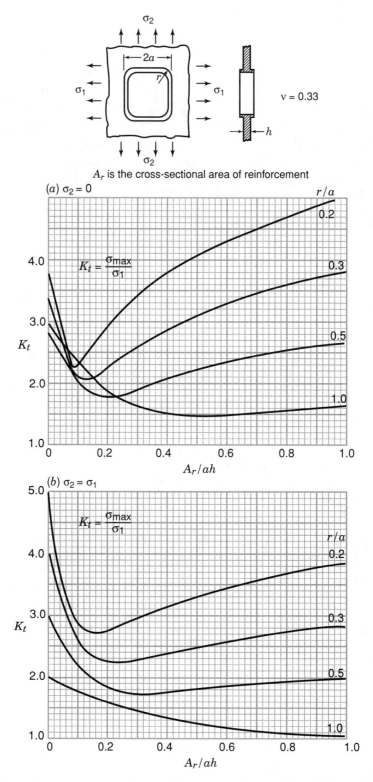

Chart 4.64a,b Stress concentration factors of round-cornered square holes with bead reinforcement in an infinite panel under uniaxial or biaxial stresses (Sobey 1968; ESDU 1981). (*a*) $\sigma_2 = 0$; (*b*) $\sigma_2 = \sigma_1$.

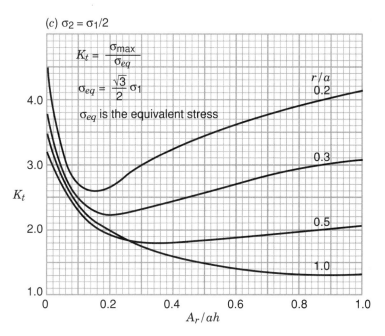

Chart 4.64c Stress concentration factors of round-cornered square holes with bead reinforcement in an infinite panel under uniaxial or biaxial stresses (Sobey 1968; ESDU 1981). (c) $\sigma_2 = \sigma_1/2$.

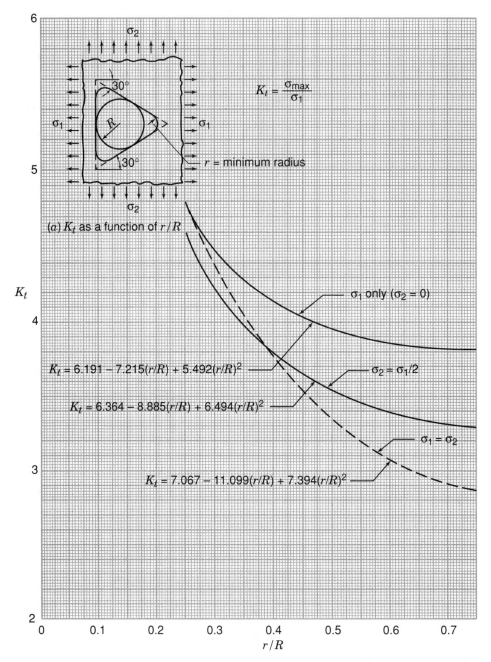

Chart 4.65a Stress concentration factors K_t for an equilateral triangular hole with rounded corners in an infinite thin element (Wittrick 1963). (*a*) K_t as a function of r/R.

(b) K_t as a function of σ_2/σ_1

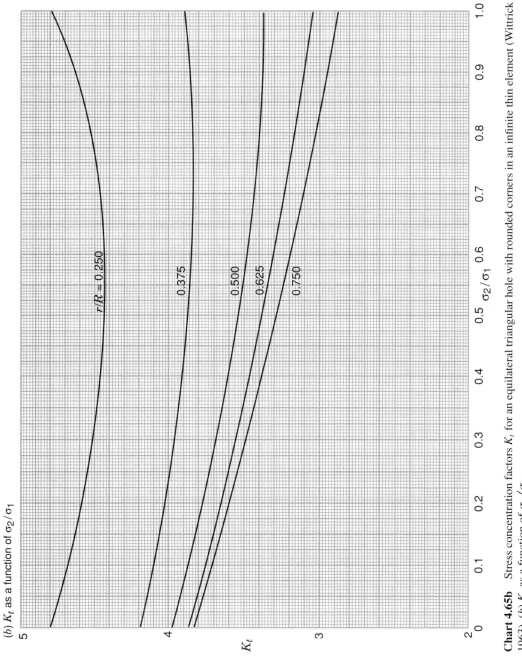

Chart 4.65b Stress concentration factors K_t for an equilateral triangular hole with rounded corners in an infinite thin element (Wittrick 1963). (b) K_t as a function of σ_2/σ_1.

342 HOLES

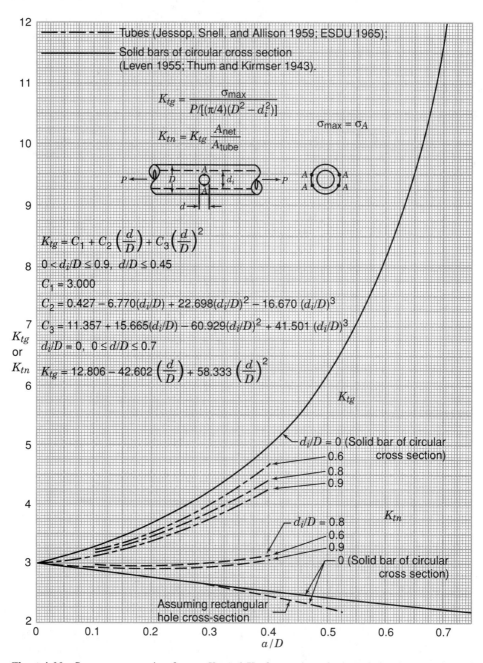

Chart 4.66 Stress concentration factors K_{tg} and K_{tn} for tension of a bar of circular cross section or tube with a transverse hole. Tubes (Jessop, Snell, and Allison 1959; ESDU 1965); Solid bars of circular cross section (Leven 1955; Thum and Kirmser 1943).

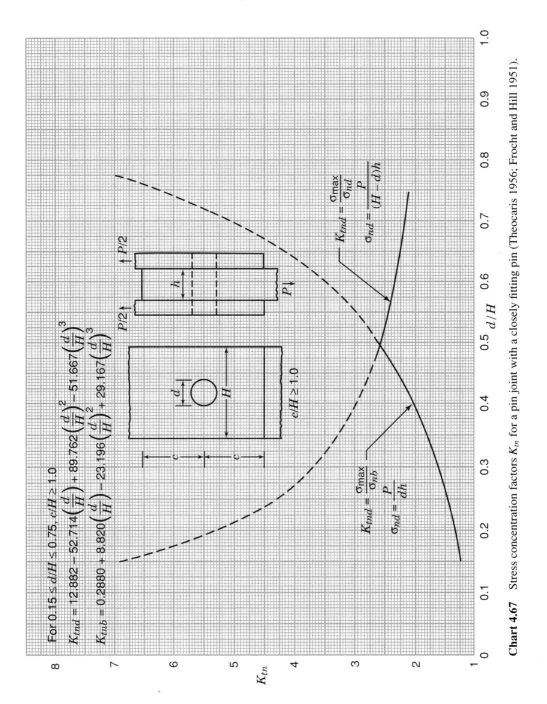

Chart 4.67 Stress concentration factors K_{tn} for a pin joint with a closely fitting pin (Theocaris 1956; Frocht and Hill 1951).

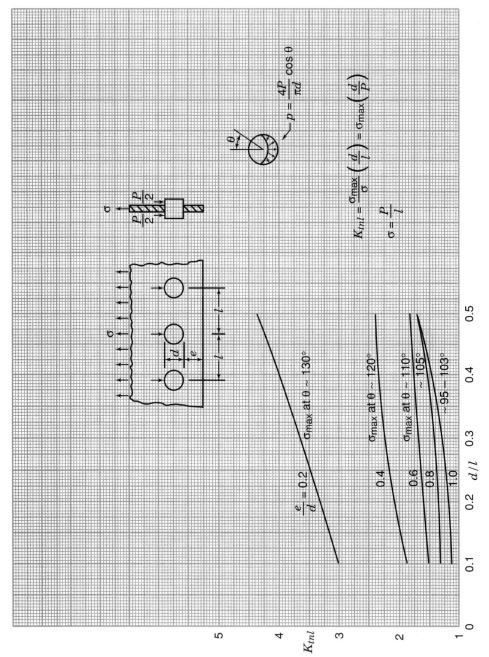

Chart 4.68 Stress concentration factors K_{tn} for a pinned or riveted joint with multiple holes (from data of Mori 1972).

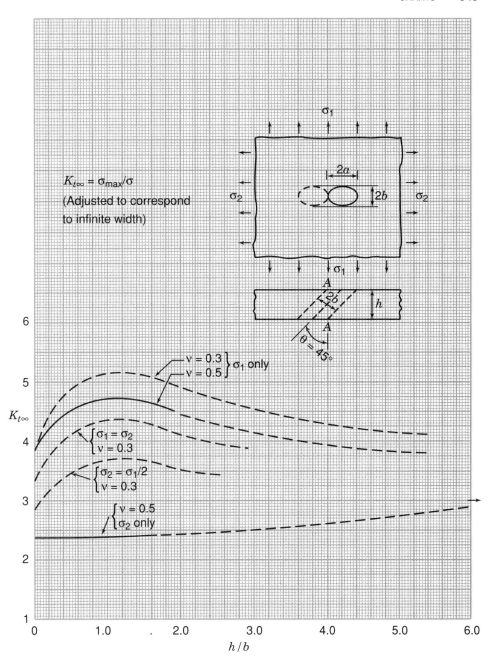

Chart 4.69 Stress concentration factors $K_{t\infty}$ for a circular hole inclined 45° from perpendicularity to the surface of an infinite panel subjected to various states of tension (based on Leven 1970; Daniel 1970; McKenzie and White 1968; Ellyin 1970).

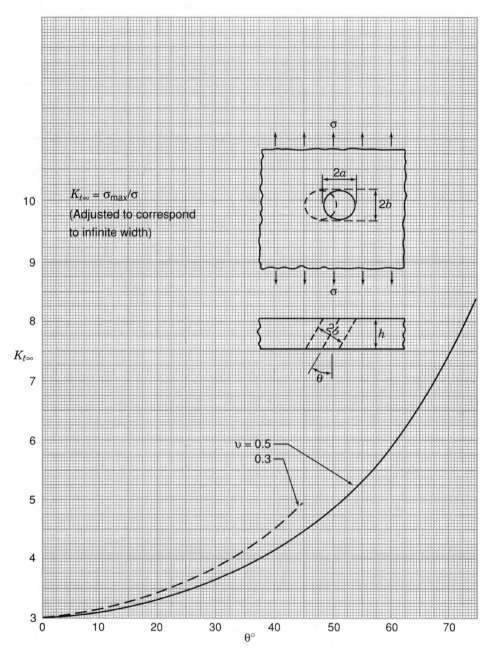

Chart 4.70 Stress concentration factors $K_{t\infty}$ for a circular hole inclined $\theta°$ from perpendicularity to the surface of an infinite panel subjected to tension, $h/b = 1.066$ (based on McKenzie and White 1968; Ellyin 1970).

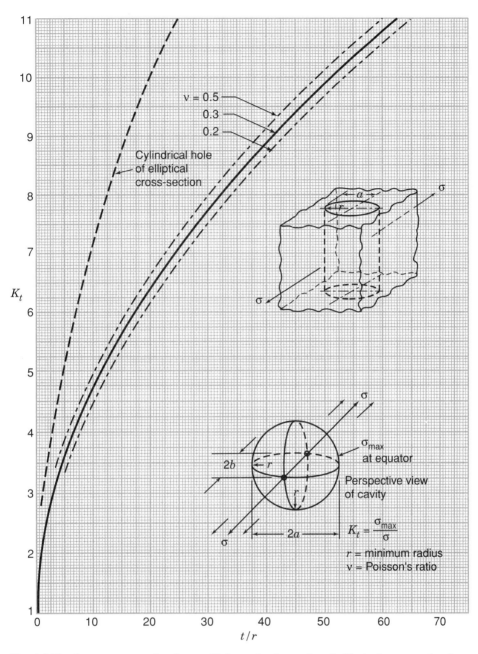

Chart 4.71 Stress concentration factors K_t for a circular cavity of elliptical cross section in an infinite body in tension (Neuber 1958).

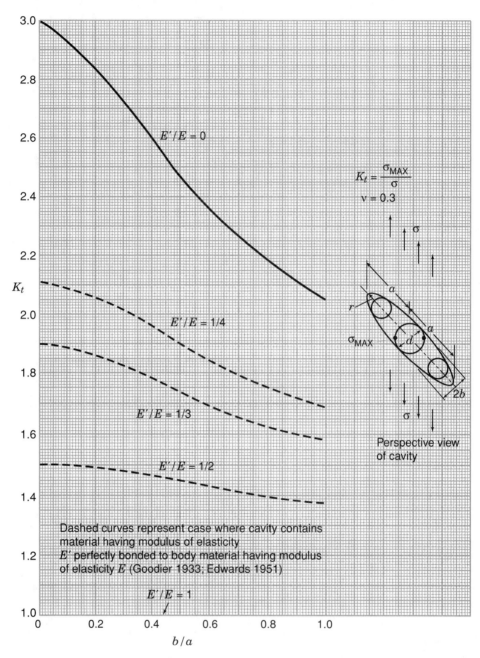

Chart 4.72 Stress concentration factors K_t for an ellipsoidal cavity of circular cross section in an infinite body in tension (mathematical analysis of Sadowsky and Sternberg 1947).

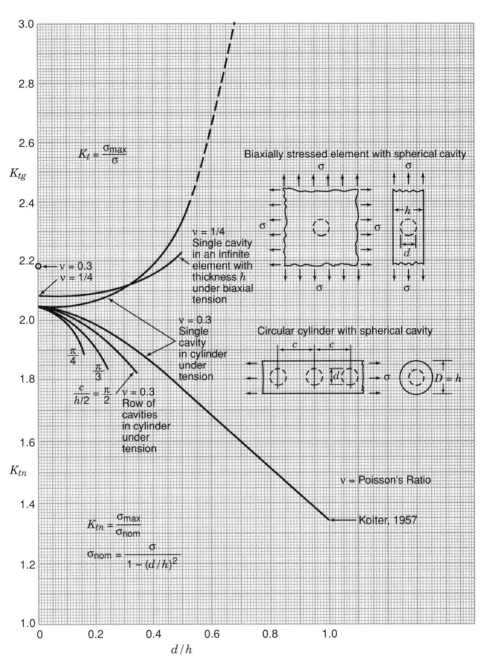

Chart 4.73 Stress concentration factors K_{tg} and K_{tn} for spherical cavities in finite-width flat elements and cylinders (Ling 1959; Atsumi 1960).

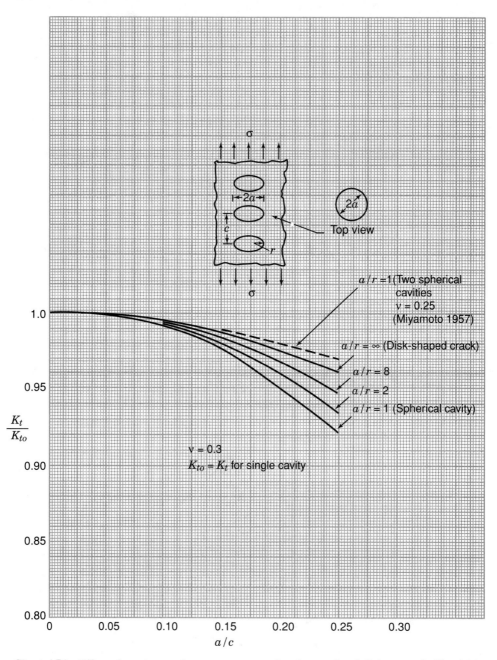

Chart 4.74 Effect of spacing on the stress concentration factor of an infinite row of ellipsoidal cavities in an infinite tension member (Miyamoto 1957).

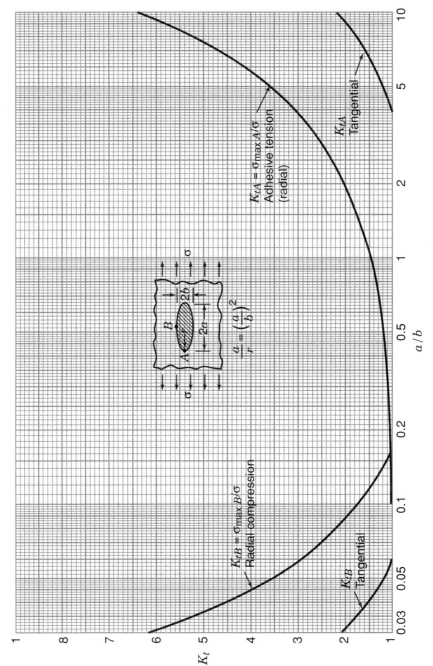

Chart 4.75 Stress concentration factors K_t for an infinite member in tension having a rigid elliptical cylindrical inclusion (Goodier 1933; Donnell 1941; Nisitani 1968).

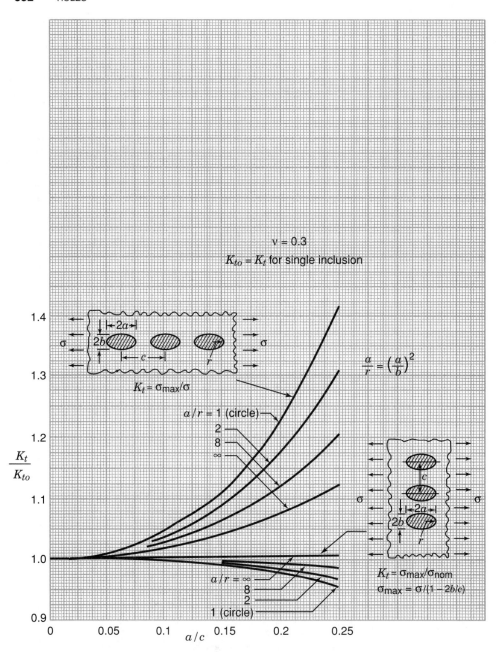

Chart 4.76 Effect of spacing on the stress concentration factor of an infinite tension panel with an infinite row of rigid cylindrical inclusions (Nisitani 1968).

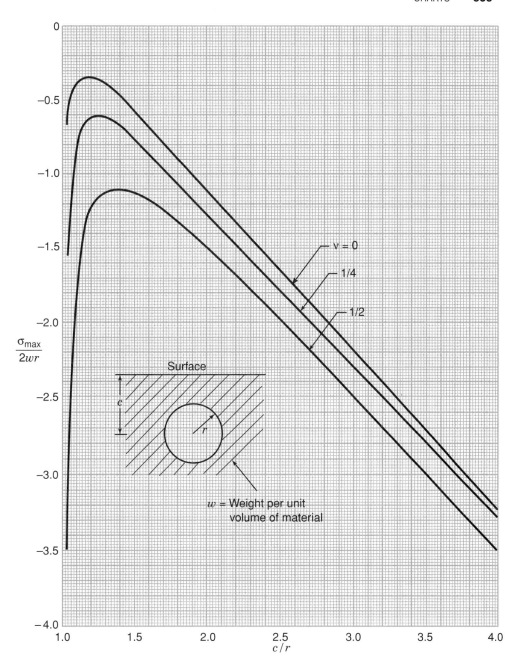

Chart 4.77 Maximum peripheral stress in a cylindrical tunnel subjected to hydrostatic pressure due to weight of material (Mindlin 1939).

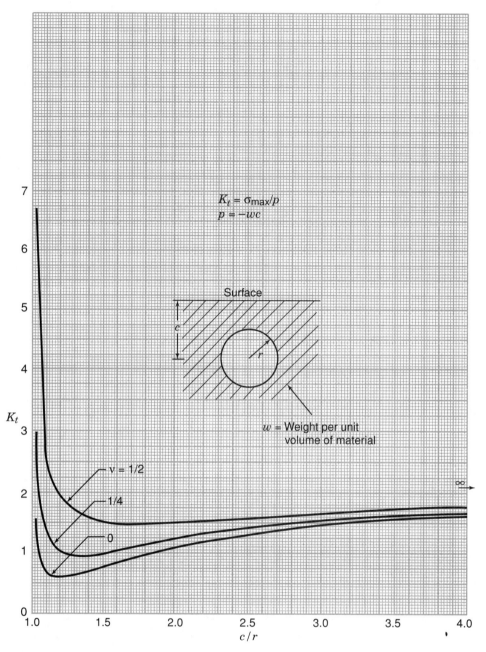

Chart 4.78 Stress concentration factors K_t for a cylindrical tunnel subjected to hydrostatic pressure due to weight of material (based on Chart 4.77).

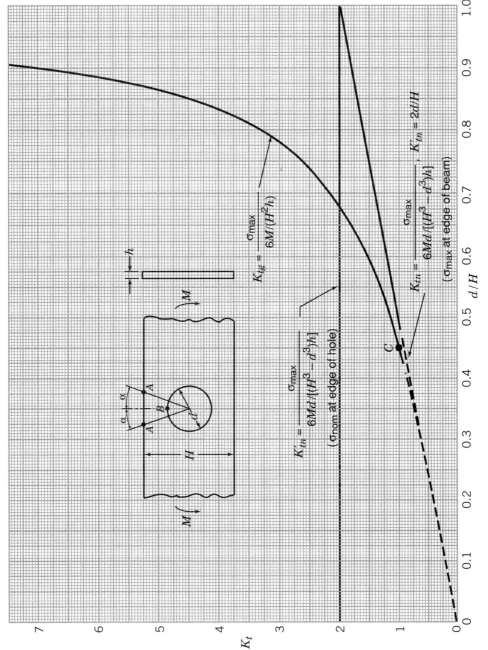

Chart 4.79 Stress concentration factors K_t for bending of a thin beam with a central circular hole (Howland and Stevenson 1933; Heywood 1952).

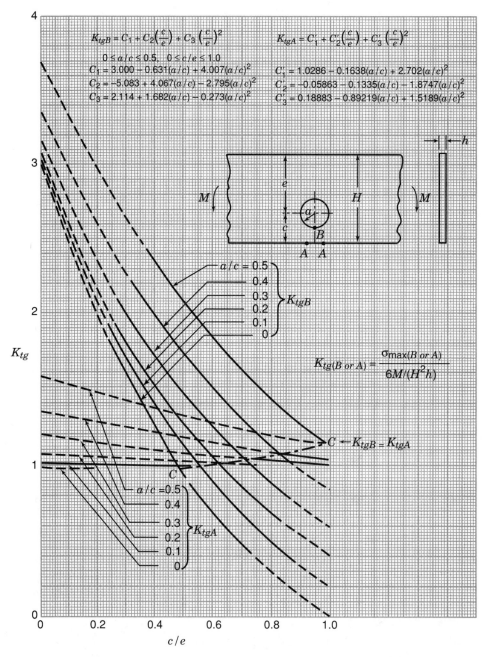

Chart 4.80 Stress concentration factors K_{tg} for bending of a thin beam with a circular hole displaced from the center line (Isida 1952).

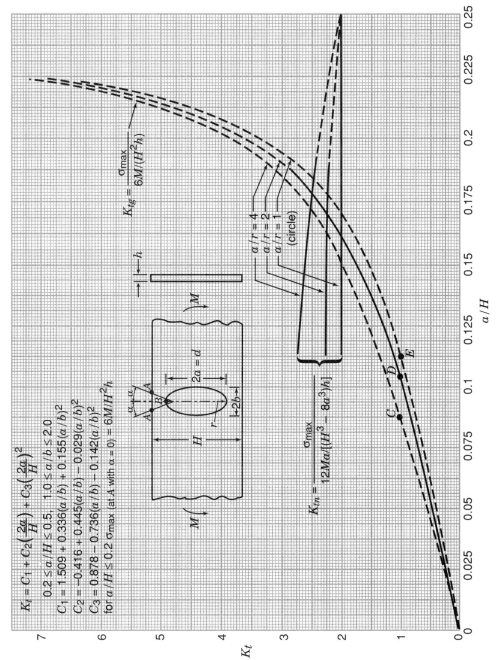

Chart 4.81 Stress concentration factors K_{tg} and K_{tn} for bending of a beam with a central elliptical hole (from data of Isida 1953).

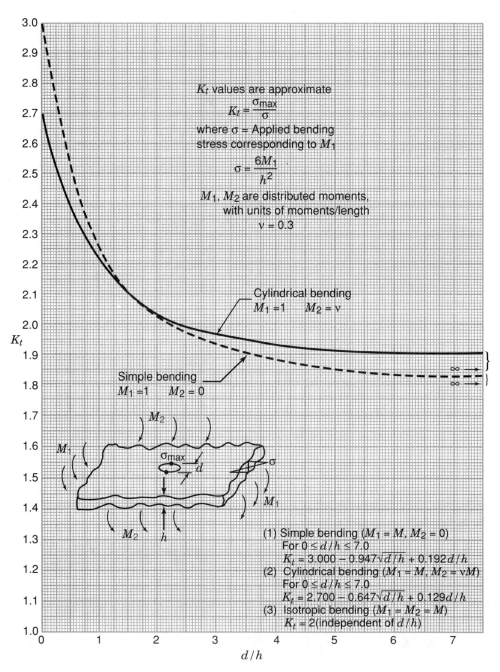

Chart 4.82 Stress concentration factors K_t for bending of an infinite plate with a hole (Goodier 1936; Reissner 1945).

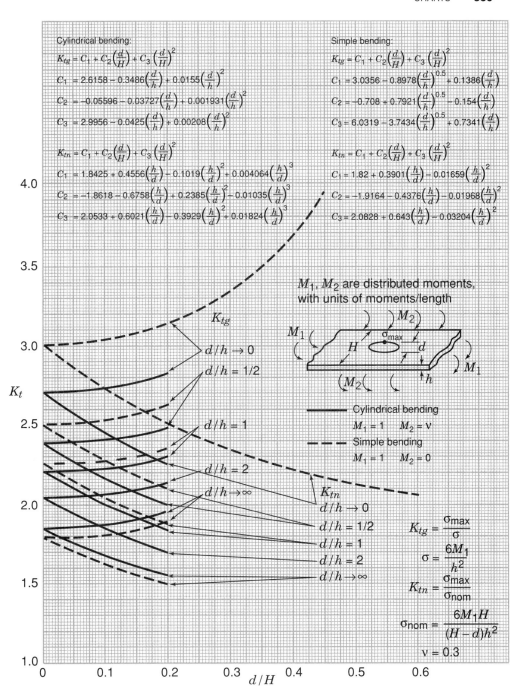

Chart 4.83 Stress concentration factors K_{tg} and K_{tn} for bending of a finite-width plate with a circular hole.

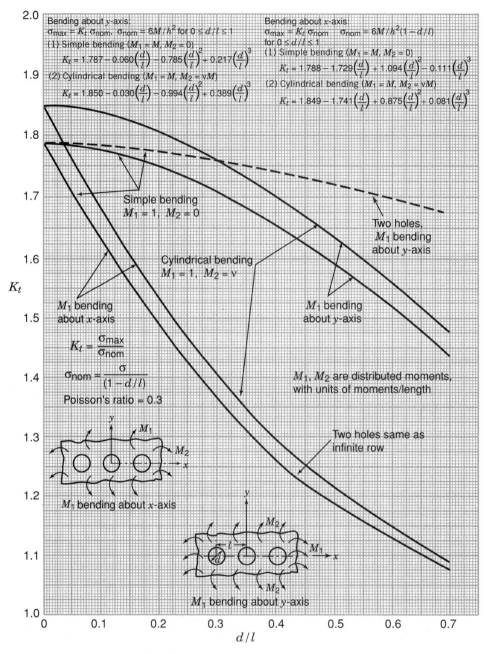

Chart 4.84 Stress concentration factors for an infinite row of circular holes in an infinite plate in bending (Tamate 1957 and 1958).

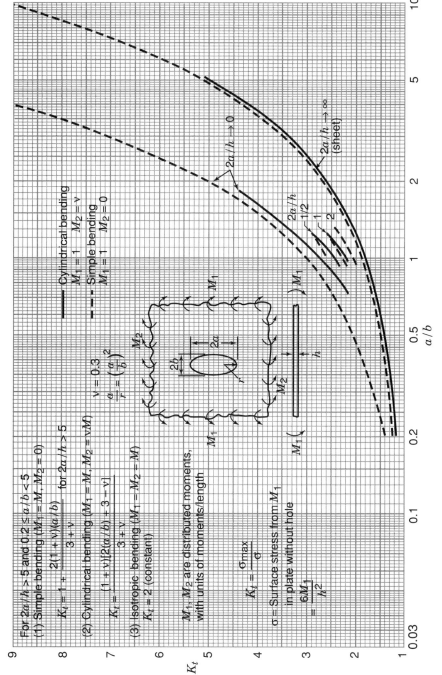

Chart 4.85 Stress concentration factors K_t for bending of an infinite plate having an elliptical hole (Neuber 1958; Nisitani 1968).

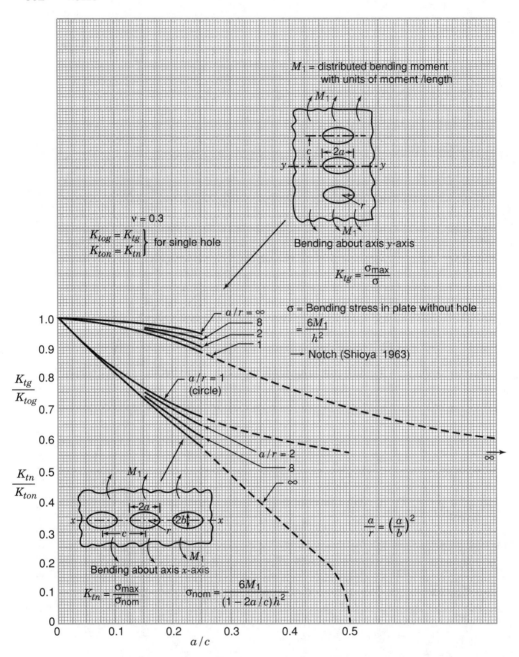

Chart 4.86 Effects of spacing on the stress concentration factors of an infinite row of elliptical holes in an infinite plate in bending (Nisitani 1968; Tamate 1958).

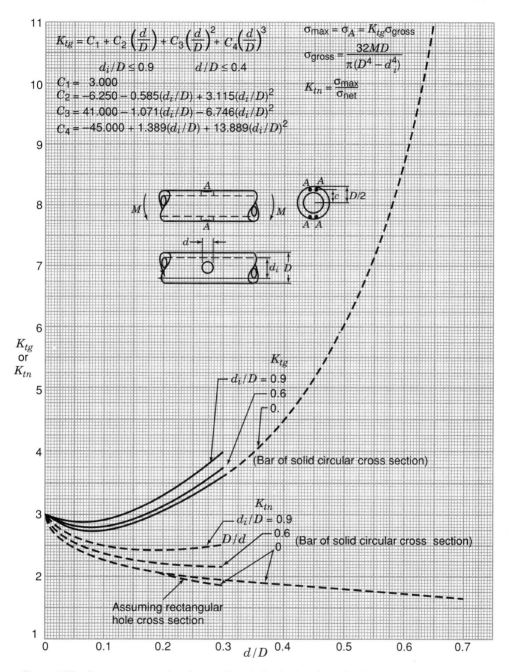

Chart 4.87 Stress concentration factors K_{tg} and K_{tn} for bending of a bar of solid circular cross section or tube with a transverse hole (Jessop, Snell, and Allison 1959; ESDU 1965); bar of solid circular cross section (Thum and Kirmser 1943).

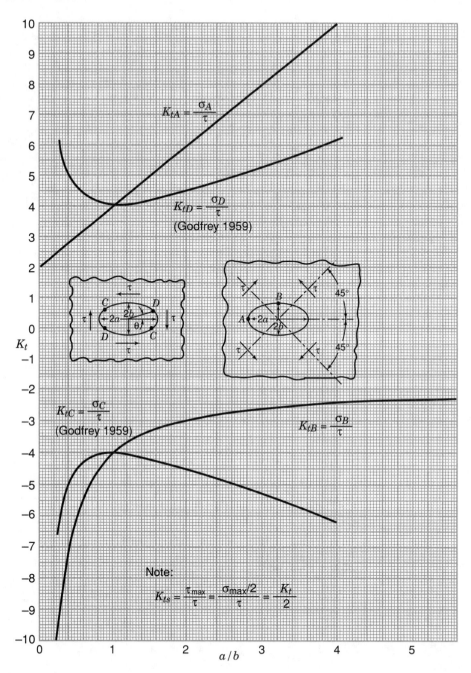

Chart 4.88 Stress concentration factors for an elliptical hole in an infinite thin element subjected to shear stress.

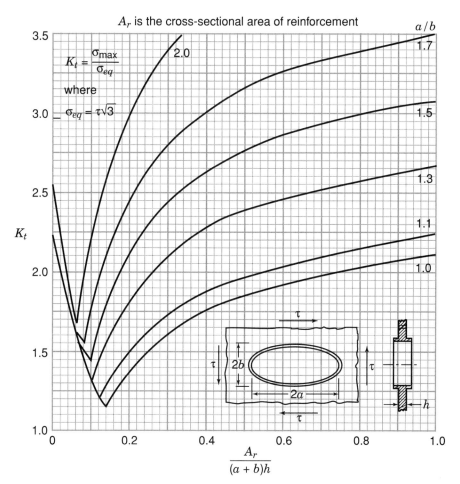

Chart 4.89 Stress concentration factors K_t for elliptical holes with symmetrical reinforcement in an infinite thin element subjected to shear stress (Wittrick 1959a, b; Houghton and Rothwell 1961; ESDU 1981).

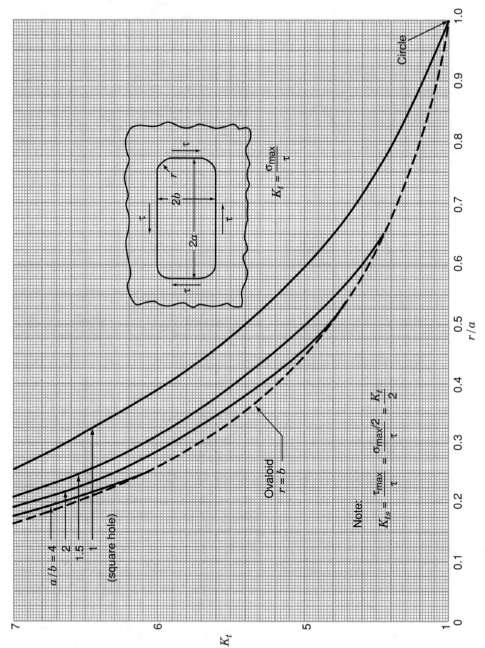

Chart 4.90 Stress concentration factors K_t for a rectangular hole with rounded corners in an infinitely wide thin element subjected to shear stress (Sobey 1963; ESDU 1970).

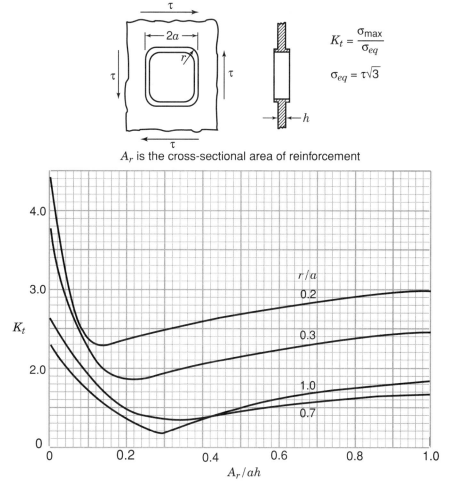

Chart 4.91 Stress concentration factor K_t for square holes with symmetrical reinforcement in an infinite thin element subject to pure shear stress (Sobey 1968; ESDU 1981).

368 HOLES

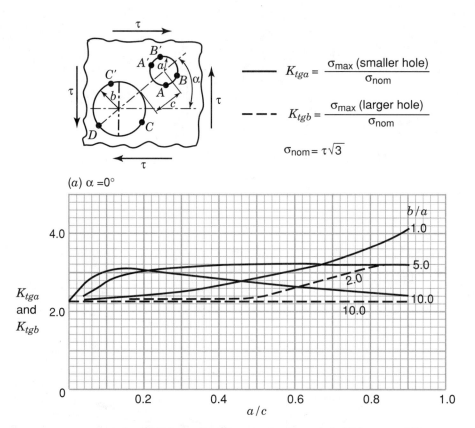

σ_{max} (smaller hole) is located close to A when a/c is high; σ_{max} shifts close to point B at a/c low.
σ_{max} (larger hole) is located close to point C.
Stresses at points A', B' and C' are equal in magnitude, but opposite in sign to those at A, B, and C.

Chart 4.92a Stress concentration factors K_{tg} for pure shear in an infinite thin element with two circular holes of unequal diameter (Haddon 1967; ESDU 1981). (a) $\alpha = 0°$.

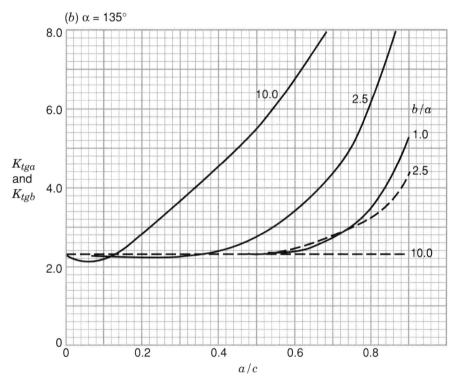

σ_{max} (smaller hole) is located close to A for all values of b/a and a/c.
σ_{max} (larger hole) is located between the points B and C when $a/c > 0.6$, at points B and D when $a/c < 0.6$.

Chart 4.92b Stress concentration factors K_{tg} for pure shear in an infinite thin element with two circular holes of unequal diameter (Haddon 1967; ESDU 1981). (b) $\alpha = 135°$.

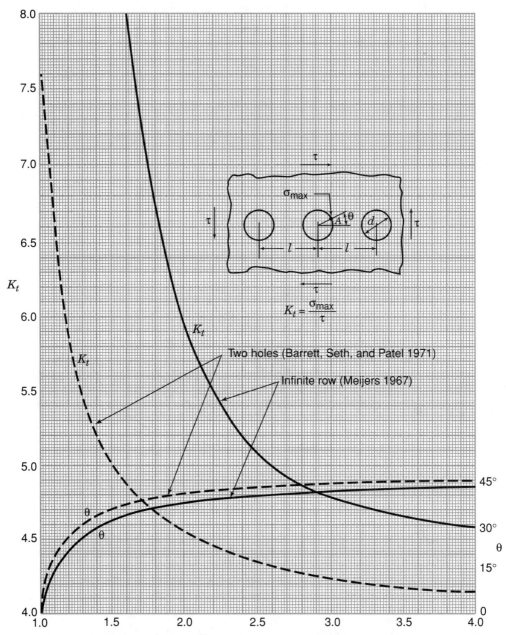

Chart 4.93 Stress concentration factors K_t for single row of circular holes in an infinite thin element subjected to shear stress.

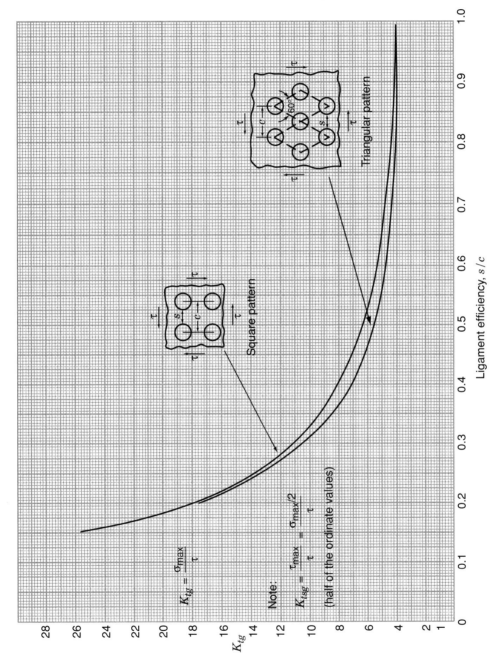

Chart 4.94 Stress concentration factors K_{tg} for an infinite pattern of holes in a thin element subjected to shear stress (Sampson 1960; Bailey and Hicks 1960; Hulbert and Niedenfuhr 1965; Meijers 1967).

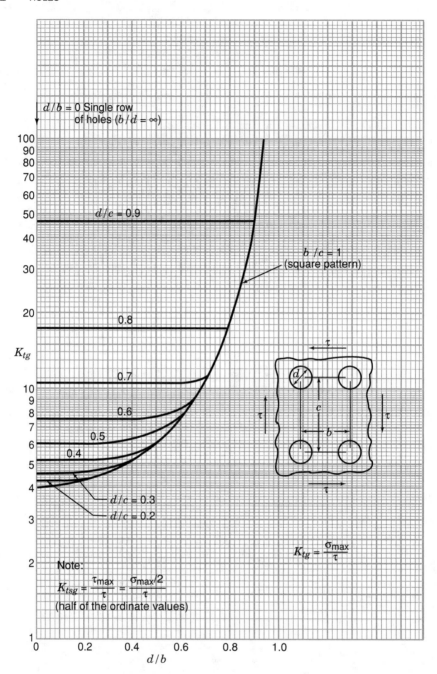

Chart 4.95 Stress concentration factors K_{tg} for an infinite rectangular pattern of holes in a thin element subjected to shear stress (Meijers 1967).

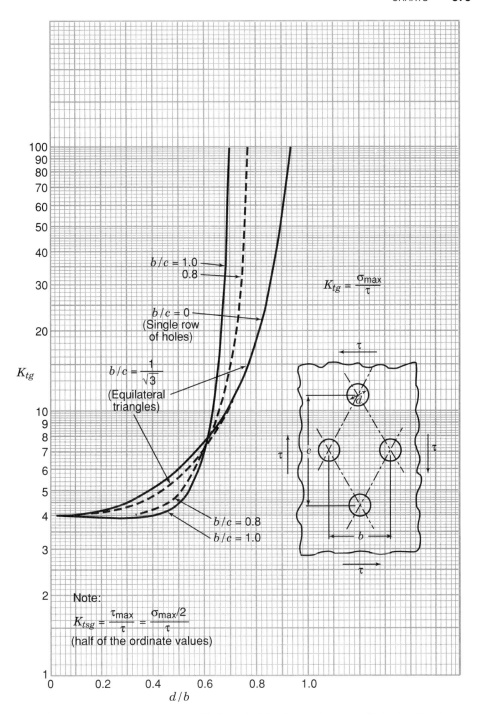

Chart 4.96 Stress concentration factors K_{tg} for an infinite diamond pattern of holes in a thin element subjected to shear stress (Meijers 1967).

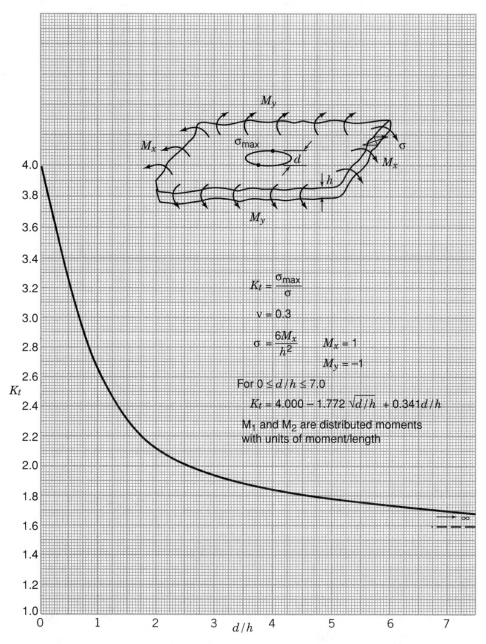

Chart 4.97 Stress concentration factors K_t for a twisted plate with a circular hole (Reissner 1945).

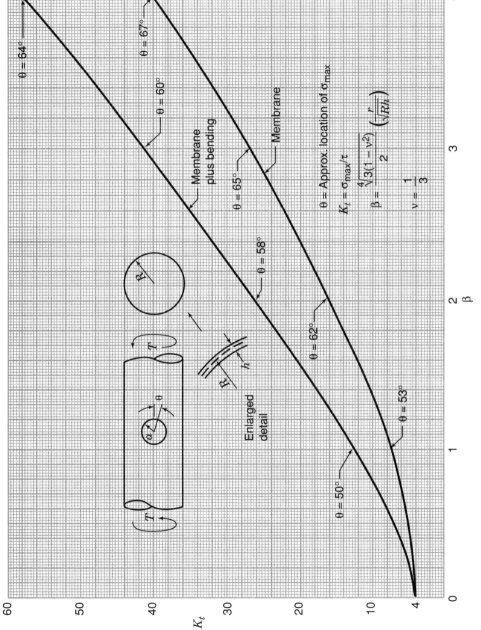

Chart 4.98 Stress concentration factors for a circular hole in a cylindrical shell stressed in torsion (based on data of Van Dyke 1965).

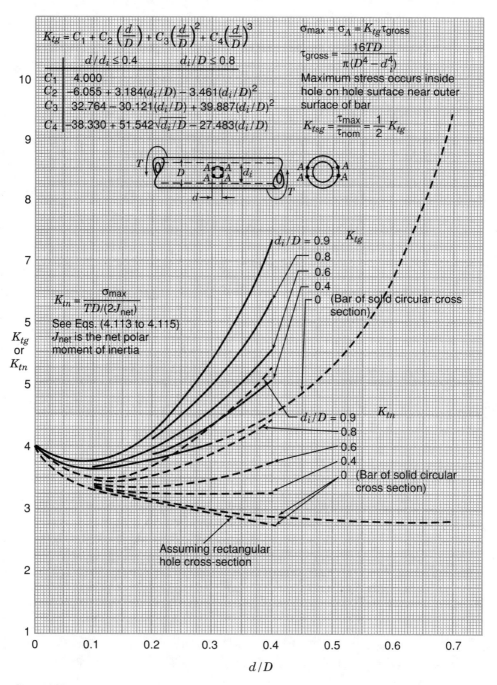

Chart 4.99 Stress concentration factors K_{tg} and K_{tn} for torsion of a tube or bar of circular cross section with transverse hole (Jessop, Snell, and Allison 1959; ESDU 1965; bar of circular cross section Thum and Kirmser 1943).

CHAPTER 5

MISCELLANEOUS DESIGN ELEMENTS

This chapter provides stress concentration factors for various elements used in machine design. These include shafts with keyseats, splined shafts, gear teeth, shrink-fitted members, bolts and nuts, lug joints, curved bars, hooks, rotating disks, and rollers.

5.1 NOTATION

Symbols:

a = wire diameter

b = keyseat width; tooth length; width of cross section

c = distance from center of hole to lug end for a lug-pin fit; distance from centroidal axis to outer fiber of a beam cross section; spring index

D = larger shaft diameter

d = shaft diameter; diameter of hole; mean coil diameter

E_{lug} = moduli of elasticity of lug

E_{pin} = moduli of elasticity of pin

e = pin-to-hole clearance as a percentage of d, the hole diameter; gear contact position height

h = thickness

I = moment of inertia of beam cross section

K_t = stress concentration factor

K_{te} = stress concentration factor for axially loaded lugs at a pin-to-hole clearance of e percent of the pin hole ($h/d < 0.5$)

377

K'_{te} = stress concentration factor for axially loaded lugs at a pin-to-hole clearance of e percent of the pin hole ($h/d > 0.5$)

K_f = fatigue notch factor

l = length

L = length (see Chart 5.10)

M = bending moment

m = thickness of bolt head

N = number of teeth

P = axial load

P_d = diametral pitch

p = pressure

R_1 = inner radius of cylinder

R_2 = outer radius of cylinder

r = radius

r_f = minimum radius

r_s = shaft shoulder fillet radius

r_t = tip radius

T = torque

t = keyseat depth; width of tooth

W = width of lug

w = gear tooth horizontal load

w_n = gear tooth normal load

η_e = correction factor for the stress concentration factor of lug-pin fits

σ = stress

σ_{nom} = nominal stress

σ_{max} = maximum stress

ν = Poisson's ratio

5.2 SHAFT WITH KEYSEAT

The U.S. standard keyseat (keyway) (ANSI 1967) has approximate average[1] values of $b/d = 1/4$ and $t/d = 1/8$ for shaft diameters up to 6.5 in. (Fig. 5.1). For shaft diameters above 6.5 in., the approximate average values are $b/d = 1/4$ and $t/d = 0.09$. The suggested approximate fillet radius proportions are $r/d = 1/48 = 0.0208$ for shaft diameters up to 6.5 in. and $r/d = 0.0156$ for shaft diameters above 6.5 in.

In design of a keyed shaft, one must also take into consideration the shape of the end of the keyseat. Two types of keyseat ends are shown in Fig. 5.2. The end-milled keyseat is more widely used, perhaps owing to its compactness and definite key positioning longitudinally. However, in bending, the stress concentration factor is lower for the sled-runner keyseat.

---

[1] The keyseat width, depth, and fillet radius are in multiples of 1/32 in., each size applying to a range of shaft diameters.

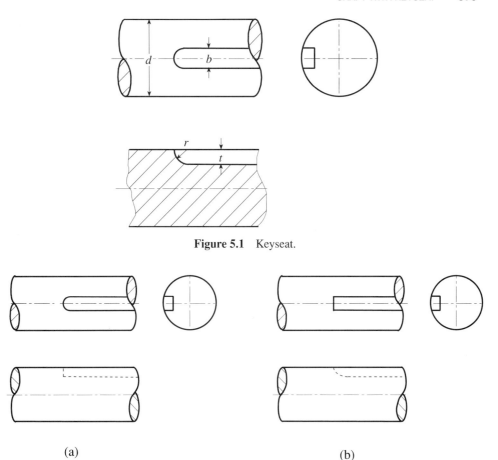

Figure 5.1 Keyseat.

Figure 5.2 Types of keyseat ends: (*a*) End milled keyseat (also, referred to as semicircular or profiled end); (*b*) sled-runner keyseat.

5.2.1 Bending

A comparison of the surface stresses of the two types of keyseat in bending ($b/d = 0.313$, $t/d = 0.156$) was made photoelastically by Hetényi (1939) who found $K_t = 1.79$ (semicircular end) and $K_t = 1.38$ (sled-runner). Fatigue tests (Peterson 1932) resulted in K_f factors (fatigue notch factors) having about the same ratio as these two K_t values.

More recently a comprehensive photoelastic investigation (Fessler et al. 1969) has been made of British standard end-milled keyseats. In Chart 5.1 corresponding to the U.S. standard the British value $K_{tA} = 1.6$ has been used for the surface, since in both cases $b/d = 1/4$ and since it appears that the surface factor is not significantly affected by moderate changes in the keyseat depth ratio (Fessler et al. 1969). The maximum stress occurs at an angle within about $10°$ from the point of tangency (see location A, Chart 5.1). Note that K_{tA} is independent of r/d.

For K_t at the fillet (Chart 5.1, location B), the British K_t values for bending have been adjusted for keyseat depth (U.S. $t/d = 1/8 = 0.125$; British $t/d = 1/12 = 0.0833$) in accordance with corresponding t/d values from extrapolations in Chart 3.1. Also note that the maximum fillet stress is located at the end of the keyway, about 15° up on the fillet.

The foregoing discussion refers to shafts below 6.5-in. diameter, for which $t/d = 0.125$. For larger diameter shafts, where $t/d = 0.09$, it would seem that the K_t factor would not differ significantly, taking account of the different values of t/d and r/d. It is suggested that for design, the K_t values for $t/d = 0.125$ and $r/d = 0.0208$ be used for all shaft diameters.

5.2.2 Torsion

For the surface at the semicircular keyseat end, Leven (1949) and Fessler et al. (1969a and b) found $K_{tA} = \sigma_{max}/\tau \approx 3.4$. The maximum normal stress, tangential to the semicircle, occurs at 50° from the axial direction and is independent of r/d. The maximum shear stress is at 45° to the maximum normal stress and is half its value: $K_{tsa} = \tau_{max}/\tau \approx 1.7$. For the stresses in the fillet of the straight part of a U.S. standard keyseat, Leven (1949) obtained K_{ts} values mathematically and also obtained confirmatory results photoelastically. The maximum shear stresses at B (Chart 5.2) are in the longitudinal and perpendicular directions. The maximum normal stresses are of the same magnitude and are at 45° to the shear stress. Therefore $K_{tsB} = \tau_{max}/\tau$ is equal to $K_{tB} = \sigma_{max}/\sigma = \sigma_{max}/\tau$, where $\tau = 16T/\pi d^3$.

Nisida (1963) made photoelastic tests of models with a keyseat having the same depth ratio ($t/d = 1/8$) but with somewhat greater width ratio ($b/d = 0.3$). Allowing for the difference in keyseat shape, the degree of agreement of K_t factors is good. Griffith and Taylor (1917–18) and Okubo (1950a) obtained results for cases with other geometrical proportions. For a semicircular groove (Timoshenko and Goodier 1970), $K_t = 2$ for $r/d \to 0$. For $r/d = 0.125$, K_t based on gross section would be somewhat higher, less than 2.1 as an estimate. This fits quite well with an extension of Leven's curve. The photoelastic results of Fessler et al. (1969a and b) are in reasonable agreement with Leven's values at $r/d = 0.0052$ and 0.0104, but as $r/d = 1/48 = 0.0208$ the K_t value (Fessler et al. 1969a and b) seems low in comparison with the previously mentioned results and the extension to $r/d = 0.125$.

The K_t values in the fillet of the semicircular keyseat end (Fessler et al. 1969a and b) appear to be lower than, or about the same as, in the straight part if the shape of Leven's curve for the straight part is accepted.

5.2.3 Torque Transmitted through a Key

The stresses in the keyseat when torque is transmitted through a key have been investigated by two-dimensional photoelasticity (Solakian and Karelitz 1932; Gibson and Gilet 1938). The results are not applicable to design, since the stresses vary along the key length.

The upper dashed curve of Chart 5.2 is an estimate of the fillet K_t when torque is transmitted by a key of length $2.5d$. The dashed curve has been obtained by use of ratios for K_t values with and without a key, as determined by an "electroplating method," with keyseats of somewhat different cross-sectional proportions (Okubo et al. 1968). In their tests with a key, shaft friction was held to a minimum. In a design application the degree of press-fit pressure is an important factor.

5.2.4 Combined Bending and Torsion

In the investigation by Fessler et al. (1969a and b), a chart was developed for obtaining a design K_t for combined bending and torsion for a shaft with the British keyseat proportions $(b/d = 0.25, t/d = 1/12 = 0.0833)$ for $r/d = 1/48 = 0.0208$. The nominal stress for the chart is defined as

$$\sigma_{\text{nom}} = \frac{16M}{\pi d^3} \left[1 + \sqrt{1 + \left(\frac{T^2}{M^2}\right)} \right] \qquad (5.1)$$

Chart 5.3 provides a rough estimate for the U.S. standard keyseat, based on use of straight lines to approximate the results of the British chart. Note that $K_t = \sigma_{\max}/\sigma_{\text{nom}}$ and $K_{ts} = \tau_{\max}/\sigma_{\text{nom}}$, with σ_{nom} defined in Eq. (5.1).

Chart 5.3 is for $r/d = 1/48 = 0.0208$. The effect of a smaller r/d is to move the middle two lines upward in accordance with the values of Charts 5.1 and 5.2; however, the top and bottom lines remain fixed.

5.2.5 Effect of Proximitiy of Keyseat to Shaft Shoulder Fillet

Photoelastic tests (Fessler et al. 1969a and b) were made of shafts with $D/d = 1.5$ (large diameter/small diameter) and with British keyseat proportions $(b/d = 0.25, t/d = 0.0833, r/d = 0.0208)$. With the keyseat end located at the position where the shaft shoulder fillet begins, Fig. 5.3a, there was no effect on the maximum keyseat fillet stress by varying the shaft shoulder fillet over a r_s/d range of 0.021 to 0.083, where r_s is the shaft shoulder fillet radius. In torsion the maximum surface stress on the semicircular keyseat end terminating at the beginning of the shoulder fillet was increased about 10% over the corresponding stress of a straight shaft with a keyseat. The increase decreases to zero as the keyseat end was moved a distance of $d/10$ away from the beginning of the shaft shoulder filler radius, Fig. 5.3b.

For keyseats cut into the shaft shoulder, Fig. 5.3c, the effect was to reduce K_t for bending (fillet and surface) and for torsion (surface). For torsion (fillet) K_t was reduced, except for an increase when the end of the keyway was located at an axial distance of $0.07d$ to $0.25d$ from the beginning of the fillet.

5.2.6 Fatigue Failures

A designer may be interested in applying the foregoing K_t factors for keyseats in attempting to prevent fatigue failure. Although the problem is a complex one, some comments may be helpful.

Fatigue is initiated by shear stress, but the eventual crack propagation is usually by normal stress. Referring to Charts 5.1 to 5.3, two locations are involved: keyseat fillet with a small radius, and the surface of the shaft at semicircular keyway end with a relatively larger radius, three or more times the fillet radius. Since both the initiation process and the eventual crack are functions of the stress gradiant, which is related mainly to the "notch" radius, this consideration must be kept in mind in attempting to predict fatigue failure. It is possible, in certain instances, to have a nonpropagating crack in a fillet having a small radius.

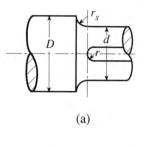

(a)

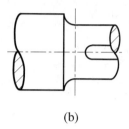

(b)

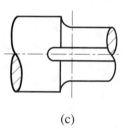

(c)

Figure 5.3 Location of end of keyseat with respect to shaft shoulder: (*a*) Keyseat end at beginning of shoulder fillet; (*b*) keyseat end away from shoulder fillet; (*c*) keyseat end cut into shoulder.

Referring to Chart 5.3, for pure torsion ($M/T = 0$), it would be expected that initiation (shear) would start in the fillet, but the stress gradiant is so steep that initial failure at the surface is also a possibility. The final crack direction will be determined by the surface stresses, where the normal stresses, associated with K_{tA}, are relatively high. For pure bending ($T/M = 0$), failure will more likely occur primarily at the surface. Laboratory fatigue tests (Peterson 1932) and service failures (Peterson 1950) appear to support the foregoing remarks. In certain instances torsional fatigue starts at the fillet and develops into a peeling type of failure. This particular type may be influenced by the key and possibly by a smaller than standard keyseat fillet radius. Predictions are difficult, owing to differing geometries, press-fit and key conditions, and ratios of steady and alternating bending and torsional stress components.

5.3 SPLINED SHAFT IN TORSION

In a three-dimensional photoelastic study by Yoshitake (1962) of a particular eight-tooth spline, the tooth fillet radius was varied in three tests. From these data a K_{ts} versus r/d curve was drawn (Chart 5.4). A test of an involute spline with a full fillet radius gave a K_{ts} value of 2.8. These values are for an open spline; that is, there is no mating member.

A test was made with a mated pair wherein the fitted length was slightly greater than the outside diameter of the spline. The maximum longitudinal bending stress of a tooth occurred at the end of the tooth and was about the same numerically as the maximum torsion stress.

A related mathematical analysis has been made by Okubo (1950b) of the torsion of a shaft with n longitudinal semicircular grooves. The results are on a strain basis. When $r/d \to 0$, $K_{ts} = 2$, as in Timoshenko and Goodier (1970) for a single groove.

5.4 GEAR TEETH

A gear tooth is essentially a short cantilever beam, with the maximum stress occurring at the fillet at the base of the tooth. Owing to the angularity of load application the stresses on the two sides are not the same. The tension side is of design interest since fatigue failures occur there. The photoelastic tests of Dolan and Broghamer (1942) provide the accepted stress concentration factors used in design (Charts 5.5 and 5.6). For gear notation see Fig. 5.4.

Ordinarily the fillet is not a constant radius but is a curve produced by the generating hob or cutter (Michalec 1966). The hobbing tool has straight sided teeth (see sketch, Chart 5.7) with a tip radius r_t which has been standardized (Baumeister 1967): $r_t = 0.209/P_d$ for 14.5° pressure angle and $r_t = 0.235/P_d$ for a 20° pressure angle, both for full-depth teeth. For stub teeth r_t has not been standardized, but $r_t = 0.3/P_d$ has been used. In Charts 5.5 and 5.6 the full curves, which are approximate in that they were obtained by interpolation of the curves corresponding to the photoelastic tests, are for these r_t values. The tool radius r_t generates a gear tooth fillet of variable radius. The minimum radius is denoted r_f. Candee

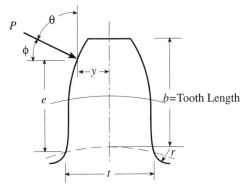

Figure 5.4 Gear notation.

(1941) has developed a relation between r_f and r_t:

$$r_f = \frac{(b-r_t)^2}{N/(2P_d) + (b-r_t)} + r_t \tag{5.2}$$

where b is the dedendum, N is the number of teeth, and P_d is the diametral pitch, (= number of teeth/pitch diameter).

The dedendum $b = 1.157/P_d$ for full-depth teeth and $1/P_d$ for stub teeth. Equation (5.2) is shown graphically in Chart 5.7. The curve for 20° stub teeth is shown dashed, since r_t is not standardized and there is uncertainty regarding application of Eq. (5.2).

The following empirical relations have been developed (Dolan and Broghamer 1942) for the stress concentration factor for the fillet on the tension side:

For 14.5° pressure angle,

$$K_t = 0.22 + \frac{1}{(r_f/t)^{0.2}(e/t)^{0.4}} \tag{5.3}$$

For 20° pressure angle,

$$K_t = 0.18 + \frac{1}{(r_f/t)^{0.15}(e/t)^{0.45}} \tag{5.4}$$

In certain instances a specific form grinder has been used to provide a semicircular fillet radius between the teeth. To evaluate the effect of fillet radius, Chart 5.8 is constructed from Eqs. (5.3) and (5.4). Note that the lowest K_t factors occur when the load is applied at the tip of the tooth. However, owing to increased moment arm, the maximum fillet stress occurs at this position—neglecting load division (Baud and Peterson 1929), which beneficial effect can be reliably taken into account only for extremely accurate gearing (Peterson 1930). Considering then the lowest curves ($e/t = 1$) and keeping in mind that r_f/t values for standard generated teeth lie in the 0.1 to 0.2 region (which depends on the number of teeth), we see that going to a semicircular fillet $r_t/t \approx 0.3$ does not result in a very large decrease in K_t. Although the decrease per se represents a definite available gain, this needs to be weighed against other factors, economic and technical, such as decreased effective rim of a pinion with a small number of teeth (DeGregorio 1968), especially where a keyway is present.

In addition to their photoelastic tests of gear tooth models, Dolan and Broghamer (1942) made tests of straight-sided short cantilever beams, varying the distance of load application and fillet radius (Chart 5.9). The following empirical formula for K_t for the tension side was developed (Dolan and Broghamer 1942):

$$K_t = 1.25 \left[\frac{1}{(r/t')^{0.2}(e/t)^{0.3}}\right] \tag{5.5}$$

Results for the compression side (K_t in most cases higher) are also given in Chart 5.9. An empirical formula was not developed.

For presenting the effect of fillet radius in gear teeth it was found preferable (see Chart 5.8) to use Eqs. (5.3) and (5.4), owing mainly to angular application of load. Chart 5.9

is included for its application to other problems. Results of Weibul (1934) and Riggs and Frocht (1938) on longer beams are included in Chart 5.9 to represent the case of large e/t.

A subsequent photoelastic investigation of gear teeth by Jacobson (1955) resulted in stress concentration factors in good agreement with the results of Dolan and Broghamer (1942). More recently Aida and Terauchi (1962) obtained the following analytical solution for the tensile maximum stress at the gear fillet:

$$\sigma_{\max} = \left(1 + 0.08\frac{t}{r}\right)\left(0.66\sigma_{Nb} + 0.40\sqrt{\sigma_{Nb}^2 + 36\tau_N^2} + 1.15\sigma_{Nc}\right) \quad (5.6)$$

where

$$\sigma_{Nb} = \frac{6Pe\sin\theta}{bt^2} \quad \text{(see Fig. 5.4)}$$

$$\sigma_{Nc} = -\frac{P\cos\theta}{bt} - \frac{6Py\cos\theta}{bt^2}$$

$$\tau_N = \frac{P\sin\theta}{bt}$$

Some further photoelastic tests resulted in a satisfactory check of the foregoing analytical results, which also were in good agreement with the results of Dolan and Broghamer (1942).

5.5 PRESS-FITTED OR SHRINK-FITTED MEMBERS

Gears, pulleys, wheels, and similar elements are often assembled on a shaft by means of a press fit or shrink fit. Photoelastic tests of the flat models of Fig. 5.5 (Peterson and Wahl 1935) have been made with $\sigma_{\text{nom}}/p = 1.36$, where σ_{nom} is the nominal bending stress in the shaft and p is the average normal pressure exerted by the member on the shaft. These led to $K_t = 1.95$ for the plain member and $K_t = 1.34$ for the grooved member.

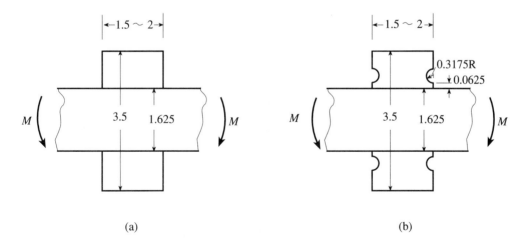

Figure 5.5 Press-fit models, with dimensions in inches: (*a*) Plain member; (*b*) grooved member.

TABLE 5.1 Stress Concentration Factors for Press Fit Shafts of 2-in. Diameter

Roller-bearing Inner Race of Case Hardened Cr–Ni–Mo Steel Pressed on the Shaft	K_f
1. No external reaction through collar	
a. 0.45% C axle steel shaft	2.3
2. External reaction through collar	
a. 0.45% C axle steel shaft	2.9
b. Cr–Ni–Mo steel, heat-treated to 310 Brinell	3.9
c. 2.6% Ni steel, 57,000 psi fatigue limit	3.3–3.8
d. Same, heat treated to 253 Brinell	3.0

Fatigue tests of the "three-dimensional" case of a collar pressed[2] on a 1.625-in.-diameter medium-carbon (0.42% C) steel shaft, the proportions being approximately the same as for the previously mentioned photoelastic models, gave the bending "fatigue-notch factors" of $K_f = 2.0$ for the plain member and $K_f = 1.7$ for the grooved member. It will be noted that the factors for the plain member seem to be in good agreement, but this is not significant since the fatigue result is due to a combination of stress concentration and "fretting corrosion" (Tomlinson 1927; Tomlinson et al. 1939; ASTM 1952; Nishioka et al. 1968), the latter producing a weakening effect over and above that produced by stress concentration. Note that the fatigue factor for the grooved member is higher than the stress concentration factor. This is no doubt due to fretting corrosion, which becomes relatively more prominent for lower-stress condition cases. The fretting corrosion effect varies considerably with different combinations of materials. Table 5.1 gives some fatigue results (Horger and Maulbetsch 1936).

Similar tests were made in Germany (Thum and Bruder 1938) which are reported in Table 5.2. Where the reaction was carried through the inner race, somewhat lower values were obtained. Some tests made with relief grooves (see Section 3.6) showed lower K_f values.

TABLE 5.2 Stress Concentration Factors for Press Fit Shafts of 0.66-in. Diameter

Reaction not carried by the inner race
1. 0.36% C axle steel shaft
a. Press fit and shoulder fillet ($r = 0.04$ in., $D/d = 1.3$)
b. Same, shoulder fillet only (no inner race present)
c. Press fit only (no shoulder)
2. 1.5% Ni–0.5% Cr steel shaft (236 Brinell)
a. Press fit and shoulder fillet ($r = 0.04$ in., $D/d = 1.3$)
b. Same, shoulder fillet (no inner race)

Note: d = Diameter of shaft; D = outer diameter of ring.

---

[2]The calculated radial pressure in this case was 16,000 lb/in.^{2} $\left(\sigma_{\text{nom}}/p = 1\right)$. However, tests (Peterson and Wahl 1935; Thum and Wunderlich 1933) indicate that over a wide range of pressures, this variable does not affect K_f, except for very light pressures which result in a lower K_f.

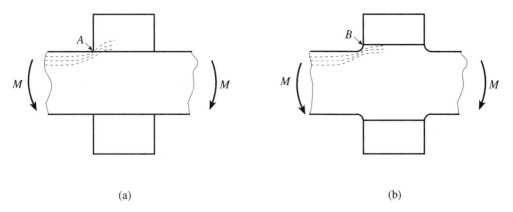

Figure 5.6 Shoulder design for fitted member, with schematic stress "flow lines": (*a*) Plain shaft; (*b*) shaft with shoulder.

Another favorable construction (Horger and Buckwalter 1940; White and Humpherson 1969), as shown in Fig. 5.6, is to enlarge the shaft at the fit and to round out the shoulders in such a way that the critical region A (Fig. 5.6*a*) is relieved as at B (Fig. 5.6*b*). The photoelastic tests (Horger and Buckwalter 1940) did not provide quantitative information, but it is clear that if the shoulder is ample, failure will occur in the fillet, in which case the design can be rationalized in accordance with Chapter 3.

As noted previously, K_f factors are a function of size, increasing toward a limiting value for increasing size of geometrically similar values. For 50-mm ($\approx$ 2-in.) diameter, $K_f = 2.8$ was obtained for 0.39% C axle steel (Nishioka and Komatsu 1967). For models $3\frac{1}{2}$ to 5 in. diameter, K_f values of the order of 3 to 4 were obtained for turbine rotor alloy steels (Coyle and Watson 1963–64). For 7- to $9\frac{1}{2}$-in. wheel fit models (Horger 1953, 1956; Association of American Railroads 1950), K_f values of the order 4 to 5 were obtained for a variety of axle steels, based on the fatigue limit of conventional specimens. Nonpropagating cracks were found, in some instances at about half of the fatigue limit of the press-fitted member. Photoelastic tests (Adelfio and DiBenedetto 1970) of a press-fitted ring on a shaft with six lands or spokes gave K_t factors on the order of 2 to 4.

The situation with regard to press fits is complicated by the simultaneous presence of stress concentration and fretting corrosion. The relations governing fretting corrosion are not well understood at the present time.

5.6 BOLT AND NUT

It has been estimated (Martinaglia 1942) that bolt failures are distributed about as follows: (1) 15% under the head, (2) 20% at the end of the thread, and (3) 65% in the thread at the nut face. By using a reduced bolt shank (Fig. 2.5*c* as compared to Fig. 2.5*b*), the situation with regard to fatigue failures of group *b* type can be improved (Staedel 1933; Wiegand 1933). With a reduced shank a larger fillet radius can be provided under the head (see Section 5.7), thereby improving design with regard to group *a* type failure.

With regard to failure in the threads at the nut face, group *c* type, Hetényi (1943) investigated various bolt-and-nut combinations by means of three-dimensional photoelastic tests. For Whitworth threads, with root radius of 0.1373 pitch (Baumeister 1967), he obtained for the designs shown in Fig. 5.7: $K_{tg} = 3.85$ for bolt and nut of standard proportions; $K_{tg} = 3.00$ for nut having lip, based on the full body (shank) nominal stress. If the factors are calculated for the area at the thread bottom (which is more realistic from a stress concentration standpoint, since this corresponds to the location of the maximum stress), then $K_{tn} = 2.7$ for the standard nut, and $K_{tn} = 2.1$ for the tapered nut.

Later tests by Brown and Hickson (1952–53) using a Fosterite model twice as large and thinner slices, resulted in $K_{tg} = 9$ for the standard nut based on body diameter (see authors' closure) (Brown and Hickson 1952–53). This corresponds to $K_{tn} = 6.7$ for the standard nut, based on root diameter. This compares with the Hetényi (1943) value of 2.7. The value of 6.7 should be used in design where fatigue or embrittling is involved, with a correction for notch sensitivity (Fig. 1.31).

In the discussion of the foregoing paper (Brown and Hickson 1952–53), Taylor reports a fatigue $K_{fn} = 7$ for a 3-in.-diameter bolt with a root contour radius/root diameter half that of the photoelastic model of Brown and Hickson. He estimates that if his fatigue test had been made on a bolt of the same geometry as the photoelastic model, the K_{fn} value might be as low as 4.2.

For a root radius of 0.023 in., a notch sensitivity factor q of about 0.67 is estimated from Fig. 1.31 for "mild steel." The photoelastic $K_{tn} = 6.7$ would then correspond to $K_{fn} = 4.8$. Although this is in fair agreement with Taylor's estimate, the basis of the estimate has some uncertainties.

A photoelastic investigation (Marino and Riley 1964) of buttress threads showed that by modifying the thread-root contour radius, a reduction of the maximum stress by 22% was achieved.

In a nut designed with a lip (Fig. 5.8b), the peak stress is relieved by the lip being stressed in the same direction as the bolt. Fatigue tests (Wiegand 1933) showed the lip design to be about 30% stronger than the standard nut design (Fig. 5.8a), which is in approximate agreement with photoelastic tests (Hetényi 1943).

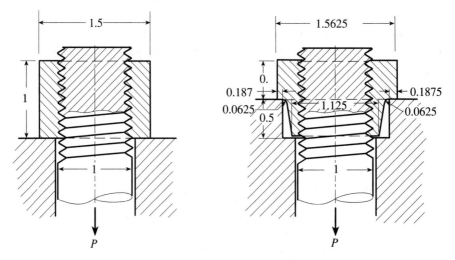

Figure 5.7 Nut designs tested photoelastically, with dimensions in inches (Hetényi 1943).

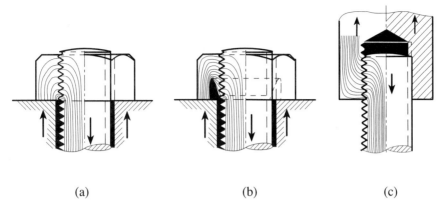

(a)   (b)   (c)

Figure 5.8 Nut designs fatigue tested (Wiegand 1933).

In the arrangement shown in Fig. 5.8c, the transmitted load is not reversed. Fatigue tests (Wiegand 1933) showed a fatigue strength more than double that of the standard bolt-and-nut combination (Fig. 5.8a).

The use of a nut of material having a lower modulus of elasticity is helpful in reducing the peak stress in the bolt threads. Fatigue tests (Wiegand 1933; Kaufmann and Jäniche 1940) have shown gains in strength of 35% to 60% depending on materials.

Other methods can be used for reduction of the K_t factor of a bolt-and-nut combination, such as tapered threads and differential thread spacing, but these methods are not as practical.

5.7 BOLT HEAD, TURBINE-BLADE, OR COMPRESSOR-BLADE FASTENING (T-HEAD)

A vital difference between the case of a bar with shoulder fillets (Fig. 5.9a) and the T-head case (Fig. 5.9b) is the manner of loading. Another difference in the above cases is the dimension L, Fig. 5.9b, which is seldom greater than d. As L is decreased, bending of the overhanging portion becomes more prominent.

Chart 5.10 presents $\sigma_{\max}/\sigma$ values as determined photoelastically by Hetényi (1939b). In this case σ is simply P/A, the load divided by the shank cross-sectional area. Thus $\sigma_{\max}/\sigma$ values express stress concentration in the simplest form for utilization in design.

However, it is also useful to consider a modified procedure for K_t, so that when comparisons are made between different kinds of fatigue tests, the resulting notch sensitivity values will have a more nearly comparable meaning, as explained in the introduction to Chapter 4. For this purpose we will consider two kinds of K_t factors, K_{tA} based on tension and K_{tB} based on bending.

For tension

$$K_{tA} = \frac{\sigma_{\max}}{\sigma} \tag{5.7}$$

where

$$\sigma = \sigma_{\text{nom }A} = \frac{P}{hd}$$

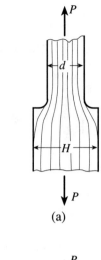

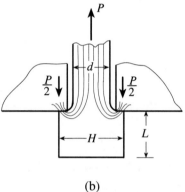

Figure 5.9 Transmittal of load (schematic): (a) Stepped tension bar; (b) T-head.

For bending

$$K_{tB} = \frac{\sigma_{max}}{\sigma_{nom B}}$$

where

$$\sigma_{nom B} = \frac{M}{I/c} = \frac{Pl}{2}\left(\frac{6}{hL^2}\right) = \frac{3}{4}\frac{Pd}{hL^2}\left(\frac{H}{d} - 1\right) \qquad (5.8)$$

with $l = (H - d)/4$. Thus

$$K_{tB} = \frac{\sigma_{max}}{\sigma\left[3(H/d - 1)/4(L/d)^2\right]} \qquad (5.9)$$

Note that for $K_{tA} = K_{tB}$,

$$\frac{(H/d - 1)}{(L/d)^2} = \frac{4}{3} \quad \text{or} \quad \frac{ld}{L^2} = \frac{1}{3} \qquad (5.10)$$

In Chart 5.10f values of K_{tA} and K_{tB} are plotted with ld/L^2 as the abscissa variable. For $(ld/L^2) > 1/3$, K_{tB} is used; for $(ld/L^2) < 1/3$, K_{tA} is used. This is a procedure similar to that used for the pinned joint (Section 5.8, and earlier Section 4.5.8) and, as in that case, not only gets away from extremely high factors but also provides a safer basis for extrapolation (in this case to smaller L/d values).

In Charts 5.10a and 5.10d the dashed line represents equal K_{tA} and K_{tB} values, Eqs. (5.7) and (5.9). Below this line the $\sigma_{\max}/\sigma$ values are the same as K_{tA}. Above the dashed line, all the $\sigma_{\max}/\sigma$ values are higher, usually much higher, than the corresponding K_{tB} values, which in magnitude are all lower than the values represented by the dashed line (i.e., the dashed line represents maximum K_{tA} and K_{tB} values as shown by the peaks in Chart 5.10f).

The effect of moving concentrated reactions closer to the fillet is shown in Chart 5.10e. The sharply increasing K_t values are due to a proximity effect (Hetényi 1939b), since the nominal bending is decreasing and the nominal tension remains the same.

The T-head factors may be applied directly in the case of a T-shaped blade fastening of rectangular cross section. In the case of the head of a round bolt, somewhat lower factors will result, as can be seen from Chapters 2 and 3. However, the ratios will not be directly comparable to those of Chapter 3, since part of the T-head factor is due to proximity effect. To be on the safe side, it is suggested to use the unmodified T-head factors for bolt heads.

Steam-turbine blade fastenings are often made as a "double T-head." In gas-turbine blades multiple projections are used in the "fir-tree" type of fastening. Some photoelastic data have been obtained for multiple projections (Heywood 1969; Durelli and Riley 1965).

5.8 LUG JOINT

The stress concentration factors at the perimeter of a hole in a lug with a pin are studied in this section (Frocht and Hill 1940; Theocaris 1956; Cox and Brown 1964; Meek 1967; Gregory 1968; Whitehead et al. 1978; ESDU 1981). See Fig. 5.10 for notation.

The pin-to-hole clearance as a percentage of the hole diameter d is designated as e. Thus, from Fig. 5.11, $e = \delta/d$. The quantity K_{te} is the stress concentration factor at e percent clearance. Thus $K_{t0.2}$ refers to a 0.2% clearance between the hole and the pin. K_{t100} is used for the limiting case of point (line) loading. In this case the load P is applied uniformly across the thickness of the lug at location C of Fig. 5.10. When h/d (the ratio of lug thickness to hole diameter) < 0.5, the stress concentration factor is designated as K_{te}. For $h/d > 0.5$, K'_{te} is used.

The stress concentration factor is influenced by lug geometry as well as the clearance of the pin in the hole. For perfectly fitting pins, $\sigma_{\max}$ should occur at the points labeled with A in Fig. 5.10. If there is a clearance between the pin and the hole, $\sigma_{\max}$ increases in value and occurs at points B, for which $10° < \theta < 35°$.

Bending stresses can be expected to occur along the C–D section if $c - d/2$ is quite small. Then $\sigma_{\max}$ will occur at point D. This phenomenon is not treated here.

Stress concentration factors K_{te} are presented in Charts 5.11 and 5.12 for $c/H > 1.5$ (ESDU 1981). Actually these charts were prepared for the limiting condition $c/H \to \infty$. In practice, however, they appear to be appropriate for any c/H greater than 1.5.

Studies (ESDU 1981) have shown that if $h/d < 0.5$ for a lug, the stress concentration factor K_{te} is not significantly affected if the pin and lug are made of different materials. More specifically, there is no significant effect if the ratio of the elastic moduli $E_{\text{pin}}/E_{\text{lug}}$ is

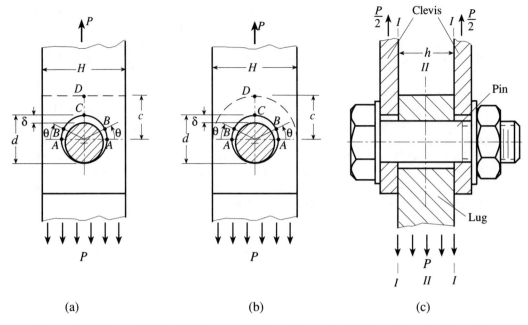

Figure 5.10 Lugs with pins: (*a*) Square-ended lug; (*b*) round-ended lug; (*c*) section through lug assembly center line.

between 1 and 3. (See Section 5.8.2 for further discussion of the effect of different materials, especially for $h/d > 0.5$.)

5.8.1 Lugs with $h/d < 0.5$

For precisely manufactured pins and lug-holes, as might occur for laboratory instrumentation, e tends to be less than 0.1%. Stress concentration factor curves for square-ended lugs for this case are given in Chart 5.11. The solid K_{te} curves of Chart 5.11 are for pin clearances in square-ended lugs of 0.2% of d. The upper limit K_{t100} curve of Chart 5.12 is probably a reasonable limiting estimate for the square-ended lugs.

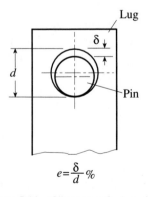

Figure 5.11 Clearance of a lug-pin fit.

For round-ended lugs Chart 5.12 gives stress concentration factor $K_{t0.2}$ and K_{t100} curves. Experimental results (ESDU 1981) validate these curves as reasonable approximations.

The stress concentration factor K_{te} for any lug-hole, pin clearance percent e can be obtained from $K_{t0.2}$ and K_{t100}. Define (ESDU 1981) a correction factor f as $f = (K_{te} - K_{t0.2})/(K_{t100} - K_{t0.2})$ so that $K_{te} = K_{t0.2} + f(K_{t100} - K_{t0.2})$. A plot of f is provided in Chart 5.12. Surface finish variations and geometric imperfections can influence stress concentration factors. Values of f should be treated as approximate, especially for $e < 0.1$.

5.8.2 Lugs with $h/d > 0.5$

The stress concentration factors K'_{te} for pin, hole joints with $h/d > 0.5$ can be obtained from Chart 5.13, using K_{te} values taken from Charts 5.11 and 5.12, as appropriate. The ratio K'_{te}/K_{te} versus h/d is shown in Chart 5.13 (ESDU 1981). This chart applies for small pin, hole clearances. The curves were actually prepared for square-ended lugs with $d/H = 0.45$ and $c/H = 0.67$. However, they probably provide reasonable estimates for square- and round-ended lugs with $0.3 \leq d/H \leq 0.6$. The upper curve in Chart 5.13 applies to $E_{pin} = E_{lug}$, where E is the modulus of elasticity. The lower curve, for $E_{pin}/E_{lug} = 3.0$, was based on a single data point of $h/d = 2.24$. As h/d increases, it leads to an increase in pin bending and an accompanying increase in loading at the hole ends, sections I–I in Fig. 5.10c. This is accompanied by a decrease in loading at the center of the hole region, section II–II in Fig. 5.10c.

If there is a nonnegligible clearance between the sides of the lug and the loading fork, faces I–I of Fig. 5.10c, pin bending may increase, as may the ratio K'_{te}/K_{te} over that shown in Chart 5.13. The effect of smoothed corners of the lug hole may be similar. Stress concentrations can also be expected to increase if loading on the fork is not symmetric.

Example 5.1 Pin, Hole Joint Determine the peak stress concentration factor for the lug of Fig. 5.12. The pin is nominally of diameter 65 mm. It can vary from $65 - 0.02$ to $65. - 0.16$ mm. The hole diameter is also nominally of diameter 65 mm and can vary between $65 + 0.10$ and $65 + 0.20$ mm.

The upper bound for the clearance is $0.20 + 0.16 = 0.36$ mm, giving

$$e = \frac{0.36}{65} \times 100 = 0.55\% \tag{1}$$

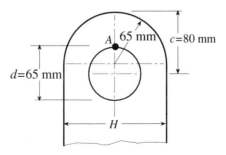

Figure 5.12 The lug of Example 5.1.

From the dimensions in Fig. 5.12,

$$\frac{d}{H} = \frac{65}{(2 \times 65)} = 0.5 \qquad (2)$$

$$\frac{c}{H} = \frac{80}{(2 \times 65)} = 0.62 \qquad (3)$$

Chart 5.12 for round-ended lugs shows values for $K_{t0.2}$ and K_{t100} of 2.68 and 3.61, respectively. We seek K_{te} for $e = 0.55\%$. From Chart 5.12 the correction factor f for $e = 0.55$ is 0.23. Then

$$K_{te} = K_{t0.2} + f(K_{t100} - K_{t0.2}) \qquad (4)$$
$$= 2.68 + 0.23(3.61 - 2.68) = 2.89$$

This is the maximum stress concentration factor for the lug. This stress occurs at point A of Fig. 5.12.

The lower bound for the clearance is $0.10 + 0.02 = 0.12$ mm, so $e = (0.12/65)100 = 0.18\%$. From Chart 5.12 the correction factor f for $e = 0.18$ is -0.02. Thus $K_{te} = K_{t0.2} + f(K_{t100} - K_{t0.2}) = 2.68 - 0.02(3.61 - 2.68) = 2.66$.

5.9 CURVED BAR

A curved bar subjected to bending will have a higher stress on the inside edge, as shown in Fig. 5.13. A discussion of the curved bar case is given in advanced textbooks (Timoshenko 1956). Formulas for typical cross sections (Pilkey 1994) and a graphical method for a general cross section have been published (Wilson and Quereau 1928). In Chart 5.14 values of K_t are given for five cross sections.

The following formula (Wilson and Quereau 1928) has been found to be reasonably accurate for ordinary cross sections, although not for triangular cross sections:

$$K_t = 1.00 + B\left(\frac{I}{bc^2}\right)\left(\frac{1}{r-c} + \frac{1}{r}\right) \qquad (5.11)$$

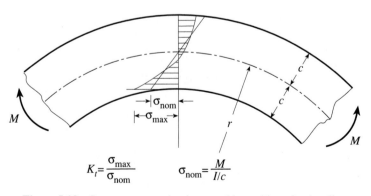

Figure 5.13 Stress concentration in curved bar subjected to bending.

where I is the moment of inertia of the cross section, b is the maximum breadth of the section, c is the distance from the centroidal axis to the inside edge, r is the radius of curvature, and $B = 1.05$ for the circular or elliptical cross section and 0.5 for other cross sections.

With regard to notch sensitivity, the q versus r curves of Fig. 1.31 do not apply to a curved bar. Use might be made of the stress gradient concept (Peterson 1938).

5.10 HELICAL SPRING

5.10.1 Round or Square Wire Compression or Tension Spring

A helical spring may be regarded as a curved bar subjected to a twisting moment and a direct shear load (Wahl 1963). The final paragraph in the preceding section applies to helical springs.

For a round wire helical compression or tension spring of small pitch angle, the *Wahl factor*, C_w, a correction factor taking into account curvature and direct shear stress, is generally used in design (Wahl 1963). See Chart 5.15.

For round wire,

$$C_w = \frac{\tau_{max}}{\tau} = \frac{4c-1}{4c-4} + \frac{0.615}{c} \tag{5.12}$$

$$\tau = \frac{T(a/2)}{J} = \frac{P(d/2)}{\pi a^3/16} = \frac{8Pd}{\pi a^3} = \frac{8Pc}{\pi a^2} \tag{5.13}$$

where T is the torque, P is the axial load, c is the spring index d/a, d is the mean coil diameter, a is the wire diameter, and J is the polar moment of inertia.

For square wire (Göhner 1932), from Fig. 5.14, for $b = h = a$ the shape correction factor α is $\alpha = 0.416$. Here a is the width and depth of the square wire. Then $1/\alpha = 2.404$ and

$$C_w = \frac{\tau_{max}}{\tau} \tag{5.14}$$

$$\tau_{max} = \tau\left(1 + \frac{1.2}{c} + \frac{0.56}{c^2} + \frac{0.5}{c^3}\right) \tag{5.15}$$

$$\tau = \frac{Pd}{\alpha a^3} = \frac{2.404 Pd}{a^3} = \frac{2.404 Pc}{a^2} \tag{5.16}$$

The corresponding stress concentration factors, which may be useful for mechanics of materials problems, are obtained by taking the nominal shear stress τ_{nom} as the sum of the torsional stress τ of Eq. (5.13) and the direct shear stress $\tau = 4P/\pi a^2$ for round wire. In the case of the wire of square cross section, τ_{nom} is the sum of the torsional stress of Eq. (5.16) and the direct shear stress $\tau = P/a^2$.

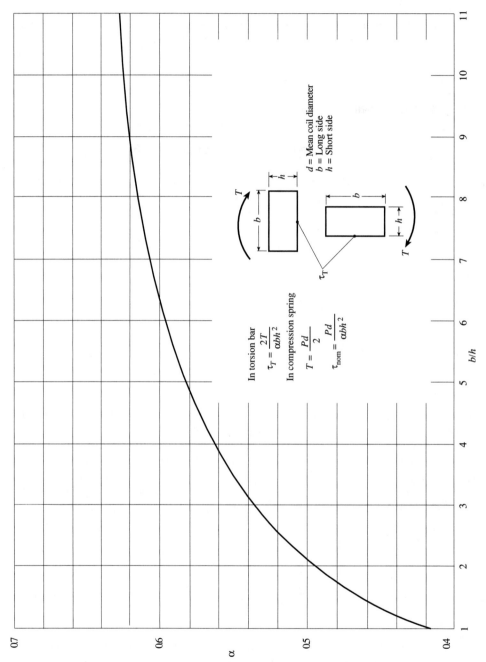

Figure 5.14 Factor α for a torsion bar of rectangular cross section.

For round wire,

$$K_{ts} = \frac{\tau_{max}}{\tau_{nom}}$$

$$= \frac{(8Pd/\pi a^3)[(4c-1)/(4c-4)] + 4.92P/\pi a^2}{8Pd/\pi a^3 + 4P/\pi a^2}$$

$$= \frac{2c[(4c-1)/(4c-4)] + 1.23}{2c+1} \tag{5.17}$$

$$\tau_{max} = K_{ts}\left[\frac{4P}{\pi a^2}(2c+1)\right] \tag{5.18}$$

For square wire,

$$K_{ts} = \frac{\tau_{max}}{\tau_{nom}}$$

$$= \frac{(2.404Pd/a^3)(1 + 1.2/c + 0.56/c^2 + 0.5/c^3)}{2.404Pd/a^3 + P/a^2}$$

$$= \frac{2.404c(1 + 1.2/c + 0.56/c^2 + 0.5/c^3)}{2.404c + 1} \tag{5.19}$$

$$\tau_{max} = K_{ts}\left[\frac{P}{a^2}(2.404c+1)\right] \tag{5.20}$$

Values of C_w and K_{ts} are shown in Chart 5.15. K_{ts} is lower than the correction factor C_w. For design calculations it is recommended that the simpler Wahl factor be used. The same value of τ_{max} will be obtained whether one uses C_w or K_{ts}.

The effect of pitch angle has been determined by Ancker and Goodier (1958). Up to 10° the effect of pitch angle is small, but at 20° the stress increases sufficiently so that a correction should be made.

5.10.2 Rectangular Wire Compression or Tension Spring

For wire of rectangular cross section, the results of Liesecke (1933) have been converted into stress concentration factors in the following way: The nominal stress is taken as the maximum stress in a straight torsion bar of the corresponding rectangular cross section plus the direct shear stress:

$$\tau_{nom} = \frac{Pd}{\alpha bh^2} + \frac{P}{bh} \tag{5.21}$$

where the shape correction factor α is given in Fig. 5.14 and b is the long side of rectangular cross section, h is the short side of rectangular cross section, and d is the mean coil diameter.

According to Liesecke (1933),

$$\tau_{max} = \frac{\beta Pd}{bh\sqrt{bh}} \tag{5.22}$$

where b is the side of rectangle perpendicular to axis of spring, h is the side of rectangle parallel to axis of spring, β is the Liesecke factor

$$K_{ts} = \frac{\tau_{\max}}{\tau_{\text{nom}}} \tag{5.23}$$

where K_{ts} is given in Chart 5.16 and α is given in Fig. 5.14.

5.10.3 Helical Torsion Spring

Torque is applied in a plane perpendicular to the axis of the spring (Wahl 1963) (see Chart 5.17):

$$K_t = \frac{\sigma_{\max}}{\sigma_{\text{nom}}}$$

For circular wire of diameter $a = h$,

$$\sigma_{\text{nom}} = \frac{32Pl}{\pi a^3} \tag{5.24}$$

For rectangular wire,

$$\sigma_{\text{nom}} = \frac{6Pl}{bh^2} \tag{5.25}$$

where Pl is the moment (torque, see Chart 5.17), b is the side of rectangle perpendicular to axis of spring, and h is the side of rectangle parallel to the axis of spring.

The effect of pitch angle has been studied by Ancker and Goodier (1958). The correction is small for pitch angles less than 15°.

5.11 CRANKSHAFT

The maximum stresses in the fillets of the pin and journal of a series of crankshafts in bending were determined by use of the strain gage by Arai (1965). Design parameters were systematically varied in a comprehensive manner involving 178 tests. The stress concentration factor is defined as $\sigma_{\max}/\sigma_{\text{nom}}$, where $\sigma_{\text{nom}} = M(d/2)/I = M/(\pi d^3/32)$. Strains were measured in the fillet in the axial plane. The smaller circumferential strain in the fillet was not measured.

It was found that the K_t values were in good agreement whether the moment was uniform or applied by means of concentrated loads at the middle of the bearing areas. The K_t values for the pin and journal fillets were sufficiently close that the average value was used.

From the standpoint of stress concentration the most important design variables are the web thickness ratio t/d and the fillet radius ratio r/d. The notation and stress concentration factors are given in Charts 5.18 and 5.19.

It was found that K_t is relatively insensitive to changes in the web width ratio b/d, and the crank "throw" as expressed[3] by s/d, over practical ranges of these parameters. It was also found that cutting the corners of the web had no effect on K_t.

Arai points out that as the web thickness t increases "extremely," K_t should agree with that of a straight stepped shaft. He refers to Fig. 65 of Peterson (1953) and an extended t/d value of 1 to 2. This is an enormous extrapolation (Chart 5.18). It seems that all that can be said is that smooth curves can be drawn to the shaft values, but this does not constitute a verification.

Referring to the sketch in Chart 5.18, it is sometimes beneficial to recess the fillet fully or partially. It was found that as δ is increased K_t increases. However, the designer should evaluate the increase against the possibility of using a larger fillet radius and increasing the bearing area or decreasing the shaft length.

An empirical formula was developed by Arai to cover the entire range of tests:

$$K_t = 4.84 C_1 \ C_2 \ C_3 \ C_4 \ C_5 \tag{5.26}$$

where

$$C_1 = 0.420 + 0.160\sqrt{[1/(r/d)] - 6.864}$$
$$C_2 = 1 + 81\{0.769 - [0.407 - (s/d)]^2\}(\delta/r)(r/d)^3$$
$$C_3 = 0.285[2.2 - (b/d)]^2 + 0.785$$
$$C_4 = 0.444/(t/d)^{1.4}$$
$$C_5 = 1 - [(s/d) + 0.1]^2/[4(t/d) - 0.7]$$

It does not appear that a corresponding investigation of the crankshaft in torsion has been published.

5.12 CRANE HOOK

A crane hook is another curved bar case. A generally applicable procedure for tensile and bending stresses in crane hooks is provided in Pilkey (1994, p. 782). Wahl (1946) has developed a simple numerical method and has applied this to a typical example of a crane hook with an approximately trapezoidal cross section, obtaining a K_t value of 1.56.

5.13 U-SHAPED MEMBER

The case of a U-shaped member subjected to a spreading type of loading has been investigated photoelastically (Mantle and Dolan 1948). This leads to Charts 5.20 and 5.21.

The location of the maximum stress depends on the proportions of the U member and the position of the load. For variable back depth d, for $b = r$, and for loads applied at distances L one to three times r from the center of curvature (Chart 5.20), the maximum

---

[3]When the inside of the crankpin and the outside of the journal are in line, $s = 0$ (see the sketch in Chart 5.18). When the crankpin is closer, s is positive (as shown in the sketch). When the crankpin's inner surface is farther away than $d/2$, s is negative.

stress occurs at position B for the smaller values of d and at A for the larger values of d. The K_t values were defined (Mantle and Dolan 1948) as follows:

For position A,

$$K_{tA} = \frac{\sigma_{max} - P/a_A}{M_A c_A/I_A} = \frac{\sigma_{max} - P/hd}{6P(L + r + d/2)/hd^2} \quad (5.27)$$

where d is the back depth (Chart 5.20), b is the arm width ($= r$), c_A is the distance of centroid of the cross section $A–A'$ to inside edge of the U-shaped member, a_A is the area of the cross section $A–A'$, M_A is the bending moment at the cross section $A–A'$, I_A is the moment of inertia of the cross section $A–A'$, h is the thickness, r is the inside radius ($= b$), and L is the distance from line of application of load to center of curvature.

For position B,

$$K_{tB} = \frac{\sigma_{max}}{M_B c_B/I_B} = \frac{\sigma_{max}}{P(L + l)c_B I_B} \quad (5.28)$$

where I_B is the moment of inertia of the cross section $B–B'$, c_B is the distance from centroid of the cross section $B–B'$ to inside edge of the U-shaped member, l is the horizontal distance from center of curvature to centroid of the cross section $B–B'$, and M_B is the bending moment at the cross section $B - B'$. In the case of position B, the angle θ (Chart 5.20) was found to be approximately $20°$.

Where the outside dimensions are constant, $b = d$ and r varies, causing b and d to vary correspondingly (Chart 5.21), the maximum stress occurs at position A, except for very large values of r/d. Values of K_t are given in Chart 5.21 for a condition where the line of load application remains the same.

5.14 ANGLE AND BOX SECTIONS

Considerable work has been done on beam sections in torsion (Lyse and Johnston 1935; Pilkey 1994). In Chart 5.22 mathematical results of Huth (1950) are given for angle and box sections. For box sections the values given are valid only when a is large compared to h, 15 to 20 times as great. An approximation of bending of angle sections can be obtained from results on knee frames (Richart et al. 1938). Pilkey (1994) contains stress formulas for numerous other cross-sectional shapes.

5.15 ROTATING DISK WITH HOLE

For a rotating disk with a central hole, the maximum stress is tangential (circumferential), occurring at the edge of the hole (Robinson 1944; Timoshenko 1956; Pilkey 1994):

$$\sigma_{max} = \frac{\gamma \Omega^2}{g} \left(\frac{3 + \nu}{4} \right) \left[R_2^2 + \left(\frac{1 - \nu}{3 + \nu} \right) R_1^2 \right] \quad (5.29)$$

where γ is the weight per unit volume, Ω is the angular velocity of rotation (rad/s), g is the gravitational acceleration, ν is Poisson's ratio, R_1 is the hole radius, R_2 is the outer radius of disk. Note that for a thin ring, $R_1/R_2 = 1$, $\sigma_{max} = (\gamma \Omega^2/g)R_2^2$.

The K_t factor can be defined in several ways, depending on the choice of nominal stress:

1. σ_{Na} is the stress at the center of a disk without a hole. At radius $(R_1 + R_2)/2$ both the radial and tangential stress reach the same maximum value.

$$\sigma_{Na} = \frac{\gamma\Omega^2}{g}\left(\frac{3+\nu}{8}\right)R_2^2 \tag{5.30}$$

Use of this nominal stress results in the top curve of Chart 5.23. This curve gives a reasonable result for a small hole; for example for $R_1/R_2 \to 0$, $K_{ta} = 2$. However, as R_1/R_2 approaches 1.0 (thin ring), the higher factor is not realistic.

2. σ_{Nb} is the average tangential stress:

$$\sigma_{Nb} = \frac{\gamma\Omega^2}{3g}\left(1 + \frac{R_1}{R_2} + \frac{R_1^2}{R_2^2}\right)R_2^2 \tag{5.31}$$

Use of this nominal stress results in a more reasonable relation, giving $K_t = 1$ for the thin ring. However, for a small hole Eq. (5.30) appears preferable.

3. The curve of σ_{Nb} (Eq. 5.31) is adjusted to fit linearly the end conditions at $R_1/R_2 = 0$ and at $R_1/R_2 = 1.0$, and σ_{Nb} becomes

$$\sigma_{Nc} = \frac{\gamma\Omega^2}{3g}\left(1 + \frac{R_1}{R_2} + \frac{R_1^2}{R_2^2}\right)\left[3\left(\frac{3+\nu}{8}\right)\left(1 - \frac{R_1}{R_2}\right) + \frac{R_1}{R_2}\right]R_2^2 \tag{5.32}$$

For a small central hole, Eq. (5.30) will be satisfactory for most purposes. For larger holes and in cases where notch sensitivity (Section 1.9) is involved, Eq. (5.32) is suggested.

For a rotating disk with a noncentral hole, photoelastic results are available for variable radial location for two sizes of hole (Barnhart et al. 1951). Here the nominal stress σ_N is taken as the tangential stress in a solid disk at a point corresponding to the outermost point (marked A, Chart 5.24) of the hole. Since the holes in this case are small relative to the disk diameter, this is a reasonable procedure.

$$\sigma_{Nc} = \frac{\gamma\Omega^2}{g}\left(\frac{3+\nu}{8}\right)\left[1 - \left(\frac{1+3\nu}{3+\nu}\right)\left(\frac{R_A}{R_2}\right)^2\right]R_2^2 \tag{5.33}$$

The same investigation (Barnhart et al. 1951) covered the cases of a disk with six to ten noncentral holes located on a common circle, the disk also containing a central hole. Hetényi (1939b) investigated the special cases of a rotating disk containing a central hole plus two or eight symmetrically disposed noncentral holes.

Similar investigations (Leist and Weber 1956; Green et al. 1964; Fessler and Thorpe 1967a and b) have been made for a disk with a large number of symmetrical noncentral holes, such as is used in gas turbine disks. The optimum number of holes was found (Fessler and Thorpe 1967a) for various geometrical ratios. Reinforcement bosses did not reduce peak stresses by a significant amount (Fessler and Thorpe 1967b), but use of a tapered disk did lower the peak stresses at the noncentral holes.

5.16 RING OR HOLLOW ROLLER

The case of a ring subjected to concentrated loads acting along a diametral line (Chart 5.25) has been solved mathematically for $R_1/R_2 = 1/2$ by Timoshenko (1922) and for $R_1/R_2 = 1/3$ by Billevicz (1931). An approximate theoretical solution is given by Case (1925). Photoelastic investigations have been made by Horger and Buckwalter (1940) and by Leven (1952). The values shown in Charts 5.25 and 5.26 represent the average of the photoelastic data and mathematical results, all of which are in good agreement. For $K_t = \sigma_{max}/\sigma_{nom}$, the maximum tensile stress is used for σ_{max}, and for σ_{nom} the basic bending and tensile components as given by Timoshenko (1956) for a thin ring are used.

For the ring loaded internally (Chart 5.25),

$$K_t = \frac{\sigma_{max\,A}[2h(R_2 - R_1)]}{P\left[1 + \dfrac{3(R_2 + R_1)(1 - 2/\pi)}{R_2 - R_1}\right]} \tag{5.34}$$

For the ring loaded externally (Chart 5.26),

$$K_t = \frac{\sigma_{max\,B}[\pi h(R_2 - R_1)^2]}{3P(R_2 + R_1)} \tag{5.35}$$

The case of a round-cornered square hole in a cylinder subjected to opposite concentrated loads has been analyzed by Seika (1958).

5.17 PRESSURIZED CYLINDER

The Lamé solution (Pilkey 1994) for a cylinder with internal pressure p is

$$\sigma_{max} = \frac{p(R_1^2 + R_2^2)}{(R_2^2 - R_1^2)} \tag{5.36}$$

where p is the pressure, R_1 is the inside radius, and R_2 is the outside radius. The two K_t relations of Chart 5.27 are

$$K_{t1} = \frac{\sigma_{max}}{\sigma_{nom}} = \frac{\sigma_{max}}{\sigma_{av}}$$

$$K_{t1} = \frac{(R_1/R_2)^2 + 1}{(R_1/R_2)^2 + R_1/R_2} \tag{5.37}$$

and

$$K_{t2} = \frac{\sigma_{max}}{p}$$

$$K_{t2} = \frac{(R_1/R_2)^2 + 1}{1 - (R_1/R_2)^2} \tag{5.38}$$

At $R_1/R_2 = 1/2$, the K_t factors are equal, $K_t = 1.666$. The branches of the curves below $K_t = 1.666$ are regarded as more meaningful when applied to analysis of mechanics of materials problems (see comments in the introduction to Chapter 4).

5.18 CYLINDRICAL PRESSURE VESSEL WITH TORISPHERICAL ENDS

Chart 5.28 for a cylindrical pressure vessel with torispherical caps is based on photoelastic data of Fessler and Stanley (1965). Since the maximum stresses used in the K_t factors are in the longitudinal (meridional) direction, the nominal stress used in Chart 5.28 is $pd/4h$, which is the stress in the longitudinal direction in a closed cylinder subjected to pressure p. Refer to the notation in Chart 5.28. Although the K_t factors for the knuckle are for $h/d = 0.05$ (or adjusted to that value), it was found (Fessler and Stanley 1965) that K_t increases only slightly with increasing thickness.

Referring to Chart 5.28, above lines ABC and CDE the maximum K_t is at the crown. Between lines ABC and FC the maximum K_t is at the knuckle. Below line FE the maximum stress is the hoop stress in the straight cylindrical portion.

Design recommendations (ASME 1971) are indicated on Chart 5.28. These include r_i/D not less than 0.06, R_i/D not greater than than 1.0, and r_i/h not less than 3.0. A critical evaluation of investigations of stresses in pressure vessels with torispherical ends has been made by Fessler and Stanley (1966).

5.19 PRESSURIZED THICK CYLINDER WITH A CIRCULAR HOLE IN THE CYLINDER WALL

Pressurized thick cylinders with circular wall holes are encountered frequently in the high-pressure equipment industry. In Chart 5.29, K_t factors (Gerdeen 1972) are defined as

$$K_t = \frac{\sigma_{\max}}{p\left[\dfrac{(R_2/R_1)^2 + 1}{(R_2/R_1)^2 - 1}\right]} \tag{5.39}$$

The denominator is the hoop stress at the inner surface of a cylinder without a hole as given by the Lamé equation (Eq. 5.36). Gerdeen (1972) also gives K_t factors for a press-fitted cylinder on an unpressurized cylinder with a sidehole or with a crosshole.

Strain gage measurements (Gerdeen and Smith 1972) on pressurized thick-walled cylinders with well rounded crossholes resulted in minimum K_t factors (1.0 to 1.1) when the holes were of equal diameter (K_t defined by Eq. 5.39). Fatigue failures in compressor heads have been reduced by making the holes of equal diameter and using larger intersection radii.

REFERENCES

Adelfio, B., and DiBenedetto, F., 1970, "Forzamento su Appoggio Discontinuo (Shrink-fitted Ring over Discontinous Support)," *Disegnio de Macchine, Palermo*, Vol. 3, p. 21.

Aida, T., and Terauchi, Y., 1962, "On the Bending Stress in a Spur Gear," *Bull. Japanese Soc. Mech. Eng.*, Vol. 5, p. 161.

Ancker, C. J., and Goodier, J. N., 1958, "Pitch and Curvature Corrections for Helical Springs," *Trans. ASME, Appl. Mech. Section*, Vol. 80, pp. 466, 471, 484.

ANSI, 1967, *Keys and Keyseats*, U.S. Standard ANSI B17.1, ASME, New York. Reaffirmed 1973 and 1987.

Arai, J., 1965, "The Bending Stress Concentration Factor of a Solid Crankshaft," *Bull. Japan Soc. Mech. Eng.*, Vol. 8, p. 322.

ASME, 1971, *Boiler and Pressure Vessel Code—Pressure Vessels*, Section VIII, ASME, New York, pp. 18, 27.

Association of American Railroads, 1950, Passenger Car Axle Tests, Fourth Progress Report, p. 26

ASTM, 1952, *Symposium on Fretting Corrosion*, STP 144, Philadelphia, Pa.

Barnhart, K. E., Hale, A. L., and Meriam, J. L., 1951, "Stresses in Rotating Disks due to Non-Central Holes," *Proc. Soc. Exp. Stress Analysis*, Vol. 9, No. 1, p. 35.

Baud, R. V., and Peterson, R. E., 1929, "Load and Stress Cycles in Gear Teeth," *Mech. Engrg.*, Vol. 51, p. 653.

Baumeister, T. ed., 1967, *Marks' Standard Handbook for Mechanical Engineers*, McGraw-Hill, New York, pp. 8, 20, 21, 133.

Billevicz, V., 1931, "Analysis of Stress in Circular Rings," PhD thesis, Univ. of Michigan.

Brown, A. F. C., and Hickson, V. M., 1952–53, "A Photoelastic Study of Stresses in Screw Threads," *Proc. IME*, Vol. 1B, p. 605. Discussion, p. 608.

Candee, A. H., 1941, *Geometrical Determination of Tooth Factor*, American Gear Manufacturers Assn., Pittsburgh, Pa.

Case, J., 1925, *Strength of Materials*, Longmans, Green, London, p. 291.

Cox, H. L., and Brown, A. F. C., 1964, Stresses around Pins in Holes," *Aeronaut. Q.*, Vol. XV, Part 4, pp. 357–372.

Coyle, M. B., and Watson, S. J., 1963-64, "Fatigue Strength of Turbine Shafts with Shrunk-on Discs," *Proc. Inst. Mech. Eng.* (London), Vol. 178, p. 147.

DeGregorio, G., 1968, "Ricerca sul Forzamento delle Corone Dentate" (Research on Shrunk-on Gear Rings), *Technica Italiana, Triesta*, Vol. 33, p. 1.

Dolan, T. J., and Broghamer, E. L., 1942, "A Photoelastic Study of Stresses in Gear Tooth Fillets," *Univ. Illinois Expt. Sta. Bull. 335*.

Durelli, A. J., and Riley, W. F., 1965, *Introduction to Photomechanics*, Prentice-Hall, Englewood Cliffs, N.J., p. 220.

ESDU (Engineering Sciences Data Unit), 1981, *Stress Concentrations*, London.

Fessler, H., and Stanley, P., 1965, "Stresses in Torispherical Drumheads: A Photoelastic Investigation," *J. Strain Anal.*, Vol. 1, p. 69.

Fessler, H., and Stanley, P., 1966, "Stresses in Torispherical Drumheads: A Critical Evaluation," *J. Strain Anal.*, Vol. 1, p. 89.

Fessler, H., and Thorpe, T. E., 1967a, "Optimization of Stress Concentrations at Holes in Rotating Discs," *J. Strain Anal.*, Vol. 2, p. 152.

Fessler, H., and Thorpe, T. E., 1967b, "Reinforcement of Non-Central Holes in Rotating Discs," *J. Strain Anal.*, Vol. 2, p. 317.

Fessler, H., Rogers, C. C., and Stanley, P., 1969a, "Stresses at End-milled Keyways in Plain Shafts Subjected to Tension, Bending and Torsion," *J. Strain Anal.*, Vol. 4, p. 180.

Fessler, H., Rogers, C. C., and Stanley, P., 1969b, "Stresses at Keyway Ends Near Shoulders," *J. Strain Anal.*, Vol. 4, p. 267.

Frocht, M. M., and Hill, H. N., 1940, "Stress-Concentration Factors around a Central Circular Hole in a Plate Loaded through Pin in the Hole," *Trans. ASME, Appl. Mech. Section,* Vol. 62, pp. A5.-A9.

Gerdeen, J. C., 1972, "Analysis of Stress Concentration in Thick Cylinders with Sideholes and Crossholes," *Trans. ASME, J. Eng. for Industry*, Vol. 94, Series B, p. 815.

Gerdeen, J. C., and Smith, R. E., 1972, "Experimental Determination of Stress Concentration Factors in Thick-walled Cylinders with Crossholes and Sideholes," *Exp. Mech.*, Vol. 12, p. 530.

Gibson, W. H., and Gilet, P. M., 1938, "Transmission of Torque by Keys and Keyways," *J. Inst. Engrs. Australia*, Vol. 10, p. 393.

Göhner, O., 1932, "Die Berechnung Zylindrischer Schraubenfedern," *Z. Ver. Deutsch. Ing.*, Vol. 76, p. 269.

Green, W. A., Hooper, G. T. J., and Hetherington, R., 1964, "Stress Distributions in Rotating Discs with Non-Central Holes," *Aeronaut. Q.*, Vol. 15, p. 107.

Gregory, R. D., 1968, "Stress Concentration around a Loaded Bolt in an Axially Loaded Bar," *Proc. Camb. Phil. Soc. No. 64*, pp. 1215–1326.

Griffith, G. I., and Taylor, A. A., 1917–1918, "Use of Soap Films in Solving Torsion Problems," *Tech. Rep. Brit. Adv. Comm. Aeronaut.*, Vol. 3, p. 910.

Hetényi, M., 1939a, "The Application of Hardening Resins in Three-Dimensional Photoelastic Studies," *J. Appl. Phys.*, Vol. 10, p. 295.

Hetényi, M., 1939b, "Some Applications of Photoelasticity in Turbine-Generator Design," *Trans. ASME, Appl. Mech. Section*, Vol. 61, p. A-151.

Hetényi, M., 1943, "The Distribution of Stress in Threaded Connections," *Proc. Soc. Exp. Stress Analysis*, Vol. 1, No. 1, p. 147.

Heywood, R. B., 1969, *Photoelasticity for Designers*, Pergamon, New York, p. 326.

Horger, O. J., 1953, "Press Fitted Assembly," *ASME Metals Engrg. Handbook—Design*, McGraw-Hill, New York, p. 178.

Horger, O. J., 1956, "Fatigue of Large Shafts by Fretting Corrosion," *Proc. Int. Conf. Fatigue, Inst. Mech. Eng.*, London, p. 352.

Horger, O. J., and Buckwalter, T. V., 1940, "Photoelasticity as Applied to Design Problems," *Iron Age*, Vol. 145, Part II, No. 21, p. 42.

Horger, O. J., and Maulbetsch, J. L., 1936, "Increasing the Fatigue Strength of Press-Fitted Axle Assemblies by Rolling," *Trans. ASME, Appl. Mech. Section*, Vol. 58, p. A-91.

Huth, J. H., 1950, "Torsional Stress Concentration in Angle and Square Tube Fillets," *Trans. ASME, Appl. Mech. Section,* Vol. 72, p. 388.

Jacobson, M. A., 1955, "Bending Stresses in Spur Gear Teeth; Proposed New Design Factors based on a Photoelastic Investigation," *Proc. Inst. Mech. Eng.*, Vol. 169, p. 587.

Kaufmann, F., and Jäniche, W., 1940, "Beitrag zur Dauerhaltbarkeit von Schraubenverbindungen," *Tech. Mitt. Krupp, Forschungsber.*, Vol. 3, p. 147.

Leist, K., and Weber, J., 1956, "Optical Stress Distributions in Rotating Discs with Eccentric Holes," Report No. 57, Institute for Jet Propulsion, German Research Institute for Aeronautics, Aachen.

Leven, M. M., 1938, 1952, Unpublished data obtained at Carnegie Inst. of Technology and Westinghouse Research Labs.

Leven, M. M., 1949, "Stresses in Keyways by Photoelastic Methods and Comparison with Numerical Solution," *Proc. Soc. Exp. Stress Analysis*, Vol. 7, No. 2, p. 141.

Liesecke, G., 1933, "Berechnung Zylindrischer Schraubenfedern mit Rechteckigem Drahtquerschnitt," *Z. VDI*, Vol. 77, pp. 425, 892.

Lyse, I., and Johnston, B. G., 1935, "Structural Beams in Torsion," Lehigh University Publication, Vol. 9, p. 477.

Mantle, J. B., and Dolan, T. J., 1948, "A Photoelastic Study of Stresses in U-Shaped Members," *Proc. Soc. Exp. Stress Analysis*, Vol. 6, No. 1, p. 66.

Marino, R. L., and Riley, W. F., 1964, "Optimizing Thread-root Contours Using Photoelastic Methods," *Exp. Mech.*, Vol. 4, p. 1.

Martinaglia, L., 1942, "Schraubenverbindungen," *Schweiz. Bauztg.*, Vol. 119, p. 107.

Meek, R. M. G., 1967, "Effect of Pin Bending on the Stress Distribution in Thick Plates Loaded through Pins," NEL Report No. 311.

Michalec, G. W., 1966, *Precision Gearing—Theory and Practice*, Wiley, New York, p. 466.

Nishioka, K., Nishimura, S., and Hirakawa, K., 1968, "Fundamental Investigations of Fretting Fatigue," *Bull. Japan Soc. Mech. Eng.*, Vol. 11, p. 437; 1969, Vol. 12, pp. 180, 397, 408.

Nishioka, K., and Komatsu, H., 1967, "Researches on Increasing the Fatigue Strength of Press-Fitted Shaft Assembly," *Bull. Japan Soc. Mech. Eng.*, Vol. 10, p. 880.

Nisida, M., 1963, "New Photoelastic Methods for Torsion Problems," *Symposium on Photoelasticity*, M. M. Frocht, Ed., Pergamon, New York, p. 109.

Okubo, H., 1950a, "On the Torsion of a Shaft with Keyways," *Q. J. Mech. Appl. Math.*, Vol. 3, p. 162.

Okubo, H., 1950b, "Torsion of a Circular Shaft with a Number of Longitudinal Notches," *Trans. ASME, Appl. Mech. Section*, Vol. 72, p. 359.

Okubo, H., Hosono, K., and Sakaki, K., 1968, "The Stress Concentration in Keyways when Torque is Transmitted through Keys," *Exp. Mech.*, Vol. 8, p. 375.

Peterson, R. E., 1930, "Load and Deflection Cycles in Gear Teeth," *Proc. 3rd Intern. Appl. Mech. Congr.*, p. 382.

Peterson, R. E., 1932, "Fatigue of Shafts Having Keyways," *Proc. ASTM*, Vol. 32, Part 2, p. 413.

Peterson, R. E., 1938, "Methods of Correlating Data from Fatigue Tests of Stress Concentration Specimens," *Stephen Timoshenko Anniversary Volume*, Macmillan, New York, p. 179.

Peterson, R. E., 1950, "Interpretation of Service Fractures," *Handbook of Experimental Stress Analysis*, Chapter 13, M. Hetényi, Ed., Wiley, New York, pp. 603, 608, 613.

Peterson, R. E., 1953, *Stress Concentration Design Factors*, John Wiley, New York.

Peterson, R. E., and Wahl, A. M., 1935, "Fatigue of Shafts at Fitted Members, with a Related Photoelastic Analysis," *Trans. ASME, Appl. Mech. Section*, Vol. 57, p. A-1.

Pilkey, W. D., 1994, *Formulas for Stress, Strain, and Structural Matrices*, Wiley, New York.

Richart, F. E., Olson, T. A., and Dolan, T. J., 1938, "Tests of Reinforced Concrete Knee Frames and Bakelite Models," *Univ. Illinois Expt. Sta. Bull. 307*.

Riggs, N. C., and Frocht, M. M., 1938, *Strength of Materials*, Ronald Press, New York, p. 389.

Roark, R. J., 1989, *Formulas for Stress and Strain*, 6th ed., McGraw-Hill, New York.

Robinson, E. L., 1944, "Bursting Tests of Steam-Turbine Disk Wheels," *Trans. ASME, Appl. Mech. Section*, Vol. 66, p. 380.

Seika, M., 1958, "The Stresses in a Thick Cylinder Having a Square Hole Under Concentrated Loading," *Trans. ASME, Appl. Mech. Section,* Vol. 80, p. 571.

Solakian, A. G., and Karelitz, G. B., 1932, "Photoelastic Studies of Shearing Stresses in Keys and Keyways," *Trans. ASME, Appl. Mech. Section*, Vol. 54, p. 97.

Staedel, W., 1933, "Dauerfestigkeit von Schrauben," *Mitt. der Materialprüfungsanstalt an der Technischen Hochschule Darmstadt*, No. 4, VDI Verlag, Berlin.

Theocaris, P. S., 1956, "The Stress Distribution in a Strip Loaded in Tension by Means of a Central Pin," *Trans. ASME*, Vol. 78, pp. 85–90.

Thum, A., and Bruder, E., 1938, "Dauerbruchgefahr an Hohlkehlen von Wellen und Achsen und ihre Minderung," *Deutsche Kraftfahrtforschung im Auftrag des Reichs-Verkehrsministeriums*, No. 11, VDI Verlag, Berlin.

Thum, A., and Wunderlich, F., 1933, "Der Einfluss von Einspann-und Kantandgriffsstellen auf die Dauerhaltbarkeit der Konstruktionen," *Z. VDI*, Vol. 77, p. 851.

Timoshenko, S., 1922, "On the Distribution of Stresses in a Circular Ring Compressed by Two Forces along a Diameter," *Phil. Mag.*, Vol. 44, p. 1014.

Timoshenko, S., 1956, *Strength of Materials,* Part II, 3rd ed., Van Nostrand, Princeton, N.J., pp. 214, 362, 380, 444.

Timoshenko, S., and Goodier, J. N., 1970, *Theory of Elasticity*, 3rd Ed., McGraw Hill, New York, p. 398.

Tomlinson, G. A., 1927, "The Rusting of Steel Surface on Contact," *Proc. Roy. Soc. (London) A*, Vol. 115, p. 472.

Tomlinson, G. A., Thorpe, P. L., and Gough, H. J., 1939, "An Investigation of the Fretting Corrosion of Closely Fitting Surfaces," *Proc. Inst. Mech. Engrs.* (London), Vol. 141, p. 223.

Yoshitake, H., 1962, "Photoelastic Stress Analysis of the Spline Shaft," *Bull. Japan Soc. Mech. Eng.*, Vol. 5, p. 195.

Wahl, A. M., 1946, "Calculation of Stresses in Crane Hooks," *Trans. ASME, Appl. Mech. Section*, Vol. 68, p. A-239.

Wahl, A. M., 1963, *Mechanical Springs*, 2nd ed., McGraw-Hill, New York.

Weibel, E. E., 1934, "Studies in Photoelastic Stress Determination," *Trans. ASME, Appl. Mech. Section*, Vol. 56, p. 637.

White, D. J., and Humpherson, J., 1969, "Finite-Element Analysis of Stresses in Shafts Due to Interference-Fit Hubs," *J. Strain Anal.*, Vol. 4, p. 105.

Whitehead, R. S., Abbey, A. J., and Glen-Bott, M. G., 1978, "Analytical Determination of Stress Intensity Factors for Attachment Lugs," British Aerospace, Aircraft Group, Warton Division, Report No. SON(P) 199.

Wiegand, H., 1933, "Über die Dauerfestigkeit von Schraubenwerkstoffen und Schraubenverbindungen," Thesis, *Technische Hochshule Darmstadt*. Also published as No. 14 (1934), *Wissenschaftliche Veröffentlichungen der Firma*, Bauer & Schaurte A. G., Neuss.

Wilson, B. J., and Quereau, J. F., 1928, "A Simple Method of Determining Stress in Curved Flexural Members," *Univ. Illinois Expt. Sta. Circ. 16*.

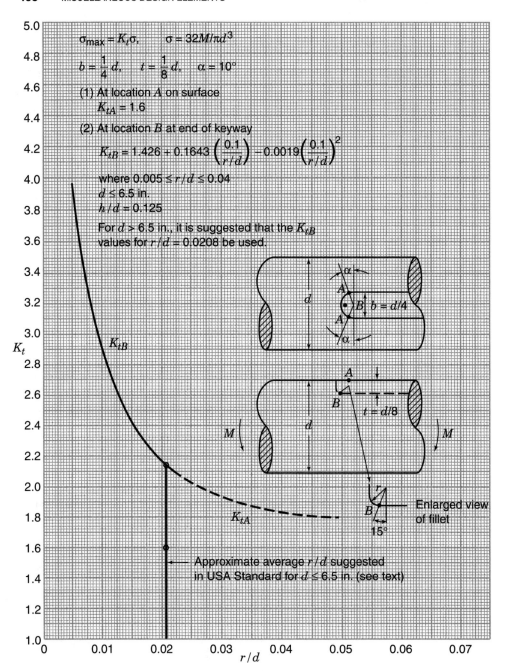

Chart 5.1 Stress concentration factors K_t for bending of a shaft of circular cross section with a semicircular end keyseat (based on data of Fessler et al. 1969).

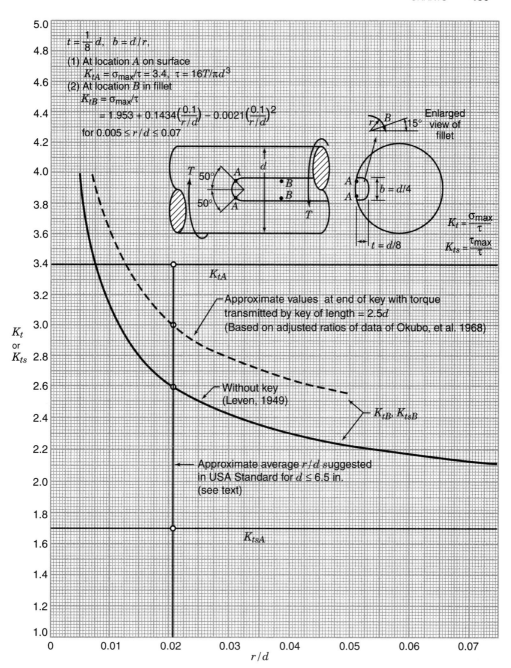

Chart 5.2 Stress concentration factors K_t, K_{ts} for a torsion shaft with a semicircular end keyseat (Leven 1949; Okubo et al. 1968).

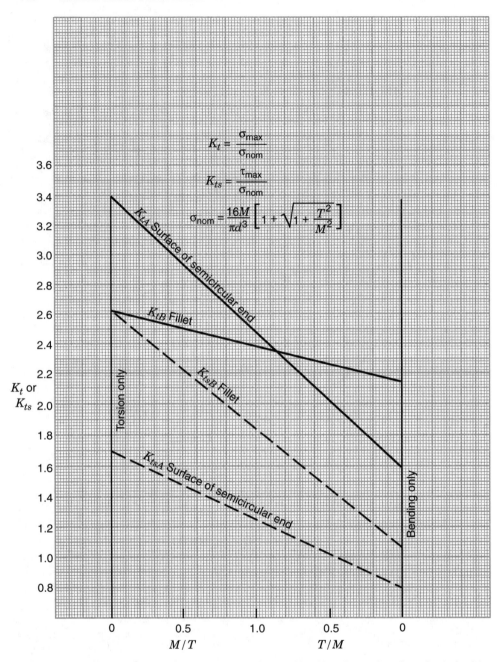

Chart 5.3 Stress concentration factors K_t and K_{ts} for combined bending and torsion of a shaft with a semicircular end keyseat. $b/d = 1/4$, $t/d = 1/8$, $r/d = 1/48 = 0.0208$ (approximate values based on method of Fessler et al. 1969).

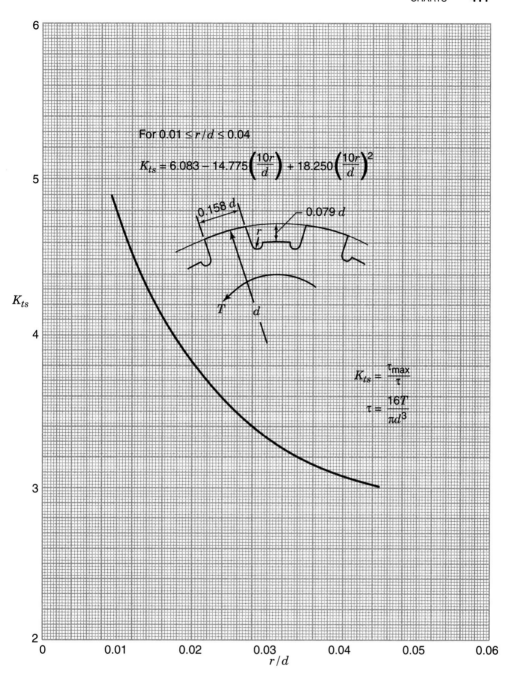

Chart 5.4 Stress concentration factors K_{ts} for torsion of a splined shaft without mating member (photoelastic tests of Yoshitake, et al. 1962). Number of teeth = 8.

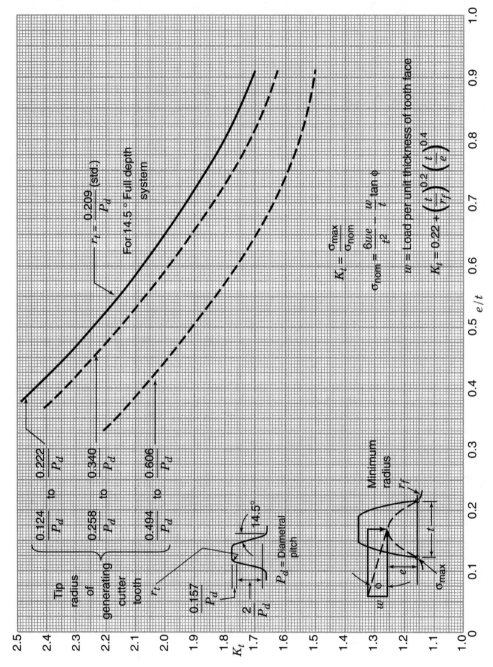

Chart 5.5 Stress concentration factors K_t for the tension side of a gear tooth fillet with 14.5° pressure angle (from photoelastic data of Dolan and Broghamer 1942).

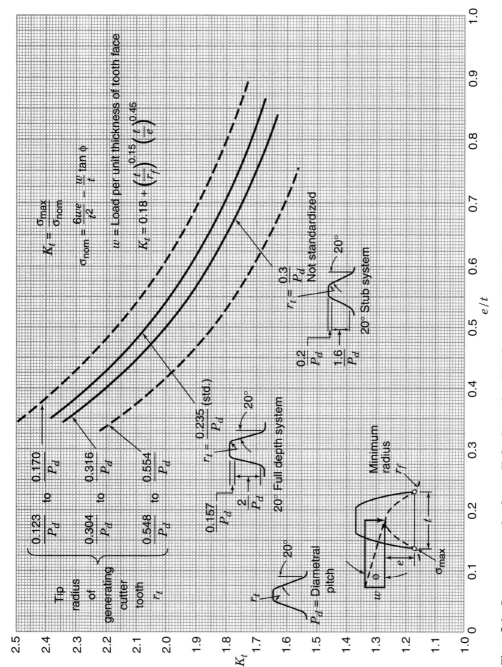

Chart 5.6 Stress concentration factors K_t for the tension side of a gear tooth fillet, 20° pressure angle (from photoelastic data of Dolan and Broghamer 1942).

414 MISCELLANEOUS DESIGN ELEMENTS

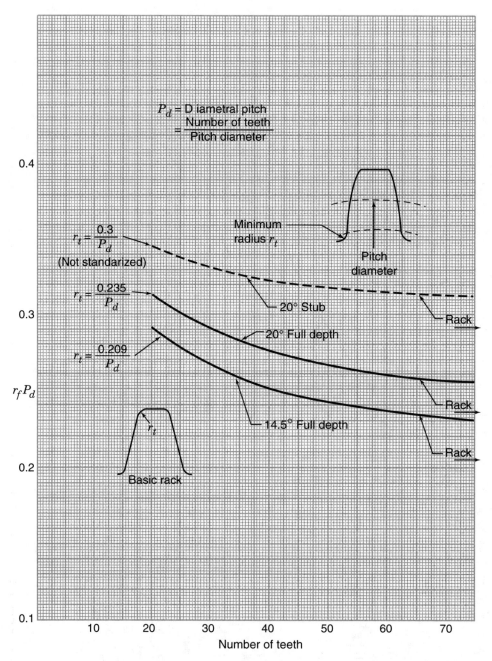

Chart 5.7 Minimum fillet radius r_f of gear tooth generated by basic rack (formula of Candee 1941).

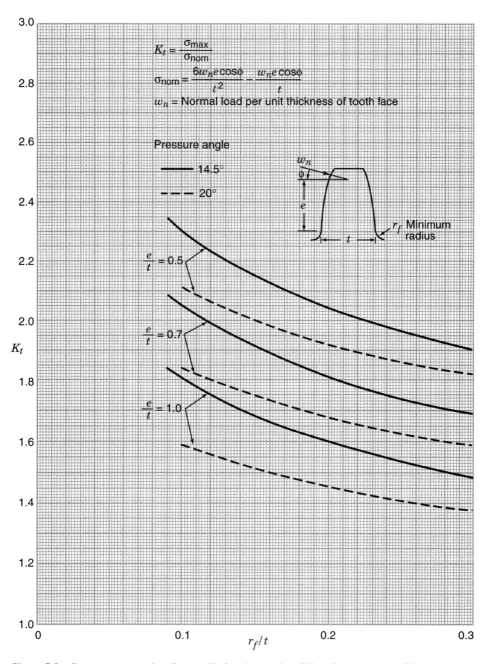

Chart 5.8 Stress concentration factors K_t for the tension Side of a gear tooth fillet (empirical formula of Dolan and Broghamer 1942).

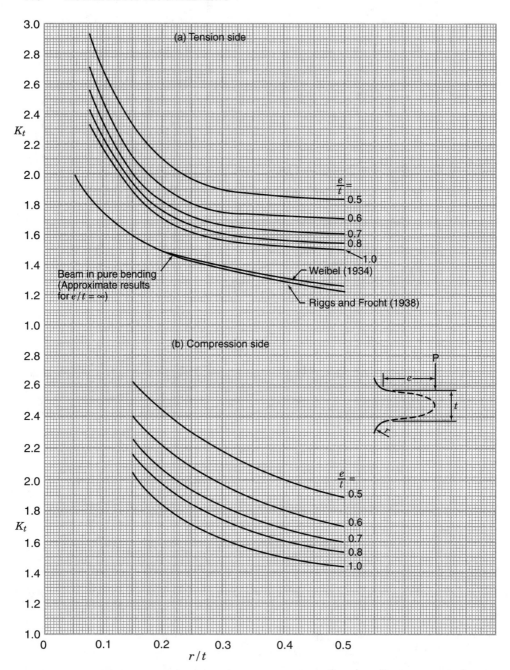

Chart 5.9 Stress concentration factors K_t for bending of a short beam with a shoulder fillet (photoelastic tests of Dolan and Broghamer 1942). (*a*) Tension side; (*b*) compression side.

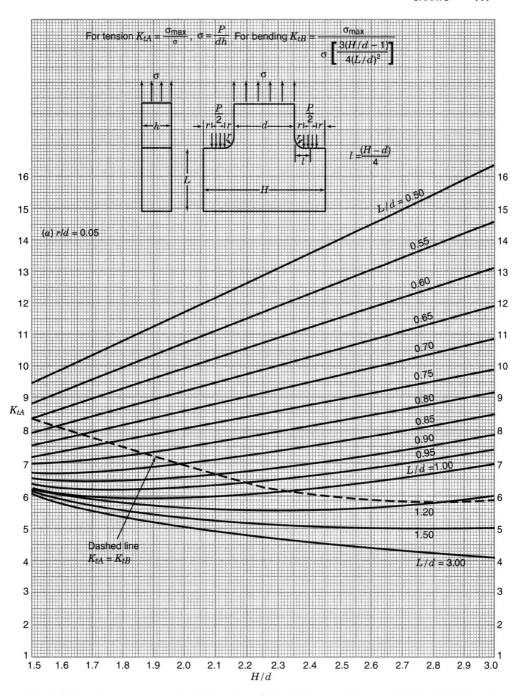

Chart 5.10a Stress concentration factors for a T-Head (photoelastic tests of Hetényi 1939b and 1943). (a) $r/d = 0.05$.

418 MISCELLANEOUS DESIGN ELEMENTS

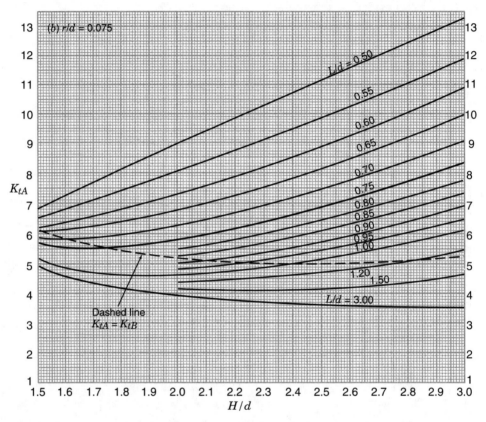

Chart 5.10b Stress concentration factors for a T-Head (photoelastic tests of Hetényi 1939b and 1943). (*b*) $r/d = 0.075$.

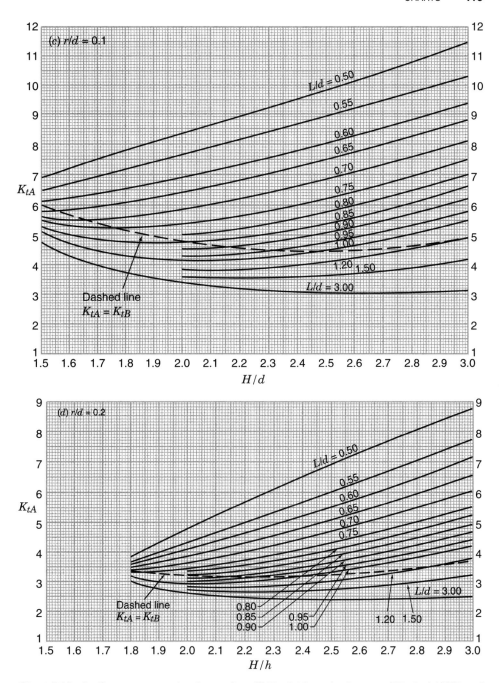

Chart 5.10c,d Stress concentration factors for a T-Head (photoelastic tests of Hetényi 1939b and 1943). (c) $r/d = 0.1$; (d) $r/d = 0.2$.

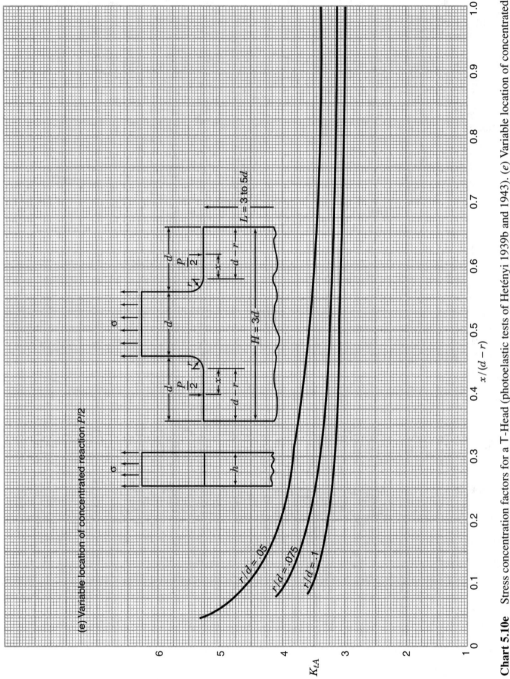

Chart 5.10e Stress concentration factors for a T-Head (photoelastic tests of Hetényi 1939b and 1943). (*e*) Variable location of concentrated reaction $P/2$.

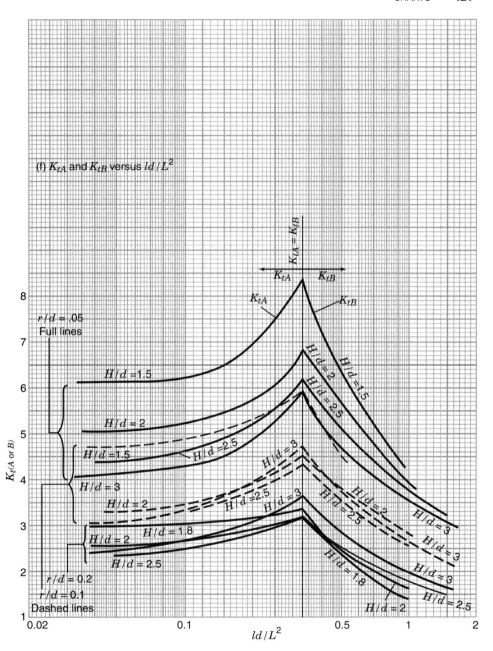

Chart 5.10f Stress concentration factors for a T-Head (photoelastic tests of Hetényi 1939b and 1943). (*f*) K_{tA} and K_{tB} versus ld/L^2.

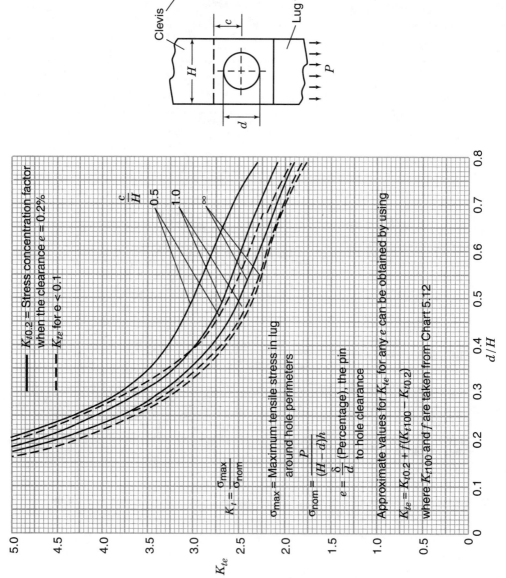

Chart 5.11 Stress concentration factors K_{te} for square-ended lugs, $c/H > 1.5$, $h/d < 0.5$ (Whitehead et al. 1978; ESDU 1981).

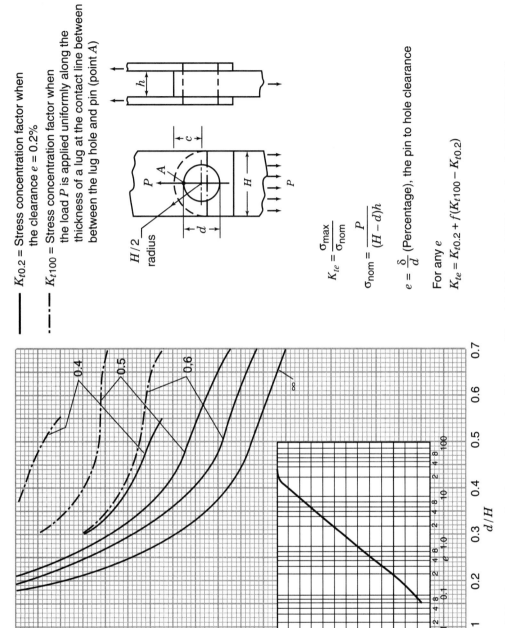

— $K_{t0.2}$ = Stress concentration factor when the clearance $e = 0.2\%$

−·−·− K_{t100} = Stress concentration factor when the load P is applied uniformly along the thickness of a lug at the contact line between between the lug hole and pin (point A)

$$K_{te} = \frac{\sigma_{max}}{\sigma_{nom}}$$

$$\sigma_{nom} = \frac{P}{(H-d)h}$$

$e = \frac{\delta}{d}$ (Percentage), the pin to hole clearance

For any e
$$K_{te} = K_{t0.2} + f(K_{t100} - K_{t0.2})$$

Chart 5.12 Stress concentration factors K_{te} round-ended lugs, $c/H > 1.5$, $h/d < 0.5$ (Whitehead et al. 1978; ESDU 1981).

424 MISCELLANEOUS DESIGN ELEMENTS

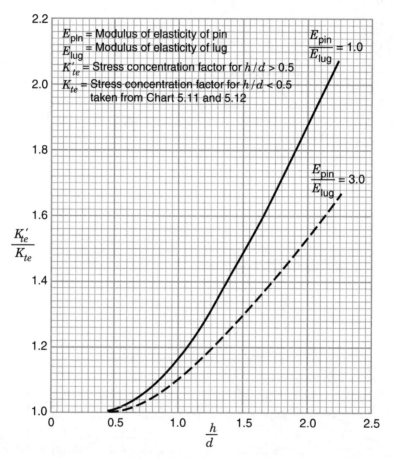

Chart 5.13 Stress concentration factors K'_{te} for thick lugs. Square or round ended lugs with $h/d > 0.5$ and $0.3 \leq d/H \leq 0.6$ (Whitehead et al. 1978; ESDU 1981).

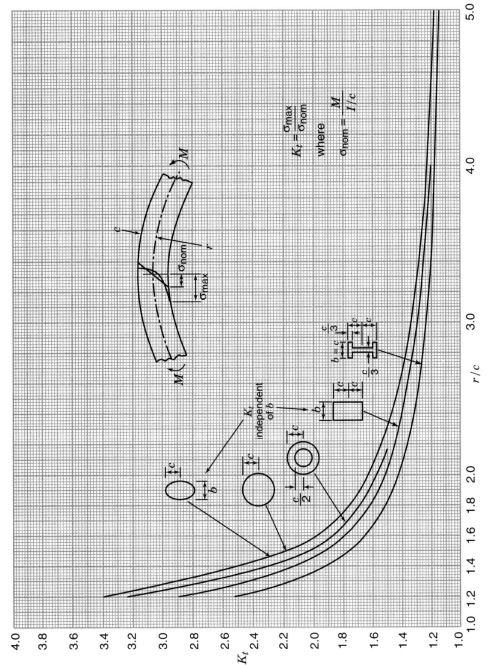

Chart 5.14 Stress concentration factors K_t for a curved bar in bending.

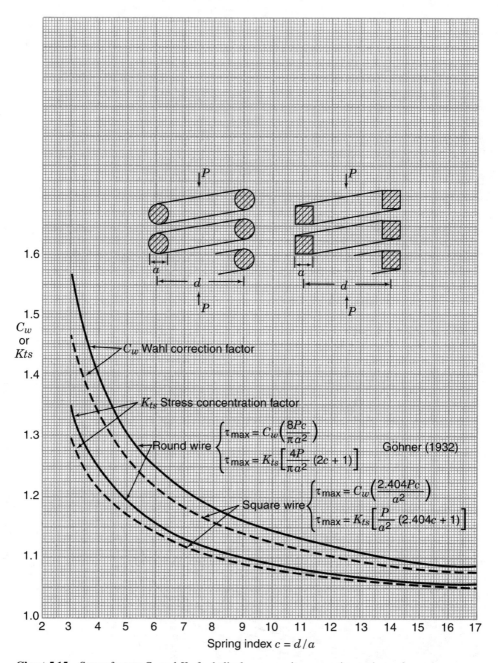

Chart 5.15 Stress factors C_w and K_{ts} for helical compression or tension springs of round or square wire (from mathamatical relations of Wahl 1963).

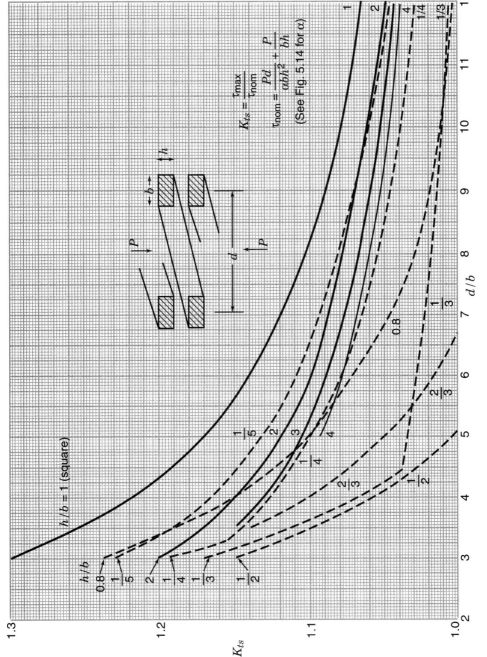

Chart 5.16 Stress concentration factors K_{ts} for a helical compression or tension spring of rectangular wire cross section (based on Liesecke 1933).

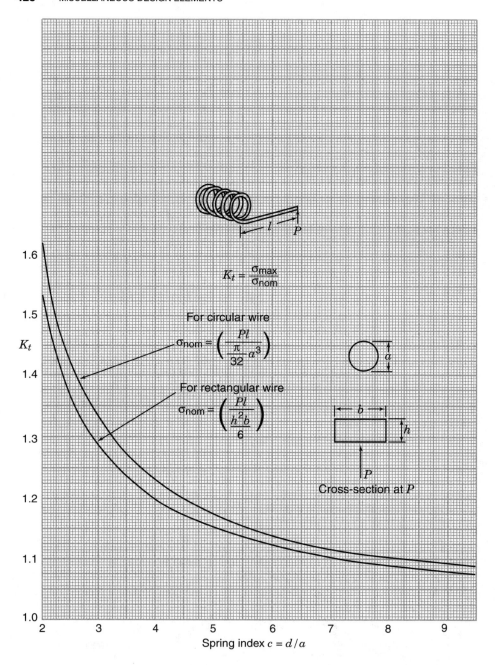

Chart 5.17 Stress concentration factors K_t for a helical torsion spring (Wahl 1963).

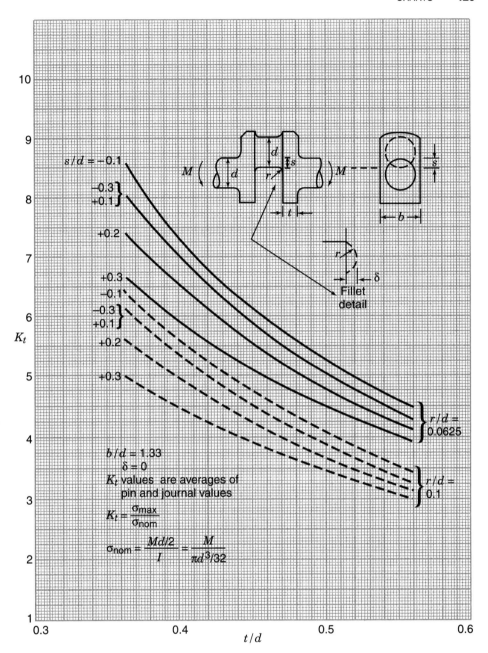

Chart 5.18 Stress concentration factors K_t for a crankshaft in bending (from strain gage values of Arai 1965).

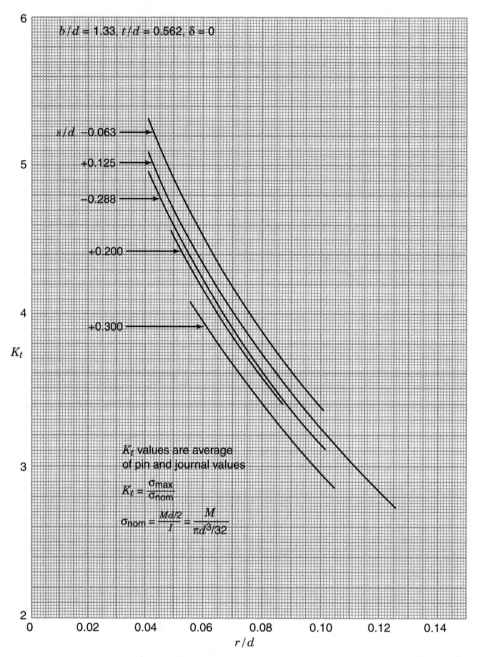

Chart 5.19 Stress concentration factors K_t for a crankshaft in bending (strain gage values of Arai 1965). See figure of Chart 5.18 for notation.

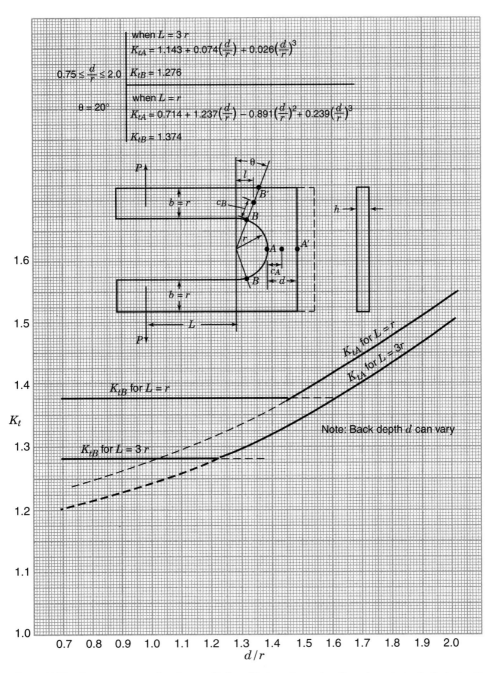

Chart 5.20 Stress concentration factors K_t for a U-shaped member (based on photoelastic tests of Mantle and Dolan 1948).

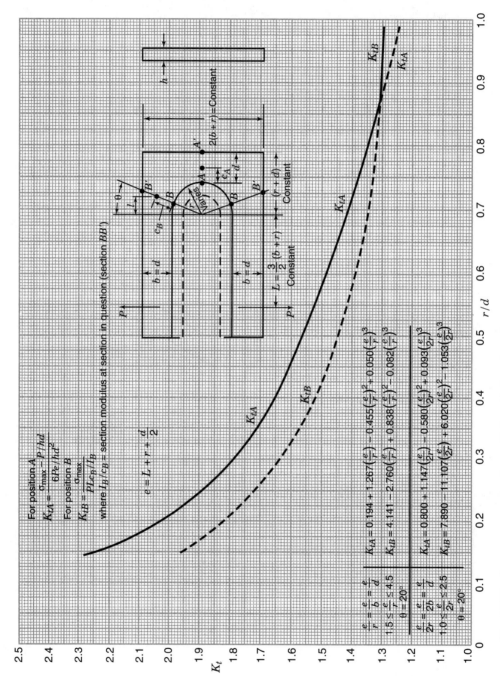

Chart 5.21 Stress concentration factors K_t for a U-shaped member (based on photoelastic tests of Mantle and Dolan 1948).

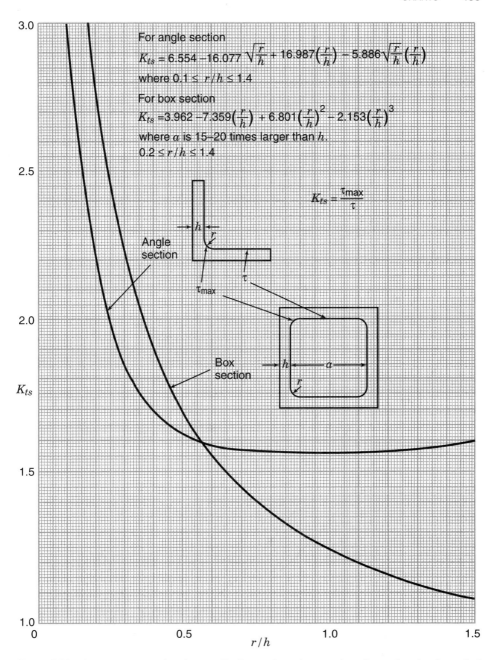

Chart 5.22 Stress concentration factors K_{ts} for angle or box sections in torsion (mathematical determination by Huth 1950).

434 MISCELLANEOUS DESIGN ELEMENTS

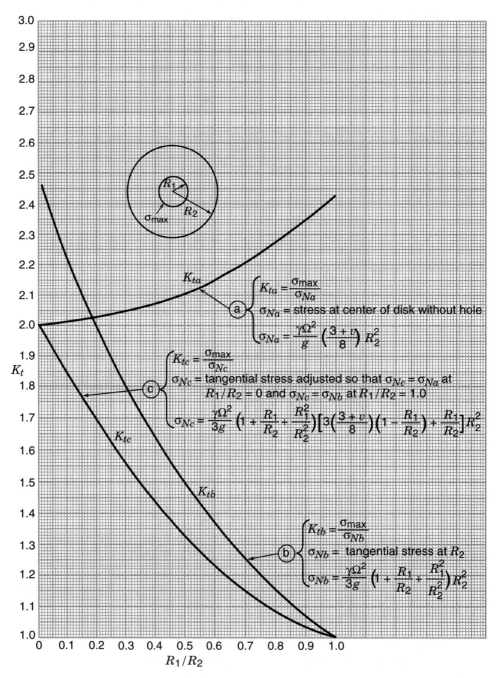

Chart 5.23 Stress concentration factors K_t for a rotating disk with a central hole.

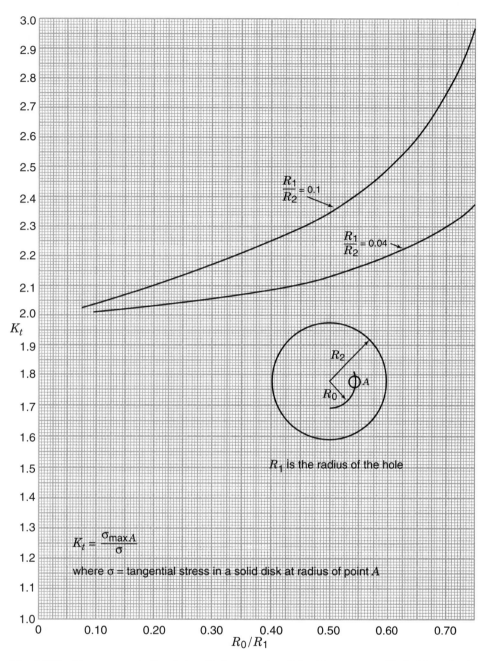

Chart 5.24 Stress concentration factors K_t for a rotating disk with a noncentral hole (photoelastic tests of Barnhart, Hale, and Meriam 1951).

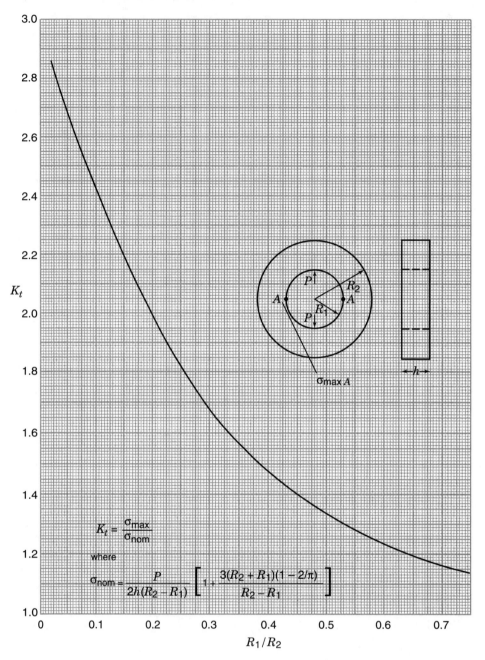

Chart 5.25 Stress concentration factors K_t for a ring or hollow roller subjected to diametrically opposite internal concentrated loads (Timoshenko 1922; Horger and Buckwalter 1940; Leven 1952).

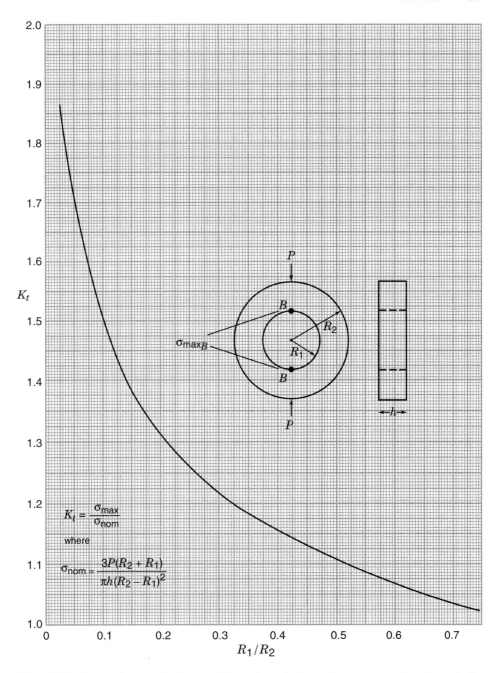

Chart 5.26 Stress concentration factors K_t for a ring or hollow roller compressed by diametrically opposite external concentrated loads (Timoshenko 1922; Horger and Buckwalter 1940; Leven 1952).

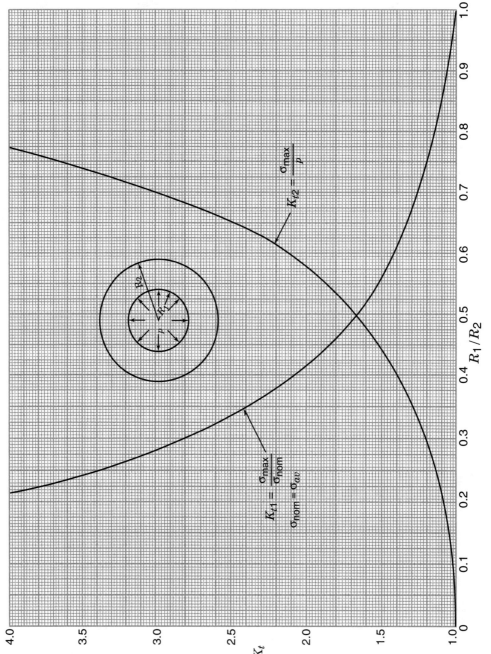

Chart 5.27 Stress concentration factors K_t for a cylinder subject to internal pressure (based on Lamé solution, Pilkey 1994).

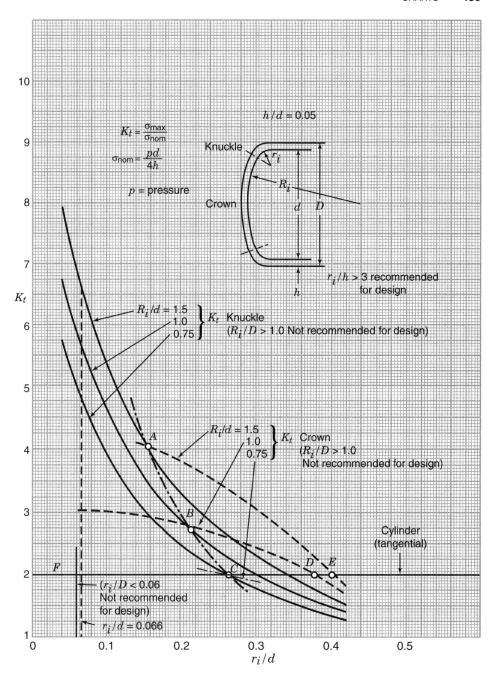

Chart 5.28 Stress concentration factors K_t for a cylindrical pressure vessel with torispherical ends (from data of Fessler and Stanley 1965).

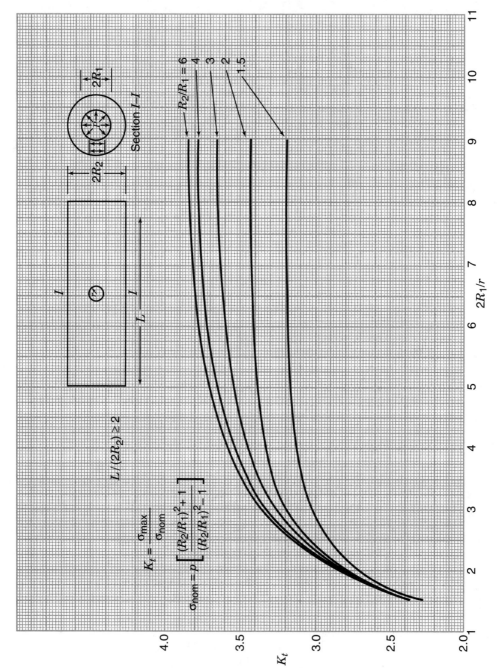

Chart 5.29 Stress concentration factors K_t for a pressurized thick cylinder with a circular hole in the cylinder wall (Gerdeen 1972).

CHAPTER 6

STRESS CONCENTRATION ANALYSIS AND DESIGN

6.1 COMPUTATIONAL METHODS

Computers have revolutionized the study of stress concentrations. Today the analysis of structures with the help of computational methods is ubiquitous. Powerful algorithms have been developed. The most acknowledged and the most flexible computational method for structural analysis is the finite element method. The analysis of stress concentration and the design to avoid harmful stress concentrations can be efficiently accomplished using this numerical tool. The universality of the finite element method allows the analysis of even complicated geometries.

High-quality finite element software for the solution of elasticity problems including those of stress concentrations is available. These codes are often very complex, and they can provide solutions to a wide range of problems. Linear static analysis in two and three dimensions which is necessary to analyze stress concentrations is a basic part of all such programs. Some of the well-known products are NASTRAN (1994), ANSYS (1992), and ABAQUS (1995). To relieve the engineer from the expensive generation of the data and evaluation of the results, finite element codes are usually supported by easy to handle pre- and postprocessing software that have graphic user interfaces. Examples are PATRAN (1996) and HyperMesh (1995). A comprehensive account of current finite element software has been compiled by Mackerle (1995).

Example 6.1 Panel with a Circular Hole under Biaxial Tension   To illustrate the computation of stress concentration factors, the thin flat element of Fig. 6.1 will be analyzed. The example problem is a 80×80 mm square panel with a circular hole of 20 mm in diameter. The panel is loaded with a biaxial tension of $\sigma_1 = \sigma_2 = \sigma = 100$ MPa. The problem is solved using a finite element research code. The solution has been obtained for several

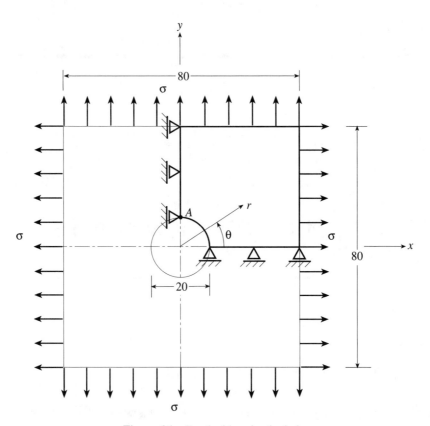

Figure 6.1 Panel with a circular hole.

different discretizations. Due to double symmetry, only one-fourth of the panel needs to be discretized in the numerical model. Figure 6.2 shows one of the discretizations used in the analysis. The maximum normal stresses $\sigma_{x\,\text{max}}$, which occur at point A, Fig. 6.1, are summarized in Table 6.1. Lines of constant σ_x, in MPa, are shown in Fig. 6.2 for mesh number 2 of Table 6.1.

The example shows that for different finite element meshes, different results are obtained. In general, it can be stated that if the number of degrees of freedom is increased, the accuracy of the results increases. As seen from the example, the rate of increase of accuracy depends on different factors. The example shows the direction dependency of the refinement. A refinement in the r direction yields a faster convergence of the results than a refinement in the θ direction. Mesh number 1 will be considered the basic discretization. Then, for example, despite a much greater effort, mesh number 3 which is a refinement in the θ direction yields only a minor improvement of the results compared with mesh number 4 which is a refinement in the r direction.

A stress concentration factor for this problem can be computed using Eq. (1.1):

$$K_t = \frac{\sigma_{\text{max}}}{\sigma_{\text{nom}}} \tag{1}$$

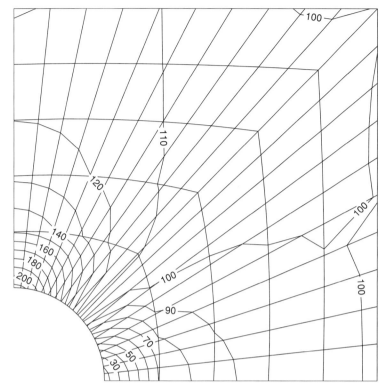

Figure 6.2 Panel with a circular hole. Stress distribution σ_x for a 20 × 5 mesh (number 2 of Table 6.1)

TABLE 6.1 Stresses in a Panel with a Circular Hole

Mesh Number	Mesh Size[a] per Quarter	DOF[b]	$\sigma_{x\,max}$[c] [MPa]
1	10 × 5	440	214.231
2	20 × 5	880	214.200
3	30 × 5	1320	214.195
4	10 × 10	840	211.990
5	10 × 15	1240	211.421
6	20 × 10	1680	211.971

[a] Mesh size—number of nine node elements in $\theta \times r$ directions (Fig. 6.1). For example, 20 × 5 indicates that there are 20 elements in the θ direction and 5 elements in the r direction as shown in Fig. 6.2.
[b] DOF—degrees of freedom (number of unknowns).
[c] $\sigma_{x\,max}$—maximum normal stress.

The nominal stress is the load, $\sigma_{nom} = \sigma$. Using the maximum stresses computed with finite elements, K_t is

$$K_t = 2.11 \tag{2}$$

To find an analytical expression for the stress concentration factor K_t, consult Chapter 4. There, the stress concentration factor for an infinite thin element with a circular hole under biaxial tension is derived. It is given by Eq. (4.18):

$$K_t = 3 - \frac{\sigma_2}{\sigma_1} = 2 \tag{3}$$

Comparison of (2) and (3) shows that the actual stress concentration in the hole is about 5% higher than obtained from the table. This is due to the fact that the panel of the example is not infinite.

Although the geometry for this example is very simple, the application of finite elements has no practical limits as to the structural geometry. Problems of geometric complexity beyond those given in the previous chapters can be solved.

In terms of design, numerical analyses permit conclusions to be drawn as to how to reduce stresses in critical regions. The results of a stress analysis can be analyzed for their sensitivity with respect to design changes. Then design changes can be determined to reduce stress concentrations, for example, by modifying the shape of the region where critical stresses occur. This is a classical design problem that can be expressed as follows: Determine the shape of a structure such that the stresses do not exceed a certain value or such that the peak stresses are minimized.

The design problem can be formulated as a *structural optimization* problem, in that the most favorable design is sought that fulfills certain *constraints* on the behavior of the structure. The *design variables* are the parameters in this process. They are to be changed in order to solve the optimization problem.

The structural optimization problem can be written as

$$
\begin{array}{lll}
\text{Objective function} & W(\mathbf{b}) \Rightarrow \min & \\
\text{Subject to} & \mathbf{D}^T \mathbf{E} \mathbf{D} \mathbf{u} + \bar{\mathbf{p}}_V = \mathbf{0} & \\
\text{Inequality constraints} & \mathbf{g}(\mathbf{u}, \mathbf{b}) \leq \mathbf{0} & (6.1)\\
\text{Equality constraints} & \mathbf{h}(\mathbf{b}) = \mathbf{0} & \\
\text{Side constraints} & \mathbf{b}^l \leq \mathbf{b} \leq \mathbf{b}^u & \\
\end{array}
$$

The objective function W is often the total or a weighted mass of the structure, but can also be any response function of the structure. Functions $\mathbf{g} = [g_i]^T$, $i = 1, \ldots, n_g$, are structural responses such as stresses or displacements. The functions $\mathbf{h} = [h_i]^T$, $i = 1, \ldots, n_h$, can be structural responses and constraints on the design variables, such as variable dependencies. The structural behavior is governed by the differential equation $\mathbf{D}^T \mathbf{E} \mathbf{D} \mathbf{u} + \bar{\mathbf{p}}_V = \mathbf{0}$ which describes the case of linear statics that will be considered here (Pilkey and Wunderlich 1993). The vector $\mathbf{u}$ contains the displacements, the matrix $\mathbf{E}$ represents the material law and contains the elastic material properties, the matrix $\mathbf{D}$ is a differential matrix operator, and the vector $\bar{\mathbf{p}}_V$ is formed of the body forces. The quantity $\mathbf{b} = [b_j]^T$, $j = 1, \ldots, n_b$, is the vector of design variables. The vectors $\mathbf{b}^l$ and $\mathbf{b}^u$ are the lower and upper bounds on the design variables, respectively. The set of all possible designs defined by these constraints is called the *design space*. A design $\mathbf{b}$ that fulfills the constraints is called a *feasible design*. Otherwise the design is called *infeasible*.

Depending on the kind of design variables, several types of structural optimization problems can be distinguished. For the first type, *sizing optimization*, the design variables are input parameters of structural elements such as beam cross-sectional properties or plate thicknesses. For the second type, the *shape optimization*, the design variables are control variables of the geometry of the structure. A third type of structural optimization problem, which will not be considered here, is known as *topology optimization*. In this case the design variables determine the material distribution in the structure.

Structural optimization capabilities are available in general finite element codes such as NASTRAN (1994) and ANSYS (1992). Special structural optimization software based on finite elements is also available. Examples are the codes GENESIS (1995) for sizing as well as for shape optimization and OptiStruct (1996) for topology optimization. Comprehensive summaries of structural optimization software can be found in Johnson (1993) and Duysinx and Fleury (1993).

This chapter begins with the foundations of the finite element technique as a method for stress concentration analysis. Second, the analysis of the sensitivity of the stresses with respect to design changes is described, and third, the computation of design improvements using optimization methods is explained. Two examples running through the chapter are used to illustrate the derivations. Although the basic equations given here are valid in general, the examples are restricted to two-dimensional problems. A somewhat selective treatment is given here of stress concentration analysis and design problems. The reader is frequently referred to textbooks and other literature for further reading. The exposition of this chapter is intended to be sufficient to give the reader a general understanding of the methodology.

6.2 FINITE ELEMENT ANALYSIS

6.2.1 Principle of Virtual Work

Figure 6.3 shows an elastic body subject to the body forces $\bar{\mathbf{p}}_V = [\bar{p}_{Vx} \; \bar{p}_{Vy} \; \bar{p}_{Vz}]^T$, the surface tractions $\bar{\mathbf{p}}_S = [\bar{p}_{Sx} \; \bar{p}_{Sy} \; \bar{p}_{Sz}]^T$, on S_p and the prescribed boundary displacements $\bar{\mathbf{u}} = [\bar{u}_x \; \bar{u}_y \; \bar{u}_z]^T$ on S_u. The basic equations of the theory of elasticity, using the matrix

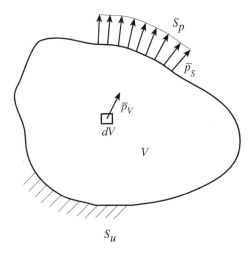

Figure 6.3 Elastic body.

calculus, are

1. The equilibrium equations

$$\mathbf{D}^T \boldsymbol{\sigma} + \bar{\mathbf{p}}_V = \mathbf{0} \tag{6.2}$$

2. The constitutive equations

$$\boldsymbol{\sigma} = \mathbf{E}\boldsymbol{\varepsilon} \tag{6.3}$$

3. The strain-displacement relationship

$$\boldsymbol{\varepsilon} = \mathbf{D}\mathbf{u} \tag{6.4}$$

In these relationships the stress vector $\boldsymbol{\sigma} = [\sigma_{xx} \quad \sigma_{yy} \quad \sigma_{zz} \quad \tau_{xy} \quad \tau_{xz} \quad \tau_{yz}]^T$ contains the components of the stress tensor, $\mathbf{E}$ is the material matrix, the strain vector $\boldsymbol{\varepsilon} = [\varepsilon_{xx} \quad \varepsilon_{yy} \quad \varepsilon_{zz} \quad \gamma_{xy} \quad \gamma_{xz} \quad \gamma_{yz}]^T$ includes the components of *Cauchy's strain tensor*, the vector $\mathbf{u} = [u_x \quad u_y \quad u_z]^T$ contains the displacements, and $\mathbf{D}$ is the differential matrix operator

$$\mathbf{D}^T = \begin{bmatrix} \partial_x & 0 & 0 & \partial_y & \partial_z & 0 \\ 0 & \partial_y & 0 & \partial_x & 0 & \partial_z \\ 0 & 0 & \partial_z & 0 & \partial_x & \partial_y \end{bmatrix} \tag{6.5}$$

with $\partial_x = \partial/\partial x$.

The relations of Eqs. (6.2) to (6.4) can be expressed as a system of differential equations for the displacements of an elastic body, called *Lamé's displacement equations*. They appear as

$$\mathbf{D}^T \mathbf{E} \mathbf{D} \mathbf{u} + \bar{\mathbf{p}}_V = \mathbf{0} \tag{6.6}$$

The corresponding boundary conditions follow from the conditions of equilibrium on the boundary of the body

$$\mathbf{A}^T \boldsymbol{\sigma} = \bar{\mathbf{p}}_S \quad \text{on} \quad S_p \tag{6.7}$$

$$\mathbf{u} = \bar{\mathbf{u}} \quad \text{on} \quad S_u \tag{6.8}$$

where $\mathbf{A}^T$ is the transformation matrix

$$\mathbf{A}^T = \begin{bmatrix} n_x & 0 & 0 & n_y & n_z & 0 \\ 0 & n_y & 0 & n_x & 0 & n_z \\ 0 & 0 & n_z & 0 & n_x & n_y \end{bmatrix} \tag{6.9}$$

The coefficients of this matrix are the components of the normal unit vector (direction cosines) $\mathbf{n} = [n_x \quad n_y \quad n_z]^T$ on the surface of the elastic body. The boundary conditions of Eq. (6.8) are called *displacement, kinematic*, or *essential* boundary conditions. These boundary conditions are limiting the *space of all admissible displacements* $\mathbf{U}$.

Statically admissible stresses $\boldsymbol{\sigma}$ satisfy the equilibrium relations of Eq. (6.2) and the stress boundary conditions of Eq. (6.7). In integral form these two sets of equations can be expressed as

$$\int_V \delta \mathbf{u}^T \left(\mathbf{D}^T \boldsymbol{\sigma} + \bar{\mathbf{p}}_V \right) dV + \int_{S_p} \delta \mathbf{u}^T (\mathbf{A}^T \boldsymbol{\sigma} - \bar{\mathbf{p}}_S) \, dS = 0 \qquad (6.10)$$

For kinematically admissible displacements, that is, displacements that satisfy the kinematic conditions of Eq. (6.4) and the displacement boundary conditions of Eq. (6.8), integration by parts (Gauss's integral theorem) leads to

$$\int_V \delta \boldsymbol{\varepsilon}^T \boldsymbol{\sigma} \, dV - \int_V \delta \mathbf{u}^T \bar{\mathbf{p}}_V \, dV - \int_{S_p} \delta \mathbf{u}^T \bar{\mathbf{p}}_S \, dS = 0 \qquad (6.11)$$

This means that for kinematically admissible displacements, the sum of the virtual work of the internal forces δW_{int} and the virtual work of the external forces δW_{ext}, is zero:

$$- \delta W_{\text{int}} - \delta W_{\text{ext}} = 0 \qquad (6.12)$$

with

$$\delta W_{\text{int}} = - \int_V \delta \boldsymbol{\varepsilon}^T \boldsymbol{\sigma} \, dV \qquad (6.13)$$

$$\delta W_{\text{ext}} = \int_V \delta \mathbf{u}^T \bar{\mathbf{p}}_V \, dV + \int_{S_p} \delta \mathbf{u}^T \bar{\mathbf{p}}_S \, dS$$

Equation (6.11) is called the *principle of virtual work*. It is often also referred to as the *weak form* of the differential equations (6.6). It is the basis for a derivation of finite element equations to solve elasticity problems.

Substitution of Eqs. (6.3) and (6.4) into Eq. (6.11) leads to

$$\int_V (\mathbf{D} \, \delta \mathbf{u})^T \mathbf{E} \mathbf{D} \mathbf{u} \, dV - \int_V \delta \mathbf{u}^T \bar{\mathbf{p}}_V \, dV - \int_{S_p} \delta \mathbf{u}^T \bar{\mathbf{p}}_S \, dS = 0 \qquad (6.14)$$

Introduction of the symmetric operator

$$\check{A}(\delta \mathbf{u}, \mathbf{u}) = \int_V (\mathbf{D} \, \delta \mathbf{u})^T \mathbf{E} \mathbf{D} \mathbf{u} \, dV \qquad (6.15)$$

and of

$$\check{p}(\delta \mathbf{u}) = \int_V \delta \mathbf{u}^T \bar{\mathbf{p}}_V \, dV + \int_{S_p} \delta \mathbf{u}^T \bar{\mathbf{p}}_S \, dS \qquad (6.16)$$

permits an abridged notation for the weak form of the displacement differential equations to be written as

$$\check{A}(\delta \mathbf{u}, \mathbf{u}) = \check{p}(\delta \mathbf{u}), \qquad \forall \delta \mathbf{u} \in \mathbf{U} \qquad (6.17)$$

which will be utilized later.

6.2.2 Element Equations

The structure that is to be analyzed is subdivided into finite elements. Such elements are defined by n_e nodes. Figure 6.4 shows, for example, a quadrilateral element for two-dimensional problems defined by four nodes ($n_e = 4$). The displacements in the element nodes are summarized in the element displacement vector $\mathbf{v}^e = [\mathbf{v}_x^e \; \mathbf{v}_y^e \; \mathbf{v}_z^e]^T$, with $\mathbf{v}_x^e = [u_{x1} \; u_{x2} \; \cdots \; u_{xn_e}]$, where $u_{x1} \; u_{x2} \; \cdots \; u_{xn_e}$ are nodal displacements. The vectors $\mathbf{v}_y^e, \mathbf{v}_z^e$ are defined similarly. The nodal applied forces $\bar{\mathbf{p}}$ are due to the body forces $\bar{\mathbf{p}}_V$ and boundary loads $\bar{\mathbf{p}}_S$. Because the element is cut out of the elastic body, there must be the unknown internal nodal forces $\mathbf{p}^e = [\mathbf{p}_x^e \; \mathbf{p}_y^e \; \mathbf{p}_z^e]^T$, with $\mathbf{p}_x^e = [p_{x1} \; p_{x2} \; \cdots \; p_{xn_e}]$, where $\mathbf{p}_y^e, \mathbf{p}_z^e$ are defined similarly and $p_{x1} \; p_{x2} \; \cdots \; p_{xn_e}$, are internal nodal forces. These forces are in equilibrium with the internal nodal forces of the adjacent elements if the body is assembled from the elements. Using these definitions, the principle of virtual work for a single element e appears as

$$0 = \int_{V^e} (\mathbf{D}\delta\mathbf{u})^T \mathbf{E}\mathbf{D}\mathbf{u} \, dV \qquad (6.18)$$

$$- \int_{V^e} \delta\mathbf{u}^T \bar{\mathbf{p}}_V \, dV - \int_{S_p^e} \delta\mathbf{u}^T \bar{\mathbf{p}}_S \, dS - \delta\mathbf{v}^{eT} \mathbf{p}^e$$

or

$$- \delta W_{\text{int}}^e - \delta W_{\text{ext}}^e = 0 \qquad (6.19)$$

The final term in Eq. (6.18) represents the work of the internal nodal forces $\mathbf{p}^e$.

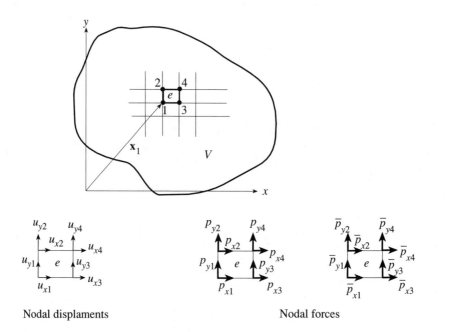

Figure 6.4 Four-node element.

The components of the displacements **u** of the continuum are approximated for each element using functions that are only dependent on the position in the element and on the unknown nodal displacements. Such functions are called *interpolation functions*. For the x component of the displacements it follows that

$$u_x \approx \hat{u}_x = \sum_{i=1}^{n_e} N_i u_{xi} = \overline{\mathbf{N}} \mathbf{v}_x^e \tag{6.20}$$

The functions N_i are called *shape functions*. In general, these functions can be any function of x, y, z. Usually, they are obtained by transforming polynomials into interpolation form. They are summarized in the vector $\overline{\mathbf{N}} = [N_1 \ \cdots \ N_i \ \cdots \ N_{n_e}]$. The selection of the shape functions will be treated later. The superscript hat of $\hat{u}_x$ indicates that $\hat{u}_x$ is an approximation of the displacement u_x.

The displacement approximation of Eq. (6.20) can be summarized for all displacement components in the form

$$\mathbf{u} \approx \hat{\mathbf{u}} = \begin{bmatrix} \overline{\mathbf{N}} & 0 & 0 \\ 0 & \overline{\mathbf{N}} & 0 \\ 0 & 0 & \overline{\mathbf{N}} \end{bmatrix} \mathbf{v}^e = \mathbf{N}\mathbf{v}^e \tag{6.21}$$

The displacement approximation is introduced into Eq. (6.18), yielding for the element e,[1]

$$0 = \delta \mathbf{v}^{eT} \left[\int_{V^e} (\mathbf{DN})^T \mathbf{EDN}\, dV \right] \mathbf{v}^e \tag{6.22}$$

$$- \delta \mathbf{v}^{eT} \int_{V^e} \mathbf{N}^T \overline{\mathbf{p}}_V\, dV - \delta \mathbf{v}^{eT} \int_{S_p^e} \mathbf{N}^T \overline{\mathbf{p}}_S\, dS - \delta \mathbf{v}^{eT} \mathbf{p}^e$$

Introduce the abbreviation

$$\mathbf{B} = \mathbf{DN} \tag{6.23}$$

Since the variations $\delta \mathbf{v}^e$ are nonzero, it follows from Eq. (6.22) that

$$\left[\int_{V^e} \mathbf{B}^T \mathbf{EB}\, dV \right] \mathbf{v}^e - \int_{V^e} \mathbf{N}^T \overline{\mathbf{p}}_V\, dV - \int_{S_p^e} \mathbf{N}^T \overline{\mathbf{p}}_S\, dS - \mathbf{p}^e = 0 \tag{6.24}$$

This equation is the equilibrium equation for the element e. In abridged form this equation reads

$$\mathbf{k}^e \mathbf{v}^e = \overline{\mathbf{p}}^e + \mathbf{p}^e \tag{6.25}$$

with the *element stiffness matrix*

$$\mathbf{k}^e = \int_{V^e} \mathbf{B}^T \mathbf{EB}\, dV \tag{6.26}$$

---
[1] Here, the matrix operation $(\mathbf{AB})^T = \mathbf{B}^T \mathbf{A}^T$ is used.

and the *element load vector*

$$\overline{\mathbf{p}}^e = \int_{V^e} \mathbf{N}^T \overline{\mathbf{p}}_V \, dV + \int_{S_p^e} \mathbf{N}^T \overline{\mathbf{p}}_S \, dS \tag{6.27}$$

6.2.3 Shape Functions

The solution of elasticity problems using the finite element method is an approximation to the actual problem solution. If polynomials are used for the shape functions $N_i \in \mathbf{N}$, the displacement approximation for the structure is a piece-wise polynomial. The discretization can be refined by two means. First, by increasing the number of elements and therefore reducing the element size, which is called *h-refinement*. Second, by increasing the polynomial degree of the shape functions, which is called *p-refinement*. Of course a combination of both refinement methods is possible, which is then called *hp-refinement*.

The results can converge to the correct solution if the discretization is refined continuously. To achieve monotonic convergence of the solution, the displacement functions of the element must be able to represent rigid body displacements. Also, if the element size approaches zero the state of constant strain must be represented. An element satisfying these two conditions is said to be *complete*. Further the element must be *compatible*; that is, the displacements within the element and across the elements edges must be continuous.

To demonstrate the fundamentals of deriving shape functions, begin with a one-dimensional problem. The function $u(\xi)$, $\xi \in \langle -1, 1 \rangle$, will be approximated using

$$u(\xi) = \sum_{i=1}^{n_e} N_i^{n_d}(\xi) u_i, \qquad n_e = n_d \tag{6.28}$$

where there are n_e nodes spaced along a line and u_i are the nodal displacements. Each shape function $N_i^{n_d}$ has the value one at the node i it corresponds to and zero at all other nodes. From this, shape functions of arbitrary order can be constructed. The superscript n_d denotes the degree of the shape function polynomial plus one. (See Pilkey and Wunderlich 1993 for details on the derivation of these shape functions.) Figure 6.5 shows the shape functions for the one-dimensional case up to the order of three. In the case of a linear approximation, $n_d = 2$, the shape functions are (Fig.6.5a)

$$N_1^2 = \frac{1}{2}(1 - \xi) \tag{6.29}$$

$$N_2^2 = \frac{1}{2}(1 + \xi)$$

For a quadratic approximation, $n_d = 3$ (Fig.6.5b),

$$N_1^3 = \frac{1}{2}\xi(\xi - 1)$$

$$N_2^3 = 1 - \xi^2 \tag{6.30}$$

$$N_3^3 = \frac{1}{2}\xi(1 + \xi)$$

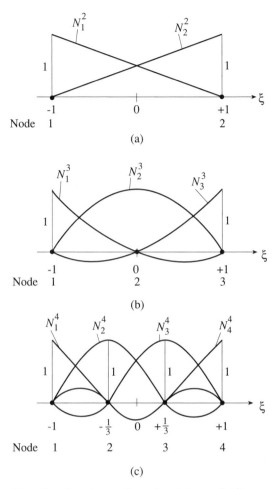

Figure 6.5 Shape functions in one dimension: (a) $n_e = 2$; (b) $n_e = 3$; (c) $n_e = 4$.

and for a cubic approximation, $n_d = 4$ (Fig. 6.5c),

$$N_1^4 = \frac{1}{16}\left(-1 + \xi + 8\xi^2 - 9\xi^3\right)$$

$$N_2^4 = \frac{1}{16}\left(9 - 27\xi - 9\xi^2 + 27\xi^3\right) \tag{6.31}$$

$$N_3^4 = \frac{1}{16}\left(9 + 27\xi - 9\xi^2 - 27\xi^3\right)$$

$$N_4^4 = \frac{1}{16}\left(-1 - \xi + 8\xi^2 + 9\xi^3\right)$$

From the functions of Eqs. (6.29) to (6.31), the shape functions for elements for two- and three-dimensional problems can be found. Figure 6.6 shows quadrilateral elements in the mathematical coordinates $\xi, \eta \in \langle -1, 1 \rangle$. There are two techniques that can be used to

452 STRESS CONCENTRATION ANALYSIS AND DESIGN

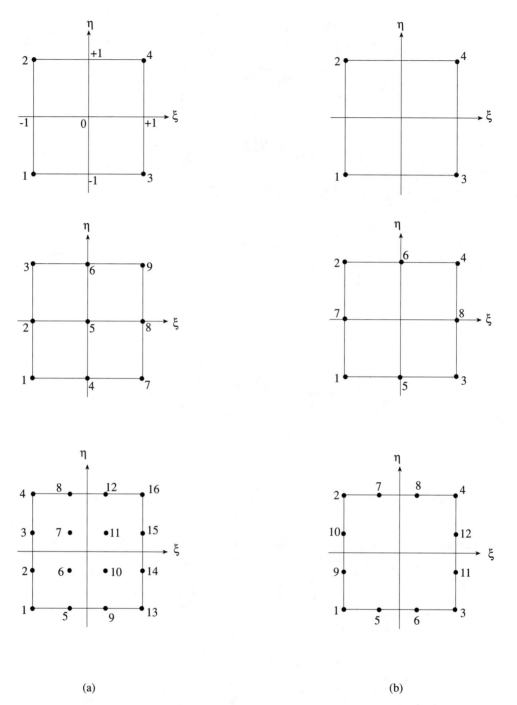

(a)   (b)

Figure 6.6 Two-dimensional elements: (*a*) Lagrange elements; (*b*) serendipity elements.

obtain these two-dimensional shape functions. The first is to use the product of the shape functions of the one-dimensional case. For an element with $n_e = n_d \times n_d$ nodes, the shape functions follow from

$$N_i = N_{IJ}^{n_d} = N_I^{n_d}(\xi)N_J^{n_d}(\eta), \qquad i = 1, \ldots, n_d^2, I, J = 1, \ldots, n_e \qquad (6.32)$$

These elements are called *Lagrange elements*. In this case internal nodes appear (Fig. 6.6a). Another technique is to have nodes only on the element boundaries. Then $n_e = 4 \times (n_d - 1)$ shape functions are to be determined. These elements are called *serendipity elements*. The shape functions for the corner nodes (Fig. 6.6b) are determined using

$$N_i = N_{IJ}^{n_d} = N_I^{n_d}(\xi)N_J^2(\eta) + N_I^2(\xi)N_J^{n_d}(\eta) - N_I^2(\xi)N_J^2(\eta) \qquad (6.33)$$
$$i = 1, \ldots, 4, I, J = 1, 2$$

and the shape functions for the nodes on the element edges follow from

$$N_i = N_{IJ}^{n_d} = N_I^{n_d}(\xi)N_J^2(\eta) \qquad (6.34)$$
$$i = 5 \ldots, 5 + n_d - 1, I = 3, n_d, J = 1, 2$$
$$N_i = N_{IJ}^{n_d} = N_I^2(\xi)N_J^{n_d}(\eta) \qquad (6.35)$$
$$i = 5 + n_d, \ldots, 4 \times (n_d - 1), I = 1, 2, J = 1, \ldots, n_d$$

In the three-dimensional case the $n_e = n_d^3$ shape functions for the hexahedral Lagrange elements of Fig. 6.7 are

$$N_i = N_{IJK}^{n_d} = N_I^{n_d}(\xi)N_J^{n_d}(\eta)N_K^{n_d}(\zeta) \qquad (6.36)$$
$$i = 1, \ldots, n_d^3, I, J, K = 1, \ldots, n_d$$

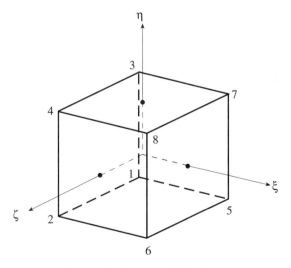

Figure 6.7 Lagrange brick elements.

For further reading on shape functions, especially for triangular and tetrahedral elements, the reader is referred to Zienkiewicz and Taylor (1989) and Bathe (1995). Shape functions can also be formed for *p*-refinement in a hierarchical manner. Then, the additional degrees of freedom for shape functions of an order higher than linear cannot be interpreted as nodal displacements and hence no additional nodes appear. For more detail on hierarchical shape functions, see Zienkiewicz and Taylor (1989), and Szabo and Babuška (1991).

For the computation of the matrix **B** of Eq. (6.23) the derivatives of the shape functions with respect to the physical coordinates x, y, z have to be calculated. However, the shape functions $\mathbf{N}(\xi, \eta, \zeta)$ are given in mathematical coordinates. Application of the chain rule of differentiation gives

$$\begin{bmatrix} \partial_\xi \\ \partial_\eta \\ \partial_\zeta \end{bmatrix} = \mathbf{J} \begin{bmatrix} \partial_x \\ \partial_y \\ \partial_z \end{bmatrix} \tag{6.37}$$

using $\partial_\xi = \partial/\partial \xi$ and the *Jacobian matrix*

$$\mathbf{J} = \begin{bmatrix} \partial_\xi x & \partial_\xi y & \partial_\xi z \\ \partial_\eta x & \partial_\eta y & \partial_\eta z \\ \partial_\zeta x & \partial_\zeta y & \partial_\zeta z \end{bmatrix} \tag{6.38}$$

From this the derivatives with respect to the x, y, z coordinates appear as

$$\begin{bmatrix} \partial_x \\ \partial_y \\ \partial_z \end{bmatrix} = \mathbf{J}^{-1} \begin{bmatrix} \partial_\xi \\ \partial_\eta \\ \partial_\zeta \end{bmatrix} \tag{6.39}$$

This relationship can be used to express Eq. (6.23).

6.2.4 Mapping Functions

To obtain the Jacobian matrix **J** of Eq. (6.38), a relationship between mathematical and physical coordinates needs to be established. The most popular method is to use the shape functions described above for the definition of the internal element geometry. Recall that the shape functions have already been used to approximate the displacements. The nodal position coordinates are assembled in the vector $\mathbf{x}^e = [\, \mathbf{x}^e_x \ \ \mathbf{x}^e_y \ \ \mathbf{x}^e_z \,]^T$, with $\mathbf{x}^e_x = [\, x_1 \ \ x_2 \ \cdots \ x_{n_e} \,]$, where n_e is the number of nodes of element e. The vectors $\mathbf{x}^e_y$, $\mathbf{x}^e_z$ are defined similarly. The nodal coordinates are input variables to the finite element analysis, since they determine the geometry of the finite element model. The position of a point of the structure $\mathbf{x} = [\, x \ \ y \ \ z \,]^T$ within the element e can be approximated using

$$\mathbf{x} \approx \hat{\mathbf{x}} = \mathbf{N}(\xi, \eta, \zeta) \mathbf{x}^e \tag{6.40}$$

Equation (6.40) is called the *mapping function* of the element. This kind of mapping is called *isoparametric mapping*, since the functions used for displacement and geometry approximations are of the same order. In contrast to the isoparametric mapping, *subparametric mapping* occurs when the mapping function is of lower order than the displacement approximation, and *superparametric mapping* is when the mapping function is of higher order then the displacement approximation. Another method to obtain mapping functions is the blending function method which is frequently applied if hierarchical shape functions are used (Szabo and Babuška 1991).

From Eqs. (6.40) and (6.38) the Jacobian matrix can be expressed as

$$\mathbf{J} = \begin{bmatrix} \partial_\xi \overline{\mathbf{N}} \mathbf{x}^e_x & \partial_\xi \overline{\mathbf{N}} \mathbf{x}^e_y & \partial_\xi \overline{\mathbf{N}} \mathbf{x}^e_z \\ \partial_\eta \overline{\mathbf{N}} \mathbf{x}^e_x & \partial_\eta \overline{\mathbf{N}} \mathbf{x}^e_y & \partial_\eta \overline{\mathbf{N}} \mathbf{x}^e_z \\ \partial_\zeta \overline{\mathbf{N}} \mathbf{x}^e_x & \partial_\zeta \overline{\mathbf{N}} \mathbf{x}^e_y & \partial_\zeta \overline{\mathbf{N}} \mathbf{x}^e_z \end{bmatrix} = \begin{bmatrix} \partial_\xi \\ \partial_\eta \\ \partial_\zeta \end{bmatrix} \mathbf{N} \mathbf{x}^{eT} \qquad (6.41)$$

To perform the integration involved in forming the element stiffness matrix $\mathbf{k}^e$ and load vector $\bar{\mathbf{p}}^e$ of Eqs. (6.26) and (6.27), the volume element dV needs to be expressed in mathematical coordinates too. Using Eq. (6.40) and the chain rule, it follows that

$$dV = dx\,dy\,dz = \det \mathbf{J}\,d\xi\,d\eta\,d\zeta \qquad (6.42)$$

Finally, the element stiffness matrix and load vector can be expressed in terms of the mathematical coordinates. The stiffness matrix of Eq. (6.26) appears as

$$\mathbf{k}^e = \int_{-1}^{+1} \int_{-1}^{+1} \int_{-1}^{+1} \mathbf{B}^T \mathbf{E} \mathbf{B} \det \mathbf{J}\,d\xi\,d\eta\,d\zeta \qquad (6.43)$$

and the load vector of Eq. (6.27)

$$\bar{\mathbf{p}}^e = \int_{-1}^{+1} \int_{-1}^{+1} \int_{-1}^{+1} \mathbf{N}^T \bar{\mathbf{p}}_V \det \mathbf{J}\,d\xi\,d\eta\,d\zeta \qquad (6.44)$$

$$+ \int_{-1}^{+1} \int_{-1}^{+1} \mathbf{N}_S^T(\xi_S, \eta_S) \bar{\mathbf{p}}_S \det \mathbf{J}_S\,d\xi_S\,d\eta_S$$

where the subscript S refers to the surface of the elastic body.

6.2.5 Numerical Integration

Since the different elements may have different geometries, the integration of Eqs. (6.43) and (6.44) is usually performed numerically. The integral over a function $\varphi(\xi)$ is written as a sum

$$\int_{-1}^{+1} \varphi(\xi)d\xi = \sum_{I}^{n_i} \varphi(\xi_I) w_I \qquad (6.45)$$

in which ξ_I is an integration point to evaluate the function $\varphi(\xi)$ (Fig. 6.8), w_I is a weighting function, and n_i is the number of integration points. The most popular and the most efficient integration scheme for finite elements is *Gaussian quadrature*. Using this scheme with n_i integration points, functions up to the degree $2n_i - 1$ are integrated exactly. For integration in two and three dimensions, the scheme is applied in each direction. Then

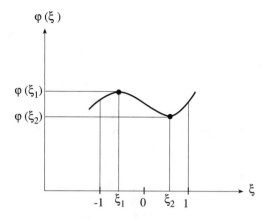

Figure 6.8 Numerical integration.

$$\int_{-1}^{+1}\int_{-1}^{+1} \varphi(\xi,\eta)d\xi\,d\eta = \sum_{I}^{n_i}\sum_{J}^{n_i} \varphi(\xi_I,\eta_J)w_I\,w_J \qquad (6.46)$$

$$\int_{-1}^{+1}\int_{-1}^{+1}\int_{-1}^{+1} \varphi(\xi,\eta,\zeta)d\xi\,d\eta\,d\zeta = \sum_{I}^{n_i}\sum_{J}^{n_i}\sum_{K}^{n_i} \varphi(\xi_I,\eta_J,\zeta_K)w_I\,w_J\,w_K \qquad (6.47)$$

Table 6.2 summarizes the collocation points ξ_I and the weights w_I for the Gaussian quadrature up to $n_i = 8$.

TABLE 6.2 Gaussian Quadrature

n_i	ξ_I	w_I
1	0.000000000000000	2.000000000000000
2	±0.577350269189626	1.000000000000000
3	±0.774596669241483	0.555555555555556
	0.000000000000000	0.888888888888889
4	±0.861136311594053	0.347854845137454
	±0.339981043584856	0.652145154862546
5	±0.906179845938664	0.236926885056189
	±0.538469310105683	0.478628670499366
	0.000000000000000	0.568888888888889
6	±0.932469514203152	0.171324492379170
	±0.661209386466265	0.360761573048139
	±0.238619186083197	0.467913934572691
7	±0.949107912342759	0.129484966168870
	±0.741531185599394	0.279705391489277
	±0.525532409916329	0.381830050505119
	0.000000000000000	0.417959183673469
8	±0.960289856497536	0.101228536290376
	±0.796666477413627	0.222381034453374
	±0.525532409916329	0.313706645877887
	±0.183434642495650	0.362683783378362

6.2.6 System Equations

To solve the elasticity problem for the structure as a whole, the element equations must be assembled to form a system of linear equations. Equation (6.12) states that for the entire domain the sum of the internal virtual work and the external virtual work is zero, while Eq. (6.19) makes the same statement for the element. Since the structure is assembled of the elements, the total work must be the sum of the work of the individual elements. Hence

$$\delta W_{\text{int}} = \sum_e \delta W_{\text{int}}^e \tag{6.48}$$

$$\delta W_{\text{ext}} = \sum_e \delta W_{\text{ext}}^e$$

To implement this summation for all elements, the element nodes must be referred to the structural (global or system) nodes.

The local (element) nodal displacements must be compatible with the global (structural) nodal displacements. The element nodal displacements $\mathbf{v}^e$ can be related to the global nodal displacements $\mathbf{v}$ using the Boolean matrix $\mathbf{a}^e$ such that

$$\mathbf{v}^e = \mathbf{a}^e \mathbf{v} \tag{6.49}$$

Matrix $\mathbf{a}^e$ is referred to by such names as the *global kinematic, connectivity, locator,* or *incidence matrix*. This matrix has $n_{\text{el}} \times n_{\text{tot}}$ components that have the values of zero or one, with n_{el} the number of element nodal displacement components, and n_{tot} the total number of structural nodal displacement components. The components of the matrix that are nonzero relate the nodal displacements of the element to the nodal displacements of the structure.

Example 6.2 Relationship between Element and System Nodes To illustrate the use of the Boolean matrix $\mathbf{a}^e$, a simple example of a system consisting of two four-node elements with one unknown displacement per node will be employed (Fig. 6.9). For this system, two relations of the type of Eq. (6.49) are necessary. They appear as

$$\mathbf{v}^e = \mathbf{a}^e \quad \mathbf{v} \tag{1}$$

$$\begin{bmatrix} u_{x1} \\ u_{x2} \\ u_{x3} \\ u_{x4} \end{bmatrix}^1 = \begin{bmatrix} 1 & 0 & 0 & 0 & 0 & 0 \\ 0 & 1 & 0 & 0 & 0 & 0 \\ 0 & 0 & 0 & 0 & 1 & 0 \\ 0 & 0 & 0 & 1 & 0 & 0 \end{bmatrix} \begin{bmatrix} u_{x1} \\ u_{x2} \\ u_{x3} \\ u_{x4} \\ u_{x5} \\ u_{x6} \end{bmatrix} \tag{2}$$

$$\begin{bmatrix} u_{x1} \\ u_{x2} \\ u_{x3} \\ u_{x4} \end{bmatrix}^2 = \begin{bmatrix} 0 & 1 & 0 & 0 & 0 & 0 \\ 0 & 0 & 1 & 0 & 0 & 0 \\ 0 & 0 & 0 & 0 & 0 & 1 \\ 0 & 0 & 0 & 0 & 1 & 0 \end{bmatrix} \begin{bmatrix} u_{x1} \\ u_{x2} \\ u_{x3} \\ u_{x4} \\ u_{x5} \\ u_{x6} \end{bmatrix} \tag{3}$$

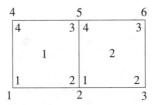

Figure 6.9 System of two elements.

Introduction of Eq. (6.49) into Eq. (6.25) and premultiplication by $\mathbf{a}^{eT}$ leads to

$$\mathbf{a}^{eT}\mathbf{k}^e\mathbf{a}^e\mathbf{v} = \mathbf{a}^{eT}[\overline{\mathbf{p}}^e + \mathbf{p}^e] \tag{6.50}$$

Since the virtual work of the system is the sum of the virtual work of the individual elements,

$$\sum_e \mathbf{a}^{eT}\mathbf{k}^e\mathbf{a}^e\mathbf{v} = \sum_e \mathbf{a}^{eT}\left[\overline{\mathbf{p}}^e + \mathbf{p}^e\right] \tag{6.51}$$

The internal nodal forces $\mathbf{p}^e$ of element e are in equilibrium with the internal nodal forces of adjacent elements. Because of this equilibrium requirement

$$\sum_e \mathbf{a}^{eT}\mathbf{p}^e = \mathbf{0} \tag{6.52}$$

Finally, the linear system equation that describes the structural behavior appears as

$$\mathbf{K}\mathbf{v} = \overline{\mathbf{p}} \tag{6.53}$$

where the *system stiffness matrix* is expressed as

$$\mathbf{K} = \sum_e \mathbf{a}^{eT}\mathbf{k}^e\mathbf{a}^e \tag{6.54}$$

and the *system load vector* is given by

$$\overline{\mathbf{p}} = \sum_e \mathbf{a}^{eT}\overline{\mathbf{p}}^e \tag{6.55}$$

Concentrated forces applied to the structure are modeled as forces applied to the nodes. From the previous derivation it follows that concentrated forces can be added to the component of the system load vector $\overline{\mathbf{p}}$ that corresponds to the node where the force is applied.

The stiffness matrix $\mathbf{K}$ is singular. Hence the system equation cannot be solved immediately. First, the rigid body motions of the structure must be suppressed by introducing the displacement boundary conditions into the system equation. The stiffness matrix with the boundary conditions introduced is often called a *reduced stiffness matrix*. Zero displacements are introduced by deleting rows and columns from the system stiffness matrix $\mathbf{K}$ and the row of the load vector $\overline{\mathbf{p}}$, respectively, that correspond to a displacement component which is zero. The deleted rows correspond to unknown reactions that occur where zero displacements are imposed. These reactions can be computed after the system displacements **v** have been calculated.

The introduction of nonzero boundary displacements as boundary conditions can be accomplished using spring elements. A spring element is an elastic boundary condition that is applied to a specific displacement component, for example, u_{yI} of a node I of the finite element model (Fig. 6.10). The stiffness of the spring e is defined by the elastic constant k^e. Then the work of Eq. (6.18) for the spring element appears as

$$0 = \delta u_{yI} k^e u_{yI} - \delta u_{yI} p_{yI} \tag{6.56}$$

which yields the element equation

$$k^e u_{yI} = p_{yI} \tag{6.57}$$

where u_{yI} is the displacement at the spring and p_{yI} is the spring force. This equation is treated like an element equation. Hence the spring constant k^e is added to the appropriate matrix element of the system stiffness matrix that corresponds to the displacement component to which the spring is applied. In the case of a prescribed displacement component $\bar{u}_{yI}$, let the spring constant be a large value, for example, $\bar{k}^e = 10^{20}$, and then apply the force $\bar{p}_{yI} = \bar{k}^e \bar{u}_{yI}$ to the node with the prescribed displacement component. This results in a displacement u_{yI} of the prescribed value.

The assembly of a system stiffness matrix using finite elements leads to some special characteristics for this matrix that are useful during the solution process. First, the matrix is symmetric, $\mathbf{K}^T = \mathbf{K}$. Second, the matrix has a band character, and third, the reduced stiffness matrix is positive definite. These characteristics can be utilized for the solution of the system equations. Due to the symmetry only half the matrix needs to be stored. The band character permits the storage of zero elements to be avoided and, if considered in the implementation of the solution algorithm, permits multiplications with zero to be avoided. This leads to a considerable reduction of the numerical effort compared to that incurred with the use of a full matrix. The positive definiteness can be utilized in some elimination methods for the solution of the system equations.

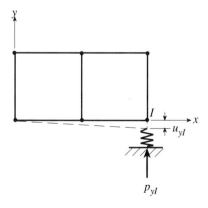

Figure 6.10 Spring element.

For the solution of the system equations, elimination or iterative methods can be used. This solution is a crucial part of a finite element program, since this step is numerically the most expensive. Many efficient algorithms are provided in the literature. The reader is referred to, for example, Zienkiewicz and Taylor (1989) and Bathe (1995).

An elimination method that takes advantage of the special characteristics of the finite element stiffness matrix is the *Gauss elimination method* using *Cholesky factorization*. During the factorization the reduced system stiffness matrix $\mathbf{K}$ is decomposed into a product of two triangular matrices

$$\mathbf{K} = \mathbf{R}^T \mathbf{R} \tag{6.58}$$

The matrix $\mathbf{R}$ has nonzero elements only above and on the main diagonal. The band character of the stiffness matrix is transmitted to the triangular matrix. From the detailed formulation of the matrix multiplication of Eq. (6.58), the elements of the triangular matrix $\mathbf{R}$ follow as

$$r_{il} r_{ii} = k_{il} - r_{1i} r_{1l} - r_{2i} r_{2l} - \cdots - r_{i-1,i} r_{i-1,l}, \quad l \geq i \tag{6.59}$$

It can be seen that for the calculation of the element r_{ii} a square root needs to be determined. The positive definiteness of the matrix $\mathbf{K}$ is required.

Using the decomposition of Eq. (6.58), the system equations of Eq. (6.53) can be solved in two steps. First, one carries out the *backward substitution*

$$\mathbf{R}^T \mathbf{f} = \overline{\mathbf{p}} \tag{6.60}$$

followed by the *forward substitution* using the vector $\mathbf{f}$, namely

$$\mathbf{R}\mathbf{v} = \mathbf{f} \tag{6.61}$$

Then one proceeds to the backward substitution using

$$r_{11} f_1 = \overline{p}_1$$

$$r_{ii} f_i = \overline{p}_i - \sum_{k=1}^{i-1} r_{ki} f_k, \quad i = 2, \ldots, n_{\text{tot}} \tag{6.62}$$

and the forward substitution

$$r_{n_{\text{tot}} n_{\text{tot}}} v_{n_{\text{tot}}} = f_{n_{\text{tot}}}$$

$$r_{ii} v_i = f_i - \sum_{k=i+1}^{n_{\text{tot}}} r_{ik} v_k, \quad i = n_{\text{tot}} - 1, \ldots, 1$$

6.2.7 Stress Computation

After solution of the system equations of Eq. (6.53), the displacements are known, and for a particular point $\mathbf{x}_c = [x_c \ y_c \ z_c]^T$ of the element e, the stresses can be obtained from Eq. (6.3):

$$\boldsymbol{\sigma} = \mathbf{E}(\mathbf{x}_c) \mathbf{B}(\mathbf{x}_c) \mathbf{v}^e \tag{6.63}$$

For the computation of stress concentration factors, the stresses of interest should be identified, and then the factors K_t, K_{ts} can be computed using Eqs. (1.1) and (1.2):

$$K_t = \frac{\sigma_{\max}}{\sigma_{\text{nom}}} \qquad (6.64)$$

$$K_{ts} = \frac{\tau_{\max}}{\tau_{\text{nom}}} \qquad (6.65)$$

One condition for the selection of the shape functions **N** was displacement continuity across the element boundaries. In general, there is no continuity imposed for the displacement derivatives. Hence the stresses at the element boundaries are discontinuous. Several methods are available to remove these discontinuities from the results. Often simple averages of the nodal stresses are calculated, and then the stresses are computed for the element using the shape functions. This method is not the most accurate. More advanced methods for stress recovery have been developed by Zienkiewicz and Zhu (1987) and Zhu and Zienkiewicz (1990).

Based on the stress computation, a posteriori estimates of the discretization error can be made. As a consequence of such an error estimate, conclusions for local mesh refinements to improve the accuracy of the stress computation can be drawn. For more detail on error estimation and adaptive mesh refinements, the reader is referred to Babuška et al. (1986), Szabo and Babuška (1991), and Zienkiewicz and Zhu (1992).

Example 6.3   Plane Stress and Plane Strain Problems   Two-dimensional elasticity has been discussed in connection with stress concentrations in Chapter 1. The *plane stress* problem is defined by the stress components in the x, y plane only (Fig. 6.11). All other stress components are zero. The equilibrium is given by Eq. (1.3). For the *plane strain* problem only the strains in the x, y plane are nonzero. Due to these assumptions the displacement vector **u** reduces to two displacements, $\mathbf{u} = [\, u_x \quad u_y \,]^T$. For both problems the vectors of stresses and strains appear as $\boldsymbol{\sigma} = [\, \sigma_x \quad \sigma_y \quad \tau_{xy} \,]^T$ and $\boldsymbol{\varepsilon} = [\, \varepsilon_x \quad \varepsilon_y \quad \gamma_{xy} \,]^T$, respectively. In the case of plane stress ($\sigma_z = 0$), the strain ε_z can be determined from the

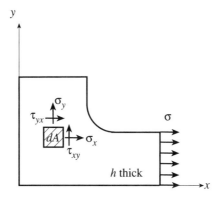

Figure 6.11   Plane stress, plane strain.

strains ε_x and ε_y,

$$\varepsilon_z = -\frac{\nu}{E}(\varepsilon_x + \varepsilon_y) \tag{1}$$

In the case of plane strain ($\varepsilon_z = 0$), the normal stress σ_z can be obtained from σ_x and σ_y:

$$\sigma_z = \nu(\sigma_x + \sigma_y) \tag{2}$$

From Eq. (1.3) it can be seen that the matrix differential operator of Eq. (6.4) must be

$$\mathbf{D}^T = \begin{bmatrix} \partial_x & 0 & \partial_y \\ 0 & \partial_y & \partial_x \end{bmatrix} \tag{3}$$

The only difference in the elasticity equations for the plane stress and plane strain problems is in the constitutive law of Eq. (6.3). For plane stress

$$\mathbf{E}^\sigma = \frac{E}{1-\nu^2}\begin{bmatrix} 1 & \nu & 0 \\ \nu & 1 & 0 \\ 0 & 0 & \frac{1-\nu}{2} \end{bmatrix} \tag{4}$$

and for plane strain

$$\mathbf{E}^\varepsilon = \frac{E}{(1+\nu)(1-2\nu)}\begin{bmatrix} 1-\nu & \nu & 0 \\ \nu & 1-\nu & 0 \\ 0 & 0 & \frac{1-2\nu}{2} \end{bmatrix} \tag{5}$$

To establish finite element equations, consider a nine-node element. At each node i the two displacement components u_{xi}, u_{yi} are defined. The displacement approximation of Eq. (6.20) involves the quadratic shape functions that are calculated from Eq. (6.32) using the functions of Eq. (6.30). Then

$$\overline{\mathbf{N}}(\xi, \eta) = [N_1 \quad N_2 \quad \cdots \quad N_9] \tag{6}$$

with

$$N_1(\xi, \eta) = \frac{1}{4}\xi(1-\xi)\eta(1-\eta)$$

$$N_2(\xi, \eta) = \frac{1}{2}\xi(1-\xi)(1-\eta^2)$$

$$N_3(\xi, \eta) = \frac{1}{4}\xi(1-\xi)\eta(1+\eta)$$

$$N_4(\xi, \eta) = \frac{1}{4}(1-\xi^2)\eta(1-\eta)$$

$$N_5(\xi, \eta) = (1-\xi^2)(1-\eta^2) \tag{7}$$

$$N_6(\xi, \eta) = \frac{1}{4}(1-\xi^2)\eta(1+\eta)$$

$$N_7(\xi, \eta) = \frac{1}{4}\xi(1 + \xi)\eta(1 - \eta)$$

$$N_8(\xi, \eta) = \frac{1}{2}\xi(1 + \xi)(1 - \eta^2)$$

$$N_9(\xi, \eta) = \frac{1}{4}\xi(1 + \xi)\eta(1 + \eta)$$

Now the displacements $\hat{\mathbf{u}}$ can be expressed in the form of Eq. (6.20)

$$\hat{\mathbf{u}} = \begin{bmatrix} \hat{u}_x \\ \hat{u}_y \end{bmatrix} = \begin{bmatrix} \overline{\mathbf{N}} & \mathbf{0} \\ \mathbf{0} & \overline{\mathbf{N}} \end{bmatrix} \begin{bmatrix} \mathbf{v}_x^e \\ \mathbf{v}_y^e \end{bmatrix} \qquad (8)$$

with $\mathbf{v}_x^e = [\, u_{x1} \quad u_{x2} \quad \cdots \quad u_{x9} \,]$ and $\mathbf{v}_y^e = [\, u_{y1} \quad u_{y2} \quad \cdots \quad u_{y9} \,]$. The shape functions are defined in mathematical coordinates. Hence, the relationship between ξ, η and x, y coordinates must be found. If an isoparametric mapping is used, the mapping function is given by

$$\begin{bmatrix} x \\ y \end{bmatrix} = \begin{bmatrix} \overline{\mathbf{N}} & \mathbf{0} \\ \mathbf{0} & \overline{\mathbf{N}} \end{bmatrix} \begin{bmatrix} \mathbf{x}_x^e \\ \mathbf{x}_y^e \end{bmatrix} \qquad (9)$$

with $\mathbf{x}_x^e = [\, x_1 \quad x_2 \quad \cdots \quad x_9 \,]^T$, and $\mathbf{x}_y^e = [\, y_1 \quad y_2 \quad \cdots \quad y_9 \,]^T$. Since the shape functions are given in two dimensions, the Jacobian matrix reduces to a 2×2 matrix $\mathbf{J} = [\, J_{lk} \,]$, $l, k = 1, 2$. The matrix $\mathbf{B}$ in Eq. (6.23) using $\mathbf{D}$ of (3) appears as

$$\mathbf{B} = \begin{bmatrix} \partial_x & 0 \\ 0 & \partial_y \\ \partial_y & \partial_x \end{bmatrix} \begin{bmatrix} \overline{\mathbf{N}} & \mathbf{0} \\ \mathbf{0} & \overline{\mathbf{N}} \end{bmatrix} \qquad (10)$$

or introducing the transformation of Eq. (6.39)

$$\mathbf{B} = \begin{bmatrix} J_{11}^* \partial_\xi \overline{\mathbf{N}} + J_{12}^* \partial_\eta \overline{\mathbf{N}} & 0 \\ 0 & J_{12}^* \partial_\xi \overline{\mathbf{N}} + J_{22}^* \partial_\eta \overline{\mathbf{N}} \\ J_{12}^* \partial_\xi \overline{\mathbf{N}} + J_{22}^* \partial_\eta \overline{\mathbf{N}} & J_{11}^* \partial_\xi \overline{\mathbf{N}} + J_{12}^* \partial_\eta \overline{\mathbf{N}} \end{bmatrix} \qquad (11)$$

where $\mathbf{J}^{-1} = [J_{lk}^*]$, $l, k = 1, 2$, and

$$\partial_\xi \overline{\mathbf{N}} = [\, \partial_\xi N_1 \quad \partial_\xi N_2 \quad \cdots \quad \partial_\xi N_9 \,] \qquad (12)$$
$$\partial_\eta \overline{\mathbf{N}} = [\, \partial_\eta N_1 \quad \partial_\eta N_2 \quad \cdots \quad \partial_\eta N_9 \,]$$

The derivatives $\partial_\xi N_i, \partial_\eta N_i, i = 1, \ldots, 9$, follow by differentiation of the shape functions of (7). Utilizing Eq. (6.43), the stiffness matrix can be calculated. If it is assumed that the thickness of the element is the constant h, then

$$\mathbf{k}^e = h \int_{-1}^{+1} \int_{-1}^{+1} \mathbf{B}^T \mathbf{E} \mathbf{B} \det \mathbf{J} \, d\xi \, d\eta \qquad (13)$$

Assemblage of the system stiffness matrix and load vector and solution of the system equations yields the nodal displacements. Then the stresses can be calculated from Eq. (6.63), and finally, from selected stresses the desired stress concentration factors can be obtained. If, for example, the von Mises stress σ_{eq} for the plane stress state at a certain point $\mathbf{x}_c$ is of interest, this can be calculated from Eq. (1.34). Adjusted to the plane stress problem,

$$\sigma_{eq} = \sqrt{\sigma_x^2 - \sigma_x\sigma_y + \sigma_y^2 + 3\tau_{xy}^2} \qquad (14)$$

where the stresses σ_x, σ_y, τ_{xy} follow from

$$\boldsymbol{\sigma} = \mathbf{E}^{\sigma}\mathbf{B}(\mathbf{x}_c)\mathbf{v}^e \qquad (15)$$

A stress concentration factor is then calculated using

$$K_t = \frac{\sigma_{eq\,\text{max}}}{\sigma_{eq\,\text{nom}}} \qquad (16)$$

The methodology described here has been applied for the solution of Example 6.1. Numerical results can be found there.

Example 6.4   Torsion Bar with Variable Diameter   Consider a body of revolution as shown in Fig. 6.12a. The shaft is loaded with the torque T. Figure 6.12b shows the stresses and the system of polar coordinates. The analytical solution to the problem using Saint Venant's semi-inverse method can be found in Timoshenko and Goodier (1951).

It is assumed that the displacements $u_x = u_r = 0$ and that the displacement u_θ does not depend on θ. Then it follows that the strains $\varepsilon_r = \varepsilon_\theta = \varepsilon_x = \gamma_{rx} = 0$ and that

$$\gamma_{r\theta} = \partial_r u_\theta - \frac{u_\theta}{r} \qquad (1)$$

$$\gamma_{\theta x} = \partial_x u_\theta \qquad (2)$$

The stress vector has two components. It appears as $\boldsymbol{\sigma} = [\,-\tau_{r\theta}\quad \tau_{\theta x}\,]^T$. Express the stresses using the stress function $\phi(r,x)$ as

$$\tau_{r\theta} = -\frac{1}{r^2}\partial_x \phi \qquad (3)$$

$$\tau_{\theta x} = \frac{1}{r^2}\partial_r \phi \qquad (4)$$

The boundary of the shaft must be traction free, which leads to

$$\partial_x \phi\, n_x + \partial_r \phi\, n_r = 0 \qquad (5)$$

For the boundaries $S_{1\ldots 4}$ in Fig. 6.12b, it follows from (5) and $T = 2\pi \int_0^R r^2 \tau_{r\theta}\, dr$ that

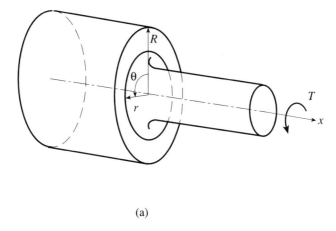

(a)

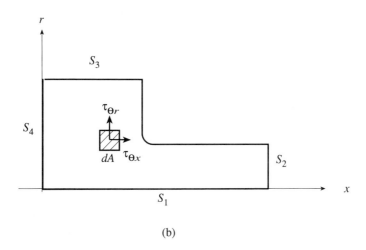

(b)

Figure 6.12 Torsion bar with variable diameter.

$$\begin{aligned}
\phi_1 &= 0 & \text{on } & S_1 \\
\phi_2 &= \frac{T}{2\pi}\left(\frac{r}{R}\right)^4 & \text{on } & S_2 \\
\phi_3 &= \frac{T}{2\pi} & \text{on } & S_3 \\
\partial_x \phi_4 &= 0 & \text{on } & S_4
\end{aligned} \qquad (6)$$

The principle of virtual work of Eq. (6.14) for this problem appears as

$$\frac{1}{G}\int_V \boldsymbol{\sigma}^T \delta\boldsymbol{\sigma}\, dV = 0 \tag{7}$$

Introduction of the stresses of (3) and (4) leads to

$$\frac{1}{G}\int_V \frac{1}{r^4}(\boldsymbol{\nabla}\phi)^T \boldsymbol{\nabla}\delta\phi\, dV = 0 \tag{8}$$

Because of the use polar coordinates, the Nabla operator $\boldsymbol{\nabla}$ appears as $\boldsymbol{\nabla} = [\partial_x \quad \partial_r]^T$. Considering that the shaft is a body of revolution, (8) can be simplified such that

$$\int_A \frac{1}{r^4}(\boldsymbol{\nabla}\phi)^T \boldsymbol{\nabla}\delta\phi\, dr\, dx = 0 \tag{9}$$

where A is the area of the solution domain (Fig. 6.12b). The stress function $\phi(x, r)$ is the only unknown for this problem. It is approximated element-wise using the same shape function as in the previous example. Then

$$\phi \approx \hat{\phi} = \overline{\mathbf{N}}\mathbf{v}^e \tag{10}$$

with $\mathbf{v}^e = [\phi_1 \quad \phi_2 \quad \cdots \quad \phi_9]^T$. An isoparametric mapping is used so that

$$\begin{bmatrix} x \\ r \end{bmatrix} = \begin{bmatrix} \overline{\mathbf{N}} & \mathbf{0} \\ \mathbf{0} & \overline{\mathbf{N}} \end{bmatrix} \begin{bmatrix} \mathbf{x}_x^e \\ \mathbf{x}_r^e \end{bmatrix} \tag{11}$$

The relationship for the matrix $\mathbf{B} = \boldsymbol{\nabla}\mathbf{N}$ can be found as

$$\mathbf{B} = \mathbf{J}^{-1}\begin{bmatrix} \overline{\mathbf{N}} & \mathbf{0} \\ \mathbf{0} & \overline{\mathbf{N}} \end{bmatrix} \tag{12}$$

with the Jacobian matrix defined analogous to Eq. (6.38). The material matrix degenerates to a unity matrix. The element stiffness matrix and load vector that follow from (9) are

$$\mathbf{k}^e = \int_{-1}^{+1}\int_{-1}^{+1} \frac{1}{r^4}\mathbf{B}^T \mathbf{B}\, d\xi\, d\eta \tag{13}$$

$$\overline{\mathbf{p}}^e = \mathbf{0}. \tag{14}$$

where $r(\xi, \eta)$ is obtained from (12).

After assembly of the system equations, the boundary conditions of (6) are introduced. The first three conditions of (6) can be treated as prescribed boundary displacements. The fourth boundary condition cannot be fulfilled due to the fact that the element vector of the unknowns $\mathbf{v}^e$ does not include the derivatives of the stress function.

Following the solution for the stress functions, the stresses for each element are calculated using

$$\boldsymbol{\sigma} = \frac{1}{r^2}\mathbf{B}\mathbf{v}^e \qquad (15)$$

A stress concentration factor K_{ts} is found from Eq. (6.65) using the equivalent shear stress

$$\tau_{eq} = \sqrt{\boldsymbol{\sigma}^T \boldsymbol{\sigma}} \qquad (16)$$

with $\boldsymbol{\sigma}$ given by (15). The nominal shear stress $\tau_{eq\,\text{nom}}$ can be selected according to Example 1.2. Some numerical results for the analysis of a shaft with variable diameter will be given later (see Example 6.6).

6.3 DESIGN SENSITIVITY ANALYSIS

The sensitivity of the structural behavior with respect to changes in the design variables needs to be determined in order to understand the effect of design changes. Here the focus is on structural responses expressed in terms of stresses or stress concentration factors. Other response functions could be the displacements or the compliance. While stresses and displacements are functions that are given at a specific point of the structure, the compliance is represented by an integral over the structural volume.

Choose an arbitrary response function

$$\psi_i = \psi_i(\mathbf{b}), \qquad i = 1, \ldots, n_\psi \qquad (6.66)$$

where n_ψ is the total number of response functions that are of interest and $\mathbf{b}$ is the vector of design variables. For a current design s, use of a Taylor series expansion provides the approximation

$$\psi_i(\mathbf{b}) \approx \hat{\psi}_i(\mathbf{b}) = \psi_i(\mathbf{b}^{(s)}) + \sum_{j=1}^{n_b} \psi_{i,j}\, \delta b_j \qquad (6.67)$$

where n_b is the number of design variables and the subscript $_{,j}$ denotes the derivative d/db_j, $j = 1, \ldots, n_b$. The quantity $\delta b_j = b_j - b_j^{(s)}$ is the variation of the design variable b_j. The superscript hat indicates that $\hat{\psi}_i$ is an approximation to ψ_i. In vector notation Eq. (6.67) appears as

$$\hat{\psi}_i(\mathbf{b}) = \psi_i(\mathbf{b}^{(s)}) + (\boldsymbol{\nabla}_b \psi_i)^T \delta \mathbf{b} \qquad (6.68)$$

where

$$\delta \mathbf{b} = \mathbf{b} - \mathbf{b}^{(s)} \qquad (6.69)$$

and $\boldsymbol{\nabla}_b \psi_i$ is the gradient of the function ψ_i with respect to the design variables b_j, $j = 1, \ldots, n_b$, that is, $\boldsymbol{\nabla}_b \psi_i = [\psi_{i,j}]^T$, $j = 1, \ldots, n_b$.

A *design sensitivity analysis* (DSA) computes the derivatives $\psi_{i,j}$ of the structural response functions. These derivatives will be called *design derivatives*. It is assumed that the response function ψ_i is differentiable with respect to the design variables and hence that the design derivatives exist. Proofs of existence for the design derivatives in the case of linear statics can be found in Haug et al. (1986).

In general, two types of sensitivity analyses can be distinguished: a finite difference analysis and an analytical sensitivity analysis. An analytical design sensitivity analysis can be based on the discretized or on the continuum model of the structure.

For further reading on design sensitivity analyses, see, for example, the books by Haug et al.(1986) and Haftka et al.(1990). A discussion on discrete versus continuum approaches to design sensitivity analyses was published by Arora (1995). Reviews on the methodology of design sensitivity analyses have been compiled by Adelmann and Haftka (1986) and Haftka and Adelmann (1989).

6.3.1 Finite Differences

The simplest, but most expensive, method to obtain design sensitivity information is to utilize finite differences. In this case the design variables b_j are changed one at a time, with the design derivatives of the response functions ψ_i computed using

$$\psi_{i,j} \approx \hat{\psi}_{i,j} = \frac{\psi_i(b_j + \Delta b_j) - \psi_i(b_j - \Delta b_j)}{2\Delta b_j} \tag{6.70}$$

The design variables change Δb_j must be small enough to obtain accurate sensitivity information but not too small, since then numerical difficulties can occur. Equation (6.70) is a central differences expression. Forward or backward differences can lead to erroneous results. In the case of shape design variables, for example, the use of forward differences would cause a loss of symmetries of the design throughout an optimization process.

6.3.2 Discrete Systems

If the design sensitivity analysis is based on the finite element discretization of the structure, the matrix equations of the numerical model are differentiated with respect to the design variables. The structural behavior is determined by the system of linear relations of Eq. (6.53). The system matrix $\mathbf{K}$ and the vector $\bar{\mathbf{p}}$ of the right-hand side depend on the design variables b_j. Hence the solution depends also on the design variables. The system equation appears as

$$\mathbf{K}(b_j)\mathbf{v}(b_j) = \bar{\mathbf{p}}(b_j). \tag{6.71}$$

The response function is given as

$$\psi_i = \psi_i(\mathbf{v}, b_j) \tag{6.72}$$

DESIGN SENSITIVITY ANALYSIS

To obtain the design sensitivity information, the function ψ_i has to be differentiated with respect to the design variables b_j, $j = 1, \ldots, n_b$. The total derivative of ψ_i appears as

$$\psi_{i,j} = \frac{\partial \psi_i}{\partial b_j} + \frac{\partial \psi_i}{\partial \mathbf{v}} \mathbf{v}_{,j}. \tag{6.73}$$

Differentiation of the system equation (6.71) with respect to the design variables b_j yields

$$\mathbf{K}\mathbf{v}_{,j} = \bar{\mathbf{p}}_{,j} - \mathbf{K}_{,j}\mathbf{v}. \tag{6.74}$$

To obtain the derivatives of the solution vector, this equation has to be solved. If an elimination method, such as described previously, is used for the solution of the system equation, the system matrix $\mathbf{K}$ does not need to be factorized again. Then, Eq. (6.74) can be solved by carrying out only the backward and forward substitution of Eqs.(6.60) and (6.61), respectively.

Another approach is to introduce Eq. (6.74) directly into the design derivative of Eq. (6.73). Then

$$\psi_{i,j} = \frac{\partial \psi_i}{\partial b_j} + \frac{\partial \psi_i}{\partial \mathbf{v}} \mathbf{K}^{-1} \left[\bar{\mathbf{p}}_{,j} - \mathbf{K}_{,j}\mathbf{v} \right]. \tag{6.75}$$

Substitution of

$$\mathbf{K}\mathbf{a}_i = \frac{\partial \psi_i^T}{\partial \mathbf{v}} \tag{6.76}$$

leads to

$$\psi_{i,j} = \frac{\partial \psi_i}{\partial b_j} + \mathbf{a}_i^T \left[\bar{\mathbf{p}}_{,j} - \mathbf{K}_{,j}\mathbf{v} \right]. \tag{6.77}$$

The solution $\mathbf{a}_i$ of Eq. (6.76) is called the *adjoint variable*. As with Eq. (6.74), no new factorization of the matrix $\mathbf{K}$ is necessary.

The use of Eq. (6.75) is called the *direct differentiation method* and the use of Eq. (6.77) is called the *adjoint variable method*. Both methods yield the same results for the design derivatives. The decision as to which method to use is based on the numerical effort. It can be seen that if the number of design variables n_b exceeds the number of response functions n_ψ, then Eq. (6.76) has fewer solution vectors than Eq. (6.74). Hence the adjoint variable method is more efficient in this case. Otherwise, direct differentiation is preferable.

To complete the derivation of the design derivatives, the derivatives of the stiffness matrix and the load vector remain to be determined. Since the Boolean matrix $\mathbf{a}^e$ in Eq. (6.49) is independent of the design variables b_j, it follows that

$$\mathbf{K}_{,j} = \sum_e \mathbf{a}^{eT} \mathbf{k}^e_{,j} \mathbf{a}^e \tag{6.78}$$

$$\bar{\mathbf{p}}_{,j} = \sum_e \mathbf{a}^{eT} \bar{\mathbf{p}}^e_{,j} \tag{6.79}$$

The derivatives of the element matrices are computed from Eq. (6.43) using the chain rule. Then

$$\mathbf{k}^e{,}_j = \int_{-1}^{+1} \int_{-1}^{+1} \int_{-1}^{+1} \mathbf{B}^T \mathbf{E}{,}_j \mathbf{B} \det \mathbf{J}\, d\xi\, d\eta\, d\zeta$$

$$+ \int_{-1}^{+1} \int_{-1}^{+1} \int_{-1}^{+1} \mathbf{B}^T \mathbf{E}\mathbf{B} (\det \mathbf{J}){,}_j\, d\xi\, d\eta\, d\zeta \qquad (6.80)$$

$$+ \int_{-1}^{+1} \int_{-1}^{+1} \int_{-1}^{+1} \left[\mathbf{B}^T{,}_j \mathbf{E}\mathbf{B} + \mathbf{B}^T \mathbf{E}\mathbf{B}{,}_j \right] \det \mathbf{J}\, d\xi\, d\eta\, d\zeta$$

In this expression, the first integral is affected if the design variables are material properties. The first and/or the second integrals are affected in the case of sizing variables, and all of the integrals are affected in the case of shape design variations. The computation of the derivative

$$\mathbf{B}{,}_j = \mathbf{D}{,}_j \mathbf{N} \qquad (6.81)$$

which is necessary for shape design variables, involves the derivative of the inverse of the Jacobian matrix, that follows from

$$\mathbf{J}^{-1}{,}_j = -\mathbf{J}^{-1} \mathbf{J}{,}_j \mathbf{J}^{-1} \qquad (6.82)$$

Differentiation of the Jacobian matrix of Eq. (6.41) with respect to the design variables yields

$$\mathbf{J}{,}_j = \begin{bmatrix} \partial_\xi \\ \partial_\eta \\ \partial_\zeta \end{bmatrix} \overline{\mathbf{N}} \mathbf{x}^e{,}_j \qquad (6.83)$$

The vector $\mathbf{x}^e{,}_j$ is the element portion of the vector $\mathbf{x}{,}_j$ which is called the *design velocity field*. It represents the sensitivity of the positions of the nodes with respect to the design changes. To refer to $\mathbf{x}{,}_j$ as a "design velocity field" is justified in the sense that one can consider the optimization process to be a time-dependent change of the structural shape where the design variable is a function of time. Figure 6.13 shows the elastic body in the configuration s and altered by a design change δb_j. Using the design velocity field $\mathbf{x}{,}_j$, the position of the point $\mathbf{x}$ is in first-order approximation

$$\mathbf{x} = \mathbf{x}^{(s)} + \delta \mathbf{x} \qquad (6.84)$$

$$= \mathbf{x}^{(s)} + \mathbf{x}{,}_j\, \delta b_j$$

The design derivatives of the load vector follow from

$$\overline{\mathbf{p}}^e{,}_j = \int_{-1}^{+1} \int_{-1}^{+1} \int_{-1}^{+1} \mathbf{N}^T \overline{\mathbf{p}}_V (\det \mathbf{J}){,}_j\, d\xi\, d\eta\, d\zeta \qquad (6.85)$$

$$+ \int_{-1}^{+1} \int_{-1}^{+1} \mathbf{N}_S^T \overline{\mathbf{p}}_S (\det \mathbf{J}){,}_j\, d\xi_S\, d\eta_S$$

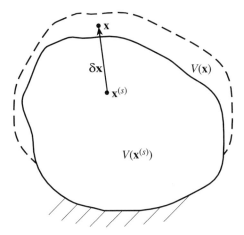

Figure 6.13 Design modification.

There is also the possibility of calculating the design derivatives of the element matrix and vector utilizing finite differences. Such an approach would be called a *semianalytical method*. There has been considerable research into this method especially in connection with shape sensitivity analyses (e.g., see Barthelemy and Haftka 1990). For certain implementations, this approach is very convenient. However, the analytical approach described here gives accurate results for the design sensitivities within the accuracy of the chosen finite element discretization. The semianalytical method is an additional approximation.

6.3.3 Continuum Systems

The derivation of a design sensitivity analysis is also possible from the weak form of Eq. (6.17) of the structural analysis problem. It is convenient here to use the response function $\psi_i, i = 1, \ldots, n_\psi$, in integral form:

$$\psi_i = \int_V \varphi_i(\mathbf{u}, \mathbf{Du}, b_j) \, dV \tag{6.86}$$

Point-wise constraints such as stresses can be expressed in such a form by using the Dirac-delta function $\delta(\mathbf{x} - \mathbf{x}_c)$. If the function of interest φ_i is defined at the point $\mathbf{x}_c$ of the structure, $\varphi_i = \varphi_i(\mathbf{x}_c)$, then the integral form of this function appears as

$$\psi_i = \int_V \varphi_i(\mathbf{u}, \mathbf{Du}, b_j) \delta(\mathbf{x} - \mathbf{x}_c) dV \tag{6.87}$$

The behavior of the structure is determined by the weak form of Eq. (6.17):

$$\check{A}(\delta \mathbf{u}, \mathbf{u}) = \check{p}(\delta \mathbf{u}) \qquad \forall \delta \mathbf{u} \in \mathbf{U} \tag{6.88}$$

where $\mathbf{U}$ is the space of all admissible solutions. The variation of the structural domain which follows from the variation of the design variables δb_j is shown in Fig. 6.13.

Before dealing with the structural analysis problem, consider an arbitrary function $\varphi(\mathbf{x}, b_j)$ which will be differentiated with respect to the design variable b_j. Application of the chain rule leads to

$$\varphi,_j = \frac{\partial \varphi}{\partial b_j} + \partial_\mathbf{x} \varphi \, \mathbf{x},_j \tag{6.89}$$

with $\partial_\mathbf{x} = \partial/\partial \mathbf{x}$. The vector $\mathbf{x},_j$ is again the design velocity field. Equation (6.89) is often called the *material derivative* of the function φ.

Also of interest are the derivatives of the volume element dV and of the surface element dS. For the volume element (Haug et al. 1986)

$$dV,_j = \partial_\mathbf{x} \mathbf{x},_j \, dV \tag{6.90}$$

and for the surface element

$$dS,_j = -\partial_\mathbf{x} \mathbf{n} \, \mathbf{n}^T \mathbf{x},_j \, dS \tag{6.91}$$

Equations (6.89) and (6.90) lead to an expression for the derivative of the function φ integrated over the structural volume

$$\left(\int_V \varphi \, dV \right),_j = \int_V \left[\frac{\partial \varphi}{\partial b_j} + \partial_\mathbf{x} \varphi \, \mathbf{x},_j + \varphi \partial_\mathbf{x} \mathbf{x},_j \right] dV \tag{6.92}$$

Using the product rule of differentiation yields

$$\left(\int_V \varphi \, dV \right),_j = \int_V \left[\frac{\partial \varphi}{\partial b_j} + \partial_\mathbf{x}(\varphi \mathbf{x},_j) \right] dV \tag{6.93}$$

Application of Gauss's integral theorem to the second term of Eq. (6.93) leads to

$$\left(\int_V \varphi \, dV \right),_j = \int_V \frac{\partial \varphi}{\partial b_j} \, dV + \int_S \varphi \mathbf{n}^T \mathbf{x},_j \, dS \tag{6.94}$$

If finite elements are employed for the evaluation of the integrals, this transformation should be avoided. Finite elements are based on the discretization of the structural volume, and hence a formulation in terms of volume integrals is more convenient. Also finite elements tend to encounter accuracy difficulties for boundary stresses, especially for stress components normal to the structural boundary.

The design derivative of an integral of the function φ over the surface follows from Eq. (6.91) and appears as

$$\left(\int_S \varphi \, dS \right),_j = \int_S \left[\frac{\partial \varphi}{\partial b_j} + \partial_\mathbf{x} \varphi \, \mathbf{x},_j - \varphi \partial_\mathbf{x} \mathbf{n} \, \mathbf{n}^T \mathbf{x},_j \right] dS \tag{6.95}$$

The material derivatives of Eq. (6.89) are now applied to find the design derivatives of the response function of Eq. (6.86):

$$\psi_{i,j} = \int_V \left[\frac{\partial \varphi}{\partial b_j} + \frac{\partial \varphi}{\partial \mathbf{u}} \mathbf{u}_{,j} + \frac{\partial \varphi}{\partial (\mathbf{Du})}(\mathbf{Du})_{,j} + \partial_{\mathbf{x}}(\varphi \mathbf{x}_{,j}) \right] dV \qquad (6.96)$$

Differentiation of the functional of Eq. (6.88) gives

$$\check{A}\left(\delta\mathbf{u}, \mathbf{u}_{,j}\right) = \check{p}(\delta\mathbf{u})_{,j} - \check{A}(\delta\mathbf{u}, \mathbf{u})_{,j} \qquad (6.97)$$

The solution of this equation can be used directly to compute the design derivatives $\mathbf{u}_{,j}$, $(\mathbf{Du})_{,j}$ for Eq. (6.96). The terms on the right-hand side of Eq. (6.97) can be obtained by differentiating Eq. (6.14) with respect to the design variables. Then

$$\check{A}(\delta\mathbf{u}, \mathbf{u})_{,j} = \int_V (\mathbf{D}\delta\mathbf{u})^T \mathbf{E}_{,j} \mathbf{Du}\, dV + \int_V \partial_{\mathbf{x}} \left[(\mathbf{D}\delta\mathbf{u})^T \mathbf{EDux}_{,j} \right] dV \qquad (6.98)$$

and

$$\check{p}(\delta\mathbf{u})_{,j} = \int_V \delta\mathbf{u}^T \bar{\mathbf{p}}_V \partial_{\mathbf{x}} \mathbf{x}_{,j}\, dV - \int_S \delta\mathbf{u}^T \bar{\mathbf{p}}_S \partial_{\mathbf{x}} \mathbf{n}\, \mathbf{n}^T \mathbf{x}_{,j}\, dS \qquad (6.99)$$

With this, a direct differentiation method based on the weak form has been found.

To obtain an adjoint variable method, evaluate Eq. (6.97) with $\delta\mathbf{u} = \mathbf{a}_i$, where $\mathbf{a}_i \in \mathbf{U}$. This leads to

$$\check{A}\left(\mathbf{a}_i, \mathbf{u}_{,j}\right) = \check{p}(\mathbf{a}_i)_{,j} - \check{A}(\mathbf{a}_i, \mathbf{u})_{,j} \qquad (6.100)$$

The function $\mathbf{a}_i$ will be determined by

$$\check{A}(\delta\mathbf{a}_i, \mathbf{a}_i) = \int_V \left[\frac{\partial \varphi}{\partial \mathbf{u}} \delta\mathbf{a}_i + \frac{\partial \varphi}{\partial (\mathbf{Du})} \mathbf{D}\delta\mathbf{a}_i \right] dV, \qquad \delta\mathbf{a}_i \in \mathbf{U} \qquad (6.101)$$

This is an adjoint equation to the problem of Eq. (6.88) with the adjoint variable $\mathbf{a}_i$. Replace $\delta\mathbf{a}_i$ by $\mathbf{u}_{,j}$ in Eq. (6.101). This is possible because $\mathbf{u}_{,j} \in \mathbf{U}$. Then

$$\check{A}\left(\mathbf{u}_{,j}, \mathbf{a}_i\right) = \int_V \left[\frac{\partial \varphi}{\partial \mathbf{u}} \mathbf{u}_{,j} + \frac{\partial \varphi}{\partial (\mathbf{Du})}(\mathbf{Du})_{,j} \right] dV \qquad (6.102)$$

Considering the symmetry of $\check{A}(\cdot, \cdot)$ in Eqs. (6.100) and (6.102), it follows that

$$\int_V \left[\frac{\partial \varphi}{\partial \mathbf{u}} \mathbf{u}_{,j} + \frac{\partial \varphi}{\partial (\mathbf{Du})}(\mathbf{Du})_{,j} \right] dV = \check{p}(\mathbf{a}_i)_{,j} - \check{A}(\mathbf{a}_i, \mathbf{u})_{,j} \qquad (6.103)$$

This equation is now introduced into Eq. (6.96), and the design derivative of the function ψ_i is obtained using the solution of the adjoint equation of Eq. (6.101):

$$\frac{d\psi_i}{db_j} = \int_V \left[\frac{\partial \varphi}{\partial b_j} + \partial_{\mathbf{x}}(\varphi \mathbf{x},_j)\right] dV + \check{p}(\mathbf{a}_i),_j - \check{A}(\mathbf{a}_i, \mathbf{u}),_j \qquad (6.104)$$

The design derivatives of the operators $\check{A}(\mathbf{a}_i, \mathbf{u})$, $\check{p}(\mathbf{a}_i)$ follow in a manner similar to Eqs. (6.98) and (6.99), respectively.

For elasticity problems, the solution of Eq. (6.97) and of the adjoint equation (6.101) may be obtained using a finite element discretization. Such a discretization is applied in a manner analogous to the finite element solution described in Section 6.2.

6.3.4 Stresses

Since stress concentration design is of special interest here, an expression for the design derivatives of the stresses will be provided. These will be treated as a special case of the functions ψ_i. Differentiation of Eq. (6.63) leads to

$$\boldsymbol{\sigma},_j = \left[\mathbf{E},_j \mathbf{B} + \mathbf{E} \mathbf{B},_j\right] \mathbf{v}^e + \mathbf{E} \mathbf{B} \mathbf{v}^e,_j \qquad (6.105)$$

The first term can be determined from the analysis results directly. The second term involves the derivatives of the element displacement vector $\mathbf{v}^e,_j$. These derivatives can be obtained from the solution $\mathbf{v},_j$ of Eq. (6.74). If the number of design variables exceeds the number of stresses that are of interest, the last term of Eq. (6.105) can be replaced in accordance with Eq. (6.77) using the solution of Eq. (6.76). Note that the number of adjoint variables is the same as the number of stress components. The design derivatives of equivalent stresses such as the von Mises stress (see Chapter 1) must be obtained using the chain rule.

If the stresses and their design derivatives are found the design derivatives of the stress concentration factors can be calculated utilizing

$$K_t,_j = \frac{1}{\sigma_{\text{nom}}^2}(\sigma_{\max},_j \sigma_{\text{nom}} - \sigma_{\max} \sigma_{\text{nom}},_j) \qquad (6.106)$$

$$K_{ts},_j = \frac{1}{\tau_{\text{nom}}^2}(\tau_{\max},_j \tau_{\text{nom}} - \tau_{\max} \tau_{\text{nom}},_j) \qquad (6.107)$$

6.3.5 Structural Volume

The sensitivity of the volume of the structure with respect to design changes is of interest. For the design optimization, the structural volume is often used as a cost or constraint function. It is calculated from

$$V = \int_V dV \qquad (6.108)$$

The volume can be computed directly from the finite element discretization.

$$V = \sum_e \int_{-1}^{+1} \int_{-1}^{+1} \int_{-1}^{+1} \det \mathbf{J} \, d\xi \, d\eta \, d\zeta \qquad (6.109)$$

Differentiation with respect to the design variables yields the design derivatives

$$V_{,j} = \sum_e \int_{-1}^{+1} \int_{-1}^{+1} \int_{-1}^{+1} (\det \mathbf{J})_{,j} \, d\xi \, d\eta \, d\zeta \tag{6.110}$$

If the continuum formulation of the sensitivity analysis is used, it can be seen from Eq. (6.90) that the design derivatives of the structural volume follow from

$$V_{,j} = \int_V \partial_{\mathbf{x}} \mathbf{x}_{,j} \, dV \tag{6.111}$$

Discretization of this equation leads again to Eq. (6.110).

6.3.6 Design Velocity Field

In the discrete and continuum approaches the design velocity field $\mathbf{x}_{,j}$ of Eqs.(6.84) and (6.89), respectively, needs to be determined. The design velocity field can also be considered to be the sensitivity of the finite element mesh to changes in the design variables. In sizing optimization, no change of the mesh is encountered if the design variables are modified. But in shape optimization the computation of the design velocity field plays an important role.

Several methods are available to obtain the design velocity field (Choi and Chang 1994). These methods are (1) finite differences method, (2) the boundary displacement method, and (3) isoparametric mapping methods.

In the finite differences approach, for each design variable a mesh with the perturbed variable is generated and the change of the nodal positions is determined. A finite difference scheme such as Eq. (6.70) gives the design velocity field.

The boundary displacement method approach is to determine the finite element mesh in the interior of the solution domain from the boundary description of the structure by solving an auxiliary structure (Yao and Choi 1989). The stiffness matrix of the structure of Eq. (6.53) is rearranged such that

$$\begin{bmatrix} \mathbf{K}_{bb} & \mathbf{K}_{bd} \\ \mathbf{K}_{db} & \mathbf{K}_{dd} \end{bmatrix} \begin{bmatrix} \mathbf{x}_{b,j} \\ \mathbf{x}_{d,j} \end{bmatrix} = \begin{bmatrix} \bar{\mathbf{p}}_b \\ \mathbf{0} \end{bmatrix} \tag{6.112}$$

where the subscript b denotes the boundary nodes and the subscript d denotes the domain nodes. The load $\bar{\mathbf{p}}_b$ is a fictitious load that would be necessary to achieve the change of the shape of the design boundary (Fig. 6.14). Also the unknowns that correspond to boundaries that do not change during the design variation are introduced as zero displacements. Finally, the second equation of Eq. (6.112) is rearranged such that

$$\mathbf{K}_{dd} \mathbf{x}_{d,j} = -\mathbf{K}_{db} \mathbf{x}_{b,j} \tag{6.113}$$

Solution of this linear system of equations yields the design velocity field $\mathbf{x}_{d,j}$ in the interior of the structure. The boundary velocity field $\mathbf{x}_{b,j}$ needs to be determined from the geometric description of the boundary shape.

In the isoparametric mapping approach the mesh is determined directly from the structural geometry (Braibant and Fleury 1984; Chang and Choi 1992; Schramm and Pilkey

476 STRESS CONCENTRATION ANALYSIS AND DESIGN

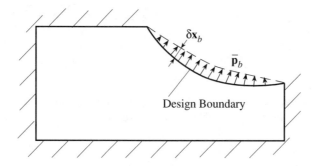

Figure 6.14 Boundary variation method.

1993). Usually the structure is subdivided into several sections or *design elements* that have analytically defined geometries. Then the positions of the finite element nodes are determined from the description of the structural geometry, for example, using parametric curves, surfaces, and solids. The design variables are the control variables of such geometries. Parametric curves, surfaces, and solids define the geometry in terms of nondimensional parameters γ_i, $\mathbf{x} = \mathbf{x}(\gamma_i)$. If the mesh is kept at the same positions in terms of γ_i, then the mesh is also modified with consecutive design changes. This method is very convenient and inexpensive. A disadvantage is that, due to the definition of the mesh, large element distortions are possible. Because of this, Lagrange elements are preferred.

In this chapter the isoparametric mapping approach will be studied more closely. In order to provide a sufficient explanation, a brief digression into the field of computer-aided geometric design (CAGD) will be made here. Farin (1992) gives a complex description of the theoretical basis of CAGD. CAGD provides the geometric elements for computer aided design (CAD). The coupling of CAD with finite element analysis and optimization codes can be accomplished through utilization of the geometry description used in CAD. From the wide variety of geometric design elements available those that are based on B-splines are the most interesting for structural shape design in connection with a finite element analysis (Braibant and Fleury 1984).

Begin with *B-spline curves*, which can be used to define the geometry of one-dimensional design elements. These curves are defined by a polygon and a knot sequence. The defining polygon is determined by the m vertices $\mathbf{w}_j = [\, w_{jx} \quad w_{jy} \quad w_{jz} \,]^T$, $j = 1, \ldots, m$. The order of the B-spline is χ. The position vector $\mathbf{x} = [\, x \quad y \quad z \,]^T$ of a point on a B-spline curve is given in terms of the dimensionless parameter γ such that $\gamma \in \langle 0, \tilde{\gamma} \rangle$, $\tilde{\gamma} = m - \chi + 1$. Then

$$\mathbf{x}(\gamma) = \begin{bmatrix} \boldsymbol{\beta}_\chi & \mathbf{0} & \mathbf{0} \\ \mathbf{0} & \boldsymbol{\beta}_\chi & \mathbf{0} \\ \mathbf{0} & \mathbf{0} & \boldsymbol{\beta}_\chi \end{bmatrix} \mathbf{w} \tag{6.114}$$

The vector $\mathbf{w}$ contains the vertex positions $\mathbf{w}_j$ of the control polygon, $\mathbf{w} = [\, \mathbf{w}_x \quad \mathbf{w}_y \quad \mathbf{w}_z \,]^T$ and $\mathbf{w}_x = [\, w_{1x} \quad w_{2x} \quad \cdots \quad w_{mx} \,]$. The vectors $\mathbf{w}_y$, $\mathbf{w}_z$ are defined similarly. The B-spline basis $\boldsymbol{\beta}_\chi = \big[\beta_{\chi j}(\gamma)\big]$, $j = 1, \ldots, m$, is obtained from the recursion formula (Rogers and

Adams 1976)

$$\beta_{kj}(\gamma) = \frac{(\gamma - \kappa_j)\beta_{k-1\,j}(\gamma)}{\kappa_{j+k-1} - \kappa_j} + \frac{(\kappa_{j+k} - \gamma)\beta_{k-1\,j+1}(\gamma)}{\kappa_{j+k} - \kappa_{j+1}} \qquad k = 2\ldots\chi \qquad (6.115)$$

with the initial values of

$$\beta_{1j}(\gamma) = \begin{cases} 1 & \text{if } \gamma \in \langle \kappa_j, \kappa_{j+1}\rangle \\ 0 & \text{otherwise} \end{cases} \qquad (6.116)$$

The integers κ_j are the elements of the knot vector $\kappa = [\kappa_j]^T$, $j = 1,\ldots,m + \chi$, that is defined by the knot sequence. The knot vector κ is obtained from

$$\kappa_j = \begin{cases} 0 & \text{if } j \leq \chi \\ j - \chi & \text{if } \chi < j \leq m \\ m - \chi + 1 & \text{if } j > m \end{cases} \qquad (6.117)$$

The basis functions $\beta_{\chi j}(\gamma)$ of Eq. (6.115) are polynomials of degree $\chi - 1$. Hence Eq. (6.114) defines a polynomial of degree of $\chi - 1$.

The characteristics of B-spline curves that are of interest for their use in finite element analysis are: (1) The curve lies in a convex hull of the vertices; (2) the curve is invariant under an affine transformation; (3) the tangents of the initial and final points are defined by the first and the last edge, respectively, of the defining polygon; (4) the position vector $\mathbf{x}(\gamma)$ and its derivatives up to order of $\chi - 2$ are continuous over the entire curve. Characteristic (3) can be used to impose C^1 continuity. C^n *continuity* of a parametric function means that the function is n times continuously differentiable. If the order χ of the curve is less than the number m of vertices, a piecewise polynomial of degree $\chi - 1$ with $C^{(\chi-2)}$ continuity will be obtained. B-spline curves can be altered by changing the number and/or position $\mathbf{w}_j$ of the vertices of the defining polygon, by changing the order χ, and by using repeated vertices. The control of a fourth-order B-spline curve, $\chi = 4$, is shown in Fig. 6.15.

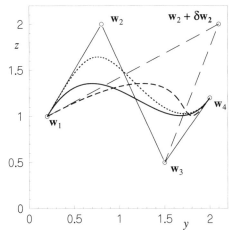

Figure 6.15 B-spline curves in the y,z plane. *Solid line:* $\chi = m = 4$. *Dashed line:* Change of the vertex position $\mathbf{w}_2$, $\delta\mathbf{w}_2 = [0 \quad 1 \quad 0]^T$. *Dotted line:* Double use of vertex $\mathbf{w}_2$, $\chi = m = 5$.

In shape sensitivity analysis and optimization, the vertex positions $\mathbf{w}_j$ are the design variables. Hence, for the determination of the design velocity field $\mathbf{x}_{,j}$, the dependence of the point vector $\mathbf{x}(\gamma)$ on the vertex positions $\mathbf{w}_j$ needs to be calculated. From Eq. (6.114) the derivatives of the position vector $\mathbf{x}(\gamma)$ with respect to the vertex $\mathbf{w}_j$ can be obtained as

$$\frac{d\mathbf{x}(\gamma)}{d\mathbf{w}_j} = \beta_{\chi j}\mathbf{I}_{(3\times 3)} \qquad (6.118)$$

B-spline surfaces are created by moving a *B*-spline curve of the order χ_1 along another *B*-spline curve of the order χ_2. The resulting surface is defined parametrically by the parameters γ_i, $i = 1, 2$. Then

$$\mathbf{x}(\gamma_1,\gamma_2) = \begin{bmatrix} \beta_{\chi_1} & 0 & 0 \\ 0 & \beta_{\chi_1} & 0 \\ 0 & 0 & \beta_{\chi_1} \end{bmatrix} \begin{bmatrix} \mathbf{W}_x & 0 & 0 \\ 0 & \mathbf{W}_y & 0 \\ 0 & 0 & \mathbf{W}_z \end{bmatrix} \begin{bmatrix} \beta_{\chi_2} & 0 & 0 \\ 0 & \beta_{\chi_2} & 0 \\ 0 & 0 & \beta_{\chi_2} \end{bmatrix}^T \qquad (6.119)$$

with the *B*-spline bases $\beta_{\chi_i} = [\beta_{\chi_i j}(\gamma_i)]^T$, $j = 1,\ldots,m_i$, $i = 1, 2$, defined by Eq. (6.115). The $m_1 \times m_2$ matrix $\mathbf{W}_x = [w_{jkx}]$, $j = 1,\ldots,m_1$, $k = 1,\ldots,m_2$, contains the x coordinate values w_{jkx} of control point positions $\mathbf{w}_{jk} = [w_{jkx} \quad w_{jky} \quad w_{jkz}]^T$ of the surface. The matrices $\mathbf{W}_y$, $\mathbf{W}_z$ are defined similarly. The vertices define a polygonal surface. The parameters γ_i, $i = 1, 2$, are defined such that $\gamma_i \in \langle 0, \tilde{\gamma}_i \rangle$, $\tilde{\gamma}_i = m_i - \chi_i + 1$. The order of the *B*-spline surface is (χ_1, χ_2). Figure 6.16 shows a *B*-spline surface defined by a third- and a fifth-order *B*-spline curve. This surface is of order $(3, 3)$.

The *B*-spline surface lies in a convex hull of the vertices, and it is invariant under an affine transformation. It can be altered by changing the number and/or position of the vertices, by changing the order χ_i, and by the use of repeated vertices. C^1 continuity between two *B*-spline surfaces can be realized using the characteristic that tangents on the surface edge are defined by the defining surfaces. More details about *B*-spline surfaces and their characteristics can be found in Farin (1992).

From Eq. (6.119) the derivatives of the point vector $\mathbf{x}(\gamma_1, \gamma_2)$ with respect to the components of the vertices $\mathbf{w}_{jk}$, which are necessary for the computation of the design velocity

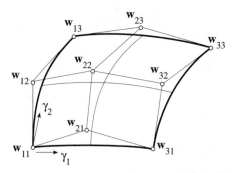

Figure 6.16 *B*-spline surface, $\chi_1 = \chi_2 = m_1 = m_2 = 3$

field $\mathbf{x}_{,j}$, are obtained as

$$\frac{d\mathbf{x}(\gamma_1, \gamma_2)}{d\mathbf{w}_{jk}} = \beta_{\chi_1 j}(\gamma_1)\beta_{\chi_2 k}(\gamma_2)\mathbf{I}_{(3\times 3)} \tag{6.120}$$

B-spline solids are generated by moving a *B*-spline surface along another *B*-spline surface or curve. Then the solid depends on three parameters: $\mathbf{x} = \mathbf{x}(\gamma_1, \gamma_2, \gamma_3)$. The vertices form a polygonal solid. The derivation of the position vector and of the design derivatives is then accomplished in the same way as for *B*-spline surfaces.

It should be noted that there is also another way to define two- and three-dimensional geometries from curves or surfaces, respectively. These geometric elements are *Gordon surfaces* and *solids*. They are based on the blending function method (Gordon 1971; Farin 1992). The principal idea for the definition of these geometries is that, surfaces are created for example, from the definition of the boundary curves of the surface.

B-spline surfaces and solids can be used to define two- and three-dimensional design elements, respectively. The finite element mesh within these design elements is then defined in terms of the parameters γ_i. Hence, an analytical definition of the nodal positions is obtained. The simplest mesh definition would be to position the nodes equidistant in terms of the parameters γ_i. Let $\tilde{\mu}_i$ be the number of finite elements in the γ_i direction. Next, the parametric nodal positions of the element number μ_i in that direction are defined by

$$\gamma_i(\xi_i) = \frac{\xi_i + 2\mu_i - 1}{2\tilde{\mu}_i}\tilde{\gamma}_i \tag{6.121}$$

where ξ_i is the mathematical coordinate of the desired node using $\xi_1 = \xi, \xi_2 = \eta, \xi_3 = \zeta$. The *B*-spline surface or solid definition is then employed to compute the nodal coordinates.

Finally, the design velocity field can be computed directly from the geometric definition of the design elements using the derivatives of the *B*-spline surface and solid definitions with respect to the vertex positions.

Example 6.5 Membrane Fillet Develop a sensitivity analysis of the fillet of Fig. 6.17a. The geometry of the fillet is defined in the x, y plane by two *B*-spline surfaces. These *B*-spline surfaces are the design elements. Section one of the fillet is defined by four vertices and hence is a *B*-spline surface of the order $(2, 2)$ using the vertices 1 to 4 (Fig. 6.17b). It is discretized by a 10×10 element mesh of nine-node elements. The second section is defined by eight vertices (3 to 10). This *B*-spline surface is of the order of $(3, 2)$. This part of the fillet is discretized by a 20×10 element mesh of nine-node elements. The fillet is under a tension load $\sigma = 100$ MPa on the free end.

The finite element analysis of plane stress problems was considered in Example 6.3. Here a sensitivity analysis for the von Mises stresses needs to be developed. The equivalent stress for a plane stress problem is given by Eq. (14) of Example 6.3,

$$\sigma_{eq} = \sqrt{\sigma_x^2 - \sigma_x\sigma_y + \sigma_y^2 + 3\tau_{xy}^2} \tag{1}$$

where the stresses $\boldsymbol{\sigma} = [\sigma_x \quad \sigma_y \quad \tau_{xy}]^T$ follow from Eq. (15) of Example 6.3,

$$\boldsymbol{\sigma} = \mathbf{E}^\sigma \mathbf{B}\mathbf{v}^e \tag{2}$$

480   STRESS CONCENTRATION ANALYSIS AND DESIGN

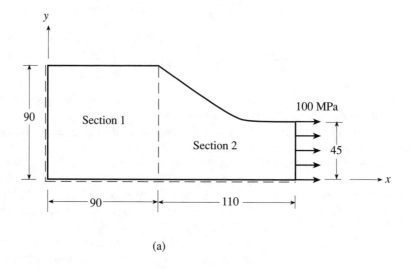

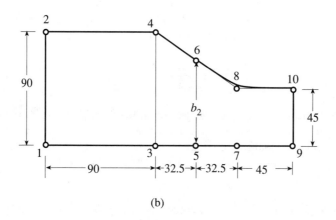

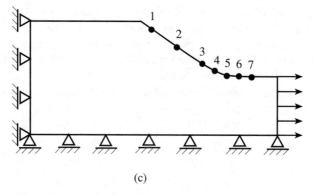

Figure 6.17 Membrane fillet: (*a*) Original geometry; (*b*) vertices; (*c*) stress sensitivity evaluation. Unit of length is millimeters.

The design derivatives of the von Mises stress of (1) appear as

$$\sigma_{eq,j} = \frac{1}{\sigma_{eq}} \begin{bmatrix} \sigma_x - \dfrac{\sigma_y}{2} \\ \sigma_y - \dfrac{\sigma_x}{2} \\ 3\tau_{xy} \end{bmatrix}^T \boldsymbol{\sigma}_{,j} \qquad (3)$$

where the derivatives of the stress components follow from Eq. (6.105). If the material matrix **E** is considered to be design independent, these derivatives appear as

$$\boldsymbol{\sigma}_{,j} = \mathbf{E}[\mathbf{B}_{,j}\mathbf{v}^e + \mathbf{B}\mathbf{v}^e_{,j}] \qquad (4)$$

The derivative $\mathbf{v}_{,j}$ follows from the solution of Eq. (6.74). The design derivative of the matrix **B**, which is needed for (4) and for the computation of the design derivatives of the stiffness matrix $\mathbf{K}_{,j}$ on the right-hand side of Eq. (6.74), is given by

$$\mathbf{B}_{,j} = \begin{bmatrix} J'_{11}\partial_\xi + J'_{12}\partial_\eta & 0 \\ 0 & J'_{12}\partial_\xi + J'_{22}\partial_\eta \\ J'_{12}\partial_\xi + J'_{22}\partial_\eta & J'_{11}\partial_\xi + J'_{12}\partial_\eta \end{bmatrix} \begin{bmatrix} \overline{\mathbf{N}} & \mathbf{0} \\ \mathbf{0} & \overline{\mathbf{N}} \end{bmatrix} \qquad (5)$$

In this expression, from Eq. (6.82), $\mathbf{J}^{-1}{}_{,j} = -\mathbf{J}^{-1}\mathbf{J}_{,j}\mathbf{J}^{-1} = [J'_{lk}], l,k = 1,2$.

The computation of the design derivatives of the Jacobian $\mathbf{J}_{,j}$ requires the knowledge of the design velocity field $\mathbf{x}_{,j}$. This is computed directly from the finite element mesh using the isoparametric method as described previously. Here a single design variable b_2 is chosen. It is the y position of the vertex number 6. The variation of the vertices changes the geometry of the design boundary. The finite element mesh is defined such that the element nodes are distributed equidistant in terms of the parameters γ_1, γ_2 of the B-spline surface.

For the stress computation, the stress components at the Gauss points are calculated using (2). Then the stress components are extrapolated to the nodes using the quadratic shape functions that are employed for the displacement approximation of the nine-node Lagrange element. In the next step the averages of the stress components at the nodes are computed. Finally, the von Mises stresses σ_{eq} of (1) at the nodes are calculated from the stress components. In the same manner the design derivatives of the stresses are obtained.

A numerical example will illustrate this. A sensitivity analysis is performed to determine the sensitivity of the stresses on the upper boundary of section two of the fillet with respect to changes of the position b_2 of the vertex number 6. The results for selected nodes (Fig. 6.17c) are summarized in Table 6.3. Note that there are differences in the accuracy of the sensitivity analysis that are due to the method used. The results for the design derivatives $\sigma^A_{eq,2}$ obtained by the analytical method are accurate within the accuracy of the finite element model.

Figure 6.18a shows the finite element mesh employed and the von Mises stress distribution. Figure 6.18b shows the distribution of design sensitivities of the von Mises stresses with respect to the design variable b_2.

Example 6.6 Torsion Bar with Variable Diameter Return to the torsion bar problem of Example 6.4, and let the response function of interest be the equivalent shear stress τ_{eq}

482 STRESS CONCENTRATION ANALYSIS AND DESIGN

TABLE 6.3 Stress Sensitivities for a Membrane Fillet

Node	x (mm)	y (mm)	σ_{eq} (MPa)	$F1$ $\sigma_{eq,2}$ (GPa/m)	$F2$ $\sigma_{eq,2}$ (GPa/m)	A $\sigma_{eq,2}$ (GPa/m)
1	99.38	83.50	20.694	−0.18256	−0.18065	−0.18060
2	119.65	69.47	44.585	−1.75863	−1.75906	−1.75911
3	140.36	55.15	86.017	−3.07039	−3.07231	−3.07252
4	149.51	49.75	120.610	−1.07792	−1.07354	−1.07306
5	161.55	45.45	160.040	2.05204	2.05445	2.05472
6	170.53	45.00	117.881	0.52566	0.52645	0.52654
7	179.47	45.00	104.304	0.17673	0.12098	0.12101

Note: $F1$—finite differences, $\Delta b_2 = 0.1$ mm; $F2$—finite differences, $\Delta b_2 = 0.01$ mm; A—analytical method.

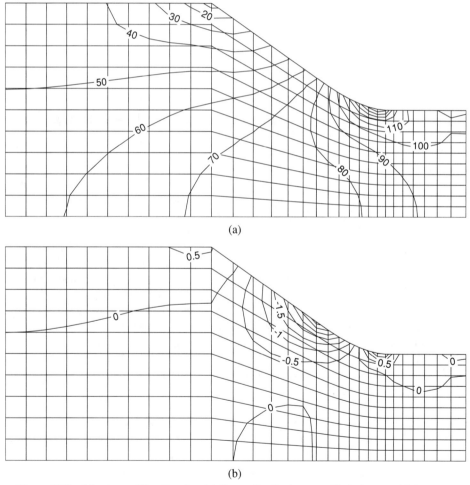

Figure 6.18 Membrane fillet. Results: (*a*) Stress distribution σ_{eq}; (*b*) design sensitivity $\sigma_{eq,2}$.

given by Eq. (16),

$$\tau_{eq} = \sqrt{\sigma^T \sigma} \tag{1}$$

with σ from Eq. (15). The design derivative of (1) appears as

$$\tau_{eq,j} = \frac{1}{\tau_{eq}} \sigma^T \sigma_{,j} \tag{2}$$

To find the design derivatives of the stresses $\sigma = [\, -\tau_{r\theta} \quad \tau_{\theta x}\,]^T$, differentiate Eq. (15) of Example 6.4. Then

$$\sigma_{,j} = \frac{1}{r^2} \left[-2\frac{r_{,j}}{r} \mathbf{B} \mathbf{v}^e + \mathbf{B}_{,j} \mathbf{v}^e + \mathbf{B} \mathbf{v}^e_{,j} \right] \tag{3}$$

Consideration of the definition of the matrix $\mathbf{B}$ of Eq. (12) of Example 6.4 and of the derivative of the Jacobi matrix $\mathbf{J}$ of Eq. (6.82) leads to

$$\mathbf{B}_{,j} = -\mathbf{J}^{-1} \mathbf{J}_{,j} \mathbf{B} \tag{4}$$

Then (3) appears as

$$\sigma_{,j} = -\frac{1}{r^2} \left[2\frac{r_{,j}}{r} \mathbf{I} + \mathbf{J}^{-1} \mathbf{J}_{,j} \right] \mathbf{B} \mathbf{v}^e \sigma + \frac{1}{r^2} \mathbf{B} \mathbf{v}^e_{,j} \tag{5}$$

The final term of (5) involves the design derivative of the state vector $\mathbf{v}_{,j}$ which needs to be obtained from Eq. (6.74). The remainder of this expression can be evaluated after the solution of the system relations of Eq. (6.53) for the problem.

The derivatives of the element matrix and vector can be found in a manner similar to the derivation of (5). They appear as

$$\mathbf{k}^e_{,j} = -\int_{-1}^{+1} \int_{-1}^{+1} \frac{1}{r^4} \mathbf{B}^T \mathbf{H} \mathbf{B} \det \mathbf{J} \, d\xi \, d\eta \tag{6}$$

$$\bar{\mathbf{p}}^e_{,j} = \mathbf{0} \tag{7}$$

with

$$\mathbf{H} = 4\frac{r_{,j}}{r} \mathbf{I} + \left[\mathbf{J}^{-1} \mathbf{J}_{,j} \right]^T + \mathbf{J}^{-1} \mathbf{J}_{,j} - \frac{(\det \mathbf{J})_{,j}}{\det \mathbf{J}} \mathbf{I}, \tag{8}$$

The matrix (6) and the vector (7) are used to assemble the right-hand side of Eq. (6.74).

As a numerical example, a shaft with a section of the shape of Fig. 6.19 will be utilized. The shaft is loaded with a torque $T = 10^4$ Nm. The geometry of the solution domain is the same as in the previous example (Fig. 6.17b). Hence the same finite element discretization can be used. The sensitivity analysis is performed to determine the sensitivity of the stresses on the surface of the shaft with respect to changes of vertex number 6.

In Table 6.4 the stresses and design sensitivities of selected points are given. The location of these points has been selected to be the same as those in Example 6.5 (Fig. 6.19).

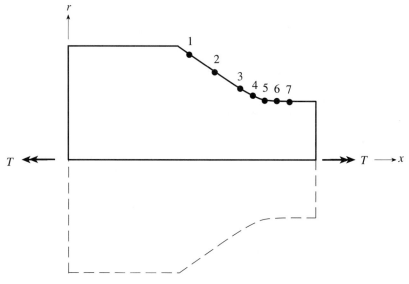

Figure 6.19 Shaft with variable diameter.

Figure 6.20a shows the finite element mesh employed and distribution of the equivalent shear stresses. Figure 6.20b shows the distribution of design sensitivities of the equivalent shear stresses with respect to the design variable b_2.

6.4 DESIGN MODIFICATION

In the previous sections the methodology for the analysis of stress concentrations was set forth and the sensitivity to design changes determined. The goal to improve the design can now be accomplished by solving the optimization problem of Eq. (6.1). This problem is nonlinear and has no special characteristics such as convexity or separability. To determine design improvements, an approximation of the optimization problem of Eq. (6.1) can be formulated using the results of the design sensitivity analysis. For a given design s, consider

TABLE 6.4 Stress Sensitivities in a Shaft of Variable Diameter

				F	A
Node	x (mm)	r (mm)	τ_{eq} (MPa)	$\tau_{eq,2}$ (GPa/m)	$\tau_{eq,2}$ (GPa/m)
1	99.38	83.50	10.970	−0.00072	−0.00196
2	119.65	69.47	20.761	−0.69620	−0.31587
3	140.36	55.15	45.377	−0.14267	−0.68389
4	149.51	49.75	66.435	−0.09772	−0.41295
5	163.36	45.17	96.355	0.34767	0.38104
6	170.53	45.00	89.140	0.08867	0.08867
7	179.47	45.00	87.031	0.01815	0.01815

F—finite differences, $\Delta b_2 = 0.01$ mm; A—analytical method.

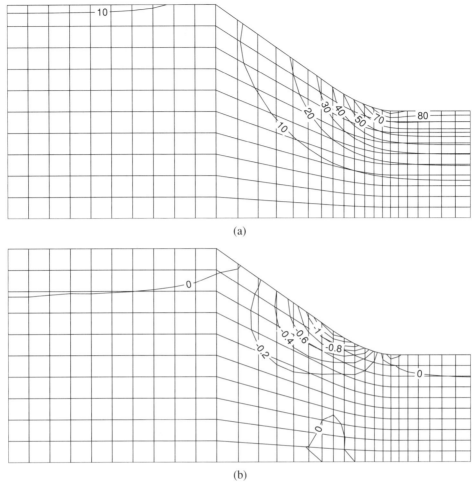

Figure 6.20 Shaft with variable diameter. Results: (*a*) Stress distribution τ_{eq}; (*b*) design sensitivity $\tau_{eq,2}$.

the solution of the approximate optimization problem

$$W(\mathbf{b}) \approx \hat{W}(\mathbf{b}, \mathbf{b}^{(s)}) \Rightarrow \min$$
$$g_i(\mathbf{b}) \approx \hat{g}_i(\mathbf{b}, \mathbf{b}^{(s)}) \leq 0, \qquad i = 1, \ldots, n_g, \qquad (6.122)$$
$$\mathbf{b}^l \leq \mathbf{b} \leq \mathbf{b}^u,$$

Superscript hats indicate approximate functions. The equality constraints of the original problem of Eq. (6.1) have been eliminated to simplify the problem. The treatment of the equality constraints will be introduced later. The approximation implied by Eq. (6.122) is called a *local approximation* (Barthelemy and Haftka 1993). Here "local" refers to the fact that the objective and constraint functions are approximated at a certain point $\mathbf{b}^{(s)}$ in the design space. Local approximations are mostly based on first-order sensitivity

486 STRESS CONCENTRATION ANALYSIS AND DESIGN

information such as derived in the previous section. Higher-order sensitivity information can be employed to improve local approximations (Fleury 1989).

Local approximations are formed in a manner such that the optimization problem of Eq. (6.122) has certain favorable characteristics that are missing from the original optimization problem of Eq. (6.1). These allow the approximate optimization problem to be solved with common mathematical programming methods. The solution of the problem of Eq. (6.122) is the updated design. Designate the updated solution as $\mathbf{b}^{(s+1)}$. Then repeat the system analysis and the sensitivity analysis (Fig. 6.21). Sequential solution of the problem of Eq. (6.122) leads ultimately to an improved structural design. The optimization iteration over s stops if

$$\left|\left(W^{(s)} - W^{(s-1)}\right)/W^{(s-1)}\right| < \epsilon_W, \qquad g_i^{(s)} < \epsilon_g, \tag{6.123}$$

where ϵ_W and ϵ_g are predetermined tolerance values.

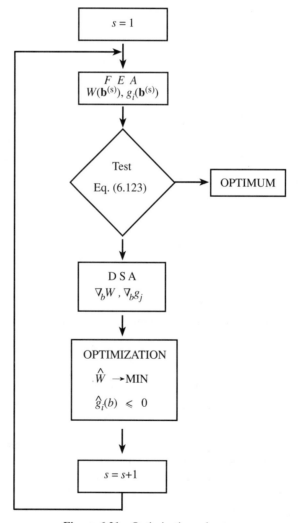

Figure 6.21 Optimization scheme.

It is assumed that in each optimization step only small changes of the structure occur and hence that the sequential solution of approximate optimization problems can lead to an overall convergence of the solution of the structural optimization problem. Although the assumption of small changes cannot always be retained, local approximation methods have been successfully applied to a variety of structural optimization problems. The design optimization process using local function approximations determines local minima only. This is caused by the nature of the structural optimization problem that, in general, has no special characteristics such as convexity. If the problem were convex, the minimum could be characterized as a global minimum.

Three methods of local approximations will be discussed here. For further reading on structural optimization methods the reader is referred to the textbooks by Morris (1982), Papalambros and Wilde (1988), and Haftka et al. (1990).

6.4.1 Sequential Linear Programming

The simplest formulation of the approximate optimization problem of Eq. (6.122) would be a linearization of the objective and constraint functions. This method of design update is called *sequential linear programming* (Zienkiewicz and Campbell 1973).

The objective function and constraints are linearized using the Taylor series expansion of Eq. (6.68). Then

$$\hat{W}(\mathbf{b}) = W(\mathbf{b}^{(s)}) + (\boldsymbol{\nabla}_b W)^T \delta\mathbf{b} \Rightarrow \min \quad (6.124)$$

$$\hat{g}_i(\mathbf{b}) = g_i(\mathbf{b}^{(s)}) + (\boldsymbol{\nabla}_b g_i)^T \delta\mathbf{b} \leq 0$$

Introduction of Eq. (6.69) leads to

$$\hat{W}(\mathbf{b}) = \overline{W} + (\boldsymbol{\nabla}_b W)^T \mathbf{b} \Rightarrow \min \quad (6.125)$$

$$\hat{g}_i(\mathbf{b}) = \overline{g}_i + (\boldsymbol{\nabla}_b g_i)^T \mathbf{b} \leq 0$$

with

$$\overline{W} = W(\mathbf{b}^{(s)}) - (\boldsymbol{\nabla}_b W)^T \mathbf{b}^{(s)} \quad (6.126)$$

$$\overline{g}_i = g_i(\mathbf{b}^{(s)}) - (\boldsymbol{\nabla}_b g_i)^T \mathbf{b}^{(s)}$$

The quantities $\overline{W}$ and $\overline{g}_i$ are constant values for the approximate optimization problem at the optimization step s. The variables for the linear program are the design variables $\mathbf{b}$. The problem of Eq. (6.125) can be solved using a linear programming method such as the simplex method.

To ensure convergence of the method, the variation of the design variables is limited for each iteration step s such that

$$\mathbf{b}^l \leq \overline{\mathbf{b}}^l \leq \mathbf{b} \leq \overline{\mathbf{b}}^u \leq \mathbf{b}^u \quad (6.127)$$

where, for example,

$$\overline{\mathbf{b}}^l = \mathbf{b}^{(s)} - c_1^{(s)} \left|\mathbf{b}^{(s)}\right| \quad (6.128)$$

$$\overline{\mathbf{b}}^u = \mathbf{b}^{(s)} + c_2^{(s)} \left|\mathbf{b}^{(s)}\right|$$

The limits $\overline{\mathbf{b}}^l$, $\overline{\mathbf{b}}^u$ are called *move limits*. During the process of the optimization iteration, the move limits are drawn tighter; that is, the constants $c_1^{(s)}$, $c_2^{(s)}$ are chosen such that

$$c_1^{(s)} \leq c_1^{(s-1)}, \quad c_2^{(s)} \leq c_2^{(s-1)} \quad (6.129)$$

An advantage of the sequential linear programming method is that it always leads to an improved design. Unfortunately, there is no mathematical proof of convergence.

6.4.2 Sequential Quadratic Programming

Similar to the sequential solution of linear optimization problems, a sequence of quadratic optimization problems can be solved. To establish the quadratic optimization problem, employ the Lagrangian multiplier method. The Lagrangian function is defined as

$$L(\mathbf{b}, \lambda) = W + \lambda^T \mathbf{g} \quad (6.130)$$

using the vector of Lagrangian multipliers $\lambda = [\lambda_i]^T$, $i = 1, \ldots, n_g$. Then the optimization problem has a solution if the Kuhn-Tucker conditions (Kuhn and Tucker 1951) for an optimum are fulfilled. These conditions are

$$\nabla_b L = \mathbf{0}, \quad \lambda^T \mathbf{g} = \mathbf{0}, \quad \lambda \geq \mathbf{0} \quad (6.131)$$

Using the Lagrangian of Eq. (6.130), a quadratic optimization problem is established to find the search direction $\mathbf{s}$ of the design improvement such that

$$\mathbf{b} = \mathbf{b}^{(s)} + \alpha \mathbf{s} \quad (6.132)$$

where α is the step length. The objective function for the quadratic optimization problem is given by

$$\hat{W}(\mathbf{s}) = W(\mathbf{b}^{(s)}) + (\nabla_b W)^T \mathbf{s} + \frac{1}{2} \mathbf{s}^T \mathbf{C}^{(s)} \mathbf{s} \Rightarrow \min \quad (6.133)$$

where the matrix $\mathbf{C}^{(s)}$ is a positive definite approximation of the Hessian matrix of the Lagrangian function L of Eq. (6.130). The *Hessian matrix* $\mathbf{C}$ of a function of multiple variables $L(b_j)$, $j = 1, \psi_b$, is the matrix of the second derivatives of the function with respect to the variables. That is, $\mathbf{C} = [d^2L/db_j db_k]$, $j, k = 1, \psi_b$. The initial approximation for the Hessian matrix in Eq. (6.133) is usually the identity matrix, $\mathbf{C}^{(1)} = \mathbf{I}$. The constraints are linearized. Then

$$\hat{g}_i(\mathbf{s}) = g_i\left(\mathbf{b}^{(s)}\right) + (\nabla_b g_i)^T \mathbf{s} \leq 0 \quad (6.134)$$

The quadratic optimization problem consisting of Eqs.(6.133) and (6.134) can be solved using the Kuhn-Tucker conditions of Eq. (6.131) to find $\mathbf{s}, \lambda$.

To find the step length α of Eq. (6.132), it is necessary to minimize a function:

$$\Phi = W(\mathbf{b}) + \sum_{i=1}^{n_g} \mu_i \max(0, g_i(\mathbf{b})) \Rightarrow \min \tag{6.135}$$

This is accomplished iteratively using the initialization $\mu_i = \left|\lambda_i^{(1)}\right|$ and subsequently

$$\mu_i = \max\left(\left|\lambda_i^{(s)}\right|, \frac{1}{2}\left(\mu_i^{(s-1)} + \left|\lambda_i^{(s-1)}\right|\right)\right) \tag{6.136}$$

After the value for α is found from the one-dimensional search for the minimum of Φ, the Hessian $\mathbf{C}^{(s)}$ is updated using the BFGS (Broyden-Fletcher-Shanno-and-Goldfarb) method (Broyden 1970; Fletcher 1970; Shanno 1970; Goldfarb 1970). The BFGS update of the Hessian matrix is given by

$$\mathbf{C}^{(s+1)} = \mathbf{C}^{(s)} - \frac{\mathbf{C}^{(s)}\mathbf{s}\mathbf{s}^T\mathbf{C}^{(s)}}{\mathbf{s}^T\mathbf{C}^{(s)}\mathbf{s}} + \frac{\pi\pi^T}{\mathbf{s}^T\pi} \tag{6.137}$$

with

$$\begin{aligned}
\pi &= a\mathbf{y} + (1-a)\mathbf{C}^{(s)}\mathbf{s} \\
\mathbf{y} &= \alpha\left[\boldsymbol{\nabla}_b L(\mathbf{b}, \lambda^{(s)}) - \boldsymbol{\nabla}_b L\left(\mathbf{b}^{(s)}, \lambda^{(s)}\right)\right] \\
a &= \begin{cases} 1 & \text{if } \mathbf{s}^T\mathbf{y} \geq 0.2\mathbf{s}^T\mathbf{Cs} \\ \dfrac{0.8\mathbf{s}^T\mathbf{Cs}}{\mathbf{s}^T\mathbf{Cs} - \mathbf{s}^T\mathbf{y}} & \text{if } \mathbf{s}^T\mathbf{y} < 0.2\mathbf{s}^T\mathbf{Cs} \end{cases}
\end{aligned} \tag{6.138}$$

Finally, let $\mathbf{b}^{(s+1)} = \mathbf{b}$, and return to the finite element analysis as shown in Fig. 6.21. In the case of the sequential quadratic programming method, a mathematical proof for convergence exists.

6.4.3 Conservative Approximation

Other methods based on first order sensitivity information include the reciprocal approximation (Storaasli and Sobieszczanski-Sobieski 1974) and the conservative approximation (Starnes and Haftka 1979). The latter technique, which is sometimes called convex linearization, forms a convex separable optimization problem. From the convex approximation, special optimization algorithms have been developed, such as methodology based on the duality of convex optimization problems (Braibant and Fleury 1985) and the method of moving asymptotes (Svanberg 1987).

The *conservative approximation* is a combination of a Taylor series expansion with respect to the design variables b_j and of a Taylor series expansion with respect to the reciprocal design variables $1/b_j$. Rewrite Eq. (6.67) as

$$\hat{\psi}_i(\mathbf{b}) = \psi_i\left(\mathbf{b}^{(s)}\right) + \sum_{j=1}^{n_b} \psi_{i,j}\, c_j\, \delta b_j \qquad (6.139)$$

The decision variable c_j is introduced to consider either the expansion with respect to the design variable or with respect to its reciprocal value. The value of c_j is determined such that the function $\hat{\psi}_i(\mathbf{b})$ assumes the largest possible value; that is, the approximation is conservative. If all design variables are chosen to have positive values,

$$\begin{aligned} c_j &= 1 & \text{if } \psi_{i,j} > 0 \\ c_j &= \frac{b_j^{(s)}}{b_j} & \text{if } \psi_{i,j} < 0 \end{aligned} \qquad (6.140)$$

which leads to

$$\hat{\psi}_i(\mathbf{b}) = \overline{\psi}_i + \sum_{\psi_{i,j}>0} \psi_{i,j}\, b_j - \sum_{\psi_{i,j}<0} \psi_{i,j}\, \frac{b_j^{(s)2}}{b_j} \qquad (6.141)$$

with

$$\overline{\psi}_i = \psi_i(\mathbf{b}^{(s)}) - \sum_j \left|\psi_{i,j}\right| b_j^{(s)} \qquad (6.142)$$

Then it can be shown that the function $\hat{\psi}_i(\mathbf{b})$ is convex. Application of Eq. (6.139) to the structural optimization problem of Eq. (6.1) leads to the optimization problem

$$\hat{W}(\mathbf{b}) = \overline{W} + \sum_{W_{,j}>0} W_{,j}\, b_j - \sum_{W_{,j}<0} W_{,j}\, \frac{b_j^{(s)2}}{b_j} \Rightarrow \min \qquad (6.143)$$

$$\hat{g}_i(\mathbf{b}) = \overline{g}_i + \sum_{g_{i,j}>0} g_{i,j}\, b_j - \sum_{g_{i,j}<0} g_{i,j}\, \frac{b_j^{(s)2}}{b_j} \leq 0.$$

where the quantities $\overline{W}$ and $\overline{g}_i$ follow from Eq. (6.142). Methods of mathematical programming can be used to solve the optimization problem of Eq. (6.143).

6.4.4 Equality Constraints

The numerical effort of the optimization process can be reduced by eliminating the equality constraints from the optimization problem. These constraints can be linearized and then eliminated. Linearization of the equality constraints in Eq. (6.1) leads to

$$\overline{\mathbf{h}} + \mathbf{H}\mathbf{b} = \mathbf{0} \qquad (6.144)$$

with

$$\bar{\mathbf{h}} = \mathbf{h}(\mathbf{b}^{(s)}) - \mathbf{H}\mathbf{b}^{(s)} \tag{6.145}$$

$$\mathbf{H} = (\nabla_b \mathbf{h})^T$$

The design variable vector **b** is separated into two parts containing the free and the dependent variables $\mathbf{b}_f$ and $\mathbf{b}_d$, respectively. Then the equality constraints in Eq. (6.144) appear as

$$\bar{\mathbf{h}} + \begin{bmatrix} \mathbf{H}_d & \mathbf{H}_f \end{bmatrix} \begin{bmatrix} \mathbf{b}_d \\ \mathbf{b}_f \end{bmatrix} = \mathbf{0}. \tag{6.146}$$

Resolution leads to

$$\mathbf{b}_f = -\mathbf{H}_f^{-1}[\bar{\mathbf{h}} + \mathbf{H}_d \mathbf{b}_d]. \tag{6.147}$$

The matrix $\mathbf{H}_f$ is invertible if the equality constraints are consistent. To introduce the equality constraints into the approximate optimization problem, the partial derivatives of the objective and the constraints have to be replaced by the total derivatives with respect to the free design variables, which for any function ψ_i are given by

$$\nabla_{b_f} \psi_i = \left(\frac{\partial \psi_i}{\partial \mathbf{b}_f} \right)^T - \mathbf{H}_f^{-1} \mathbf{H}_d \nabla_{b_d} \psi_i. \tag{6.148}$$

Another possibility to eliminate equality constraints is to replace them with two inequality constraints. For example, the constraint $h_i = 0$ can be replaced by the two conditions $h_i \leq 0$ and $-h_i \leq 0$.

6.4.5 Minimum Weight Design

In many cases structural optimization is performed to find the minimum weight design with constraints on the stresses. The optimization problem is given as

$$V \Rightarrow \min \tag{6.149}$$

$$\frac{\sigma_{eqi}}{\bar{\sigma}_{eq}} - 1 \leq 0$$

where the structural volume V is used as an equivalent measure for the structural weight. The quantity σ_{eqi} is the equivalent stress at any given point of the structure, and $\bar{\sigma}_{eq}$ is a prescribed upper limit of these stresses. Also side constraints on the design variables are considered. Of course this problem can be extended by additional constraints on displacements and other structural responses. Not all points of the structure need to be considered in the set of stress constraints. Only the most critical need to be included. The solution to the problem of Eq. (6.149) can be obtained using the algorithms described in this chapter.

6.4.6 Minimum Stress Design

To design to avoid stress concentrations, the search for a structural shape with minimum peak stresses may be of interest. The optimization problem can be expressed as

$$\max \sigma_{eq} \Rightarrow \min \tag{6.150}$$

Additionally, constraints on the design variables are formulated. To solve this optimization problem with a sequential solution of approximate optimization problems as described previously, a sequence of constrained optimization problems needs to be established. This can be done using the following formulation. At the design s, find the point $\mathbf{x}_c$ of the design with the maximum stress, and solve the problem

$$\sigma_{eq}\left(\mathbf{x}_c, \mathbf{b}^{(s)}\right) \Rightarrow \min \tag{6.151}$$

$$\frac{\sigma_{eqi}}{\overline{\sigma}_{eq}^{(s)}} - 1 \leq 0, \quad \overline{\sigma}_{eq}^{(s)} = \sigma_{eq}\left(\mathbf{x}_c, \mathbf{b}^{(s)}\right) \tag{6.152}$$

where the stresses at all points i of the structure are constrained by Eq. (6.152). Of course it is not necessary to include all possible points in the optimization problem since it is very unlikely that all stresses become critical as a result of the design update. Sequential solution of the problem of Eq. (6.151) leads to a minimum stress design.

Example 6.7 Minimum Stress Design for a Membrane Fillet Reconsider the membrane fillet of Example 6.5. Suppose that the shape of the fillet is to be designed such that the maximum von Mises stress σ_{eq} in the structure is minimized. The design optimization is performed using sequential quadratic programming.

First, the design geometry as described in Example 6.5, Fig. 6.17b, will be observed. The position b_2 of vertex number 6 is the only design variable. In a second run the design space was extended by two additional design variables b_1, b_3 (Fig. 6.22). The limits on the design variables are 10 mm $\leq b_j \leq$ 100 mm. The minimum stress design using one design variable yields a maximum von Mises stress of $\sigma_{eq\,max}$ = 130.336 MPa. The final design using three variables yields a maximum von Mises stress of $\sigma_{eq\,max}$ = 122.390 MPa. It can be seen that with a larger design space an improved solution to the optimization

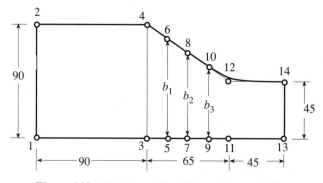

Figure 6.22 Membrane fillet. Three design variables.

TABLE 6.5 Membrane Fillet, Minimum Stresses Optimization History, One Design Variable.

s	$\sigma_{eq\,max}$ (MPa)	x_c (mm)[a]	y_c (mm)[a]	b_2 (mm)	Volume V (mm^3)
1	160.040	161.55	45.45	67.50	14,533.6
2	204.350	135.34	44.69	43.70	13,866.7
3	138.160	161.55	45.26	57.98	14,266.8
4	131.982	161.55	45.21	55.56	14,199.1
5	132.008	135.34	51.05	54.58	14,171.5
6	130.367	161.55	45.20	54.97	14,182.5
7	130.336	161.55	45.20	54.96	14,182.2
8	130.336	161.55	45.20	54.96	14,182.2

[a] $\sigma_{eq\,max} = \sigma_{eq}(x_c, y_c)$.

TABLE 6.6 Membrane Fillet Minimum Stress Optimization History, Three Design Variables

s	$\sigma_{eq\,max}$ (MPa)	b_1 (mm)	b_2 (mm)	b_3 (mm)	Volume V (mm^3)
1	159.727	78.25	67.50	56.25	14,515.9
2	270.376	43.01	36.58	41.19	13,242.8
3	148.967	71.20	61.32	53.24	14,261.3
4	147.748	70.55	60.75	52.96	14,237.8
5	168.696	52.82	44.51	42.63	13,540.1
6	130.126	63.46	54.25	48.83	13,958.7
7	123.895	61.15	52.14	47.48	13,868.1
8	124.009	60.23	51.40	47.38	13,840.7
9	122.684	60.60	51.70	47.42	13,851.7
10	122.667	60.59	51.69	47.42	13,851.4
11	122.672	60.36	51.50	47.29	13,842.6
12	122.390	60.45	51.57	47.34	13,846.1
13	122.390	60.45	51.57	47.34	13,846.1

problem can be obtained. The numerical effort for the second optimization computation was considerably higher.

The iteration histories for both examples are summarized in Tables 6.5 and 6.6. Figure 6.23 shows the shape of the final designs and the corresponding stress distributions.

Example 6.8 Minimum Weight Design for a Shaft with Variable Diameter   Return to the torsion bar of Example 6.6. The shape of the fillet will be designed such that the weight the structure is minimized. The design optimization is performed using sequential quadratic programming.

Continue to use the design geometry as described in Example 6.5, Fig. 6.17b. The position b_2 of vertex number 6 is the only design variable. In a second run the design space was extended by two design variables b_1, b_3 (Fig. 6.22). The limits on the design variables are 10 mm $\leq b_j \leq$ 100 mm. The minimum volume using one design variable is $V = 13,911$ mm^3. For the final design using three variables a minimum volume of $V = 13,474$ mm^3 was obtained.

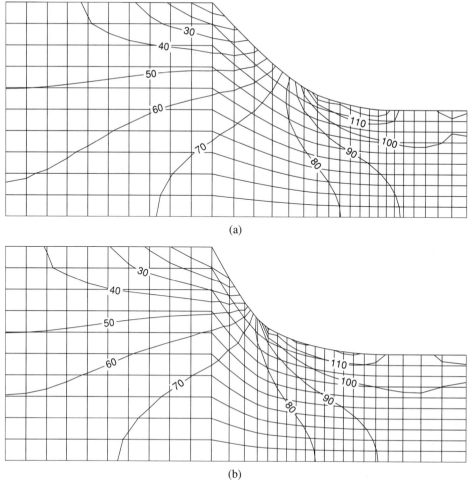

Figure 6.23 Membrane fillet. Minimum stress design. Stress distribution σ_{eq} of the final designs: (*a*) One design variable; (*b*) three design variables.

TABLE 6.7 Shaft with Variable Diameter Minimum Weight Optimization History, One Design Variable

s	Volume V (mm^2)	b_2 (mm)	$\tau_{eq\,max}$ (MPa)
1	14,533.6	67.50	96.355
2	13,733.4	38.95	133.700
3	13,884.8	44.35	103.817
4	13,854.5	43.27	108.712
5	13,826.7	42.27	113.966
6	13,903.2	45.00	101.019
7	13,911.0	45.28	99.855

TABLE 6.8 Shaft with variable diameter minimum weight optimization history, three design variables

s	Volume V (mm^2)	b_1 (mm)	b_2 (mm)	b_3 (mm)	$\tau_{eq\,max}$ (MPa)
1	14,515.9	78.25	67.50	56.25	94.438
2	12,410.5	10.00	14.38	41.96	632.272
3	14,094.9	64.60	56.88	53.39	92.647
4	13,994.6	61.35	54.35	52.71	92.174
5	13,308.0	43.80	38.80	42.18	142.411
6	13,445.3	47.31	41.91	44.29	113.992
7	13,476.6	48.11	42.62	44.77	108.609
8	13,471.9	48.44	43.73	43.06	102.933
9	13,472.9	48.38	43.51	43.40	103.311
10	13,472.5	48.40	43.59	43.28	103.124
11	13,474.0	47.98	44.01	43.30	100.411
12	13,475.1	47.88	44.07	43.38	100.117
13	13,474.8	47.90	44.06	43.36	99.991

The iteration histories for both examples are summarized in Tables 6.7 and 6.8. Figure 6.24 shows the shape of the final designs and the corresponding stress distributions.

From the iteration histories it can be seen that some of the intermediate designs are not necessarily feasible. It might be that some constraints are violated. If there is a feasible design, then the final design is also feasible. Further it can be observed that for more than one design variable, the initial guess for the Hessian matrix is a rather bad estimate. Hence the result of the first iteration step may violate the constraints (see Table 6.8).

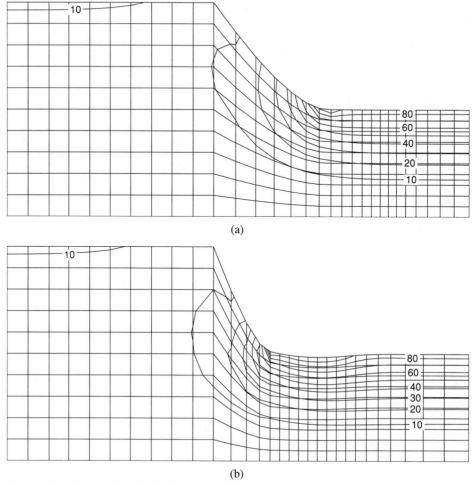

Figure 6.24 Shaft with variable diameter. Minimum weight design. Stress distribution σ_{eq} of the final designs: (*a*) One design variable. (*b*) three design variables.

REFERENCES

ABAQUS Theory Manual, V5.5, 1995, Hibbitt, Karlson & Sorenson Inc., Pawtucket, RI.

Adelmann, H.M., and Haftka, R.T., 1986, "Sensitivity Analysis of Discrete Structural Systems," *AIAA J.*, Vol. 24, pp. 823–832.

ANSYS User's Manual Rev.5.0, 1992, Swanson Analysis Systems Inc., Houston, PA.

Arora, J.S., 1995, "Structural Design Sensitivity Analysis: Continuum and Discrete Approaches," in J. Herskovits (ed.), 1995, *Advances in Structural Optimization*, Kluwer Academic Publishers, Dortrecht, The Netherlands, pp. 47–70.

Babuška, I., Zienkiewicz, O.C., Gago, J., and Oliveira, E.R.de A. (eds.), 1986, *Accuracy Estimates and Adaptive Refinements in Finite Element Computations*, Wiley, New York.

Barthelemy, B., and Haftka, R.T., 1990, "Accuracy Analysis of the Semi-Analytical Method for Shape Sensitivity Calculation," *Mech. Struct. Mach.*, Vol. 18., pp. 407–432.

Barthelemy, J.-F. M., and Haftka, R.T., 1993, "Approximation Concepts for Optimum Structural Design—A Review," *Structural Optimization*, Vol. 5., pp. 129–144.

Bathe, K.-J., 1995, *Finite Element Procedures*, Prentice Hall, Englewood Cliffs, NJ.

Braibant, V., and Fleury, C., 1984, "Shape Optimal Design using B-Splines," *Computer Methods in Appl. Mech. and Engrg.*, Vol. 44, pp. 247-267.

Braibant, V., and Fleury, C., 1985, "An Approximation Concepts Approach to Shape Optimal Design," *Computer Methods in Appl. Mech. and Engrg.*, Vol. 53, pp. 119–148.

Broyden, C.G., 1970, "The Convergence of a Class of Double-Rank Minimization Algorithms. 2: The New Algorithm," *J. Inst. for Math. Appl.*, Vol. 6, pp. 222–231.

Chang, K.-H., and Choi, K.K., 1992, "A Geometry-based Parameterization Method for Shape Design of Elastic Solids," *Mech. of Structures and Machines*, Vol. 20, pp. 215–252.

Choi, K.K., and Chang, K.-H., 1994, "A Study of Design Velocity Field Computation for Shape Optimal Design," *Finite Elements in Analysis and Design*, Vol. 15, pp. 317–342.

DOT User's Manual, V4.00, 1993, VMA Engineering, Colorado Springs, CO.

Duysinx, P., and Fleury, C., 1993, "Optimization Software: View from Europe," in M.P. Kamat (ed.), 1993, *Structural Optimization: Status and Promise*, AIAA Series in Aeronautics and Astronautics, Vol. 150, American Institute for Aeronautics and Astronautics, Washington, DC, pp. 807–850.

Farin, G., 1992, *Curves and Surfaces for Computer Aided Geometric Design, A Practical Guide*, Third Edition, Academic Press, San Diego.

Fleury, C., 1989, "Efficient Approximation Concepts using Second Order Information," *Int. J. Num. Methods in Engrg*, Vol. 28, pp. 2041–2058.

Fletcher, R., 1970, "A New Approach to Variable Metric Algorithms," *Computer J.*, Vol. 13, pp. 317–322.

Gallagher, R.H., and Zienkiewicz, O.C. (eds.), 1973, *Optimum Structural Design*, Wiley, New York.

GENESIS User Manual, V2.1, 1995, VMA Engineering, Colorado Springs, CO.

Goldfarb, D., 1970, "A Family of Variable Metric Methods Derived by Variational Means," *Math. Comp.*, Vol. 24, pp. 23–26.

Gordon, W.J., 1971, "Blending-Function Methods of Bivariate and Multivariate Interpolation and Approximation," *SIAM J. Num. Anal.*, Vol. 8, pp. 158–177.

Haftka, R.T., and Grandhi, R.V., 1986, "Structural Shape Optimization—A Survey," *Computer Methods in Appl. Mech. and Engrg.*, Vol. 57, pp. 91–106.

Haftka, R.T., and Adelmann, H.M., 1989, "Recent Developments in Structural Sensitivity Analysis," *Structural Optimization*, Vol. 1, pp. 137–152.

Haftka, R.T., Gürdal, Z., and Kamat, M.P., 1990, *Elements of Structural Optimization*, Kluwer Academic Publishers, Dortrecht, The Netherlands.

Haug, E.J., Choi, K.K., and Komkov, V., 1986, *Design Sensitivity Analysis of Structural Systems*, Academic Press, Orlando.

Herskovits, J. (ed.), 1995, *Advances in Structural Optimization*, Kluwer Academic Publishers, Dortrecht, The Netherlands.

HyperMesh Documentation, V2.0, 1995, ALTAIR Engineering, Troy, MI.

Johnson, E.H., 1993, "Tools for Structural Optimization," in M.P. Kamat (ed.), 1993, *Structural Optimization: Status and Promise*, AIAA Series in Aeronautics and Astronautics, Vol. 150, American Institute for Aeronautics and Astronautics, Washington, DC, pp. 851–863.

Kamat, M.P. (ed.), 1993, *Structural Optimization: Status and Promise*, AIAA Series in Aeronautics and Astronautics, Vol. 150, American Institute for Aeronautics and Astronautics, Washington, DC.

Kuhn, H.W., and Tucker, A.W., 1951, "Nonlinear Programming," *Proceedings of the 2nd Berkeley Symposium on Mathematical Statistics and Probability*, University of California Press, Berceley, CA.

Mackerle, J., 1995, "Linear and Nonlinear Dynamic Finite Element Analysis Codes," in W. Pilkey, and B. Pilkey (ed.), 1995, *Shock and Vibrations Computer Programs*, SAVIAC, Washington, DC., pp. 455–495.

Morris, A.J. (ed.), 1982, *Foundations of Structural Optimization: A Unified Approach*, Wiley & Sons, New York.

NASTRAN User's Manual, 1994, CSAR Corp., Agoura Hills, CA.

OptiStruct User's Manual, V2.0, 1996, ALTAIR Engineering, Troy, MI.

Papalambros, P.Y., and Wilde, D.J., 1988, *Principles of Optimal Design*, Cambridge University Press, Cambridge, UK.

PATRAN User's Guide, 1996, MSC Corp., Santa Ana, CA.

Pilkey, W., and Pilkey, B. (ed.), 1995, *Shock and Vibrations Computer Programs*, SAVIAC, Washington, DC.

Pilkey, W.D., and Wunderlich, W., 1993, *Mechanics of Structures, Variational and Computational Methods*, CRC Press, Boca Raton, FL.

Rogers, D.F., and Adams, J.A., 1976, *Mathematical Elements for Computer Graphics*, McGraw-Hill, New York.

Schramm, U., and Pilkey, W.D., 1993, "The Coupling of Geometric Descriptions and Finite Elements using NURBs - A Study in Shape Optimization," *Finite Elements in Anal. and Design*, Vol. 15, pp. 11–34.

Shanno, D.F., 1970, "Conditioning of Quasi-Newton Methods for Function Minimization," *Math. Comp.*, Vol. 24, pp. 647–656.

Starnes, J.H., and Haftka, R.T., 1979, "Preliminary Design of Composite Wings for Buckling, Strength and Displacement Constraints," *J. Aircraft*, Vol. 16, pp. 564–570.

Storaasli, O.O., and Sobieszczanski-Sobieski, J., 1974, "On the Accuracy of the Taylor Approximation for Structure Resizing," *AIAA J.*, Vol. 12, pp. 231–233.

Svanberg, K., 1987, "The Method of Moving Asymptotes - A New Method for Structural Optimization," *Int. J. Num. Meth. Engrg.*, Vol. 24, pp. 359–373.

Szabo, B.A., and Babuška, I., 1991, *Finite Element Analysis*, Wiley & Sons, New York.

Timoshenko, S.P., and Goodier, J.N., 1951, *Theory of Elasticity*, McGraw-Hill, New York.

Yao, T.M., and Choi, K.K., 1989, "3-D Shape Optimal Design and Automatic Finite Element Regridding," *Int. J. Num. Meth. Engrg.*, Vol. 28, pp. 369–384.

Zienkiewicz, O.C., and Campbell, J.S., 1973, "Shape Optimization and Sequential Linear Programming," in R.H. Gallagher, and O.C. Zienkiewicz (eds.), 1973, *Optimum Structural Design*, Wiley, New York.

Zienkiewicz, O.C., and Taylor, R.L., 1989, *The Finite Element Method. - 4th ed.*, McGraw-Hill, London.

Zienkiewicz, O.C., and Zhu, J.Z., 1987, "A Simple Error Estimation and Adaptive Procedure for Practical Engineering Analysis," *Int. J. Num. Meth. Engrg.*, Vol. 24, pp. 337–357.

Zhu, J.Z, and Zienkiewicz, O.C., 1990, "Superconvergence Patch Recovery and *A-Posteriori* Error Estimates. Part 1: The Recovery Technique," *Int. J. Num. Meth. Engrg.*, Vol. 30, pp. 1321–1339.

Zienkiewicz, O.C., and Zhu, J.Z., 1992, "The Superconvergence Recovery Technique and *A-Posteriori* Error Estimators," *Int. J. Num. Meth. Engrg.*, Vol. 33, pp. 1331–1364.

INDEX

Alternating stress
    brittle material, 45
    ductile material, 44
Analysis
    finite element method, 441
Angle section, 400, 433
Area
    gross, 6
    net, 6

B-spline
    curve, 476
    solid, 479
    surface, 478
Bar
    hole in torsion, 376
    transverse hole, 229, 342
    torsion, 245
    with hole, 242
Bar in bending
    transverse hole, 363
Beam, 105
    circular hole, 355–356
    deep hyperbolic, 70
    elliptical hole, 357
    flat bottom notch, 71
    multiple notch, 71
    semicircular, 70
    semicircular notch, 71
    semielliptical notch, 71
    shoulder fillet, 416
    U-shaped notch, 70
    V-shaped notch, 70
    with elliptical hole, 240
    with hole, 239, 355–357
    with slot, 240
    with square hole, 105, 240
Bending, 105, 143
    circumferential groove, 72
    circumferential shoulder fillet, 143
    crankshaft, 429, 430
    curved bar, 394, 425
    cylindrical, 242
    elliptical shoulder fillet, 143
    fatigue strength, 72
    fillet, 159, 163, 164, 165
    filleted bars, 160–162
    flat thin member with fillet, 143
    flat-bottom groove, 73, 125
    groove, 125
    hyperbolic groove, 72, 121
    hyperbolic notch, 72, 109
    keyseat, 379
    plate, 72
    thin element with fillet, 143
    U-shaped groove, 72, 123–124

Bending, beam, 110
  flat-bottom groove, 115, 125
  flat bottom notch, 114
  groove, 125
  notch, 104, 112
  semicircular notch, 104, 111
  semielliptical bottom notch, 113
  shallow notch, 107
  U-shaped groove, 124
  U-shaped notch, 106, 108
  V-shaped notch, 108
Bending, plate
  deep notch, 120
  elliptical notch, 118
  hyperbolic notch, 116
  rectangular notch, 117
  semicircular notch, 117, 119
  triangular notch, 117
Bending, rod
  U-shaped groove, 122
Bolt, 387
Bolt and nut, 377, 387
Bolt head
  fastening factor, 389
Box section, 400, 434
Brittle material, 25
  Maximum Stress Criterion, 43
  Mohr's theory, 43
Brown-Srawley formula
  crack, 215

Cauchy's strain tensor, 446
Cavity
  ellipsoidal, 232
  elliptical, 348
  elliptical cross-section, 347
  row of ellipsoidal, 350
  spherical, 232
  spherical in panels and cylinders, 349
Cholesky factorization, 460
Circular cavity, 347
Circular hole, 441
Combined alternating and static stresses, 45
Compatibility equation, 13
Compound
  fillet, 172–174
Compressor-blade fastening factor, 389
Computational methods, 441
Computer-aided design (CAD), 476
Computer-aided geometric design (CAGD), 476

Constitutive equations, 446
Crack, 50, 215
  Brown-Srawley formula, 215
  Fedderson formula, 215
  Koiter formula, 215
Crane hook, 399
Crankshaft, 398
  bending, 429–430
Curved bar, 377, 425
  bending, 394
  crane hook, 399
Cylinder
  circular hole, 440
  hole, 186
  hole in torsion, 375–376
  internal pressure, 438
  pressurized, 402, 440
  spherical cavity, 246
  thick, 403
  with hole, 259–260
Cylindrical bending, 242
Cylindrical pressure vessel
  torispherical cap, 403

Depression
  hemispherical, 67
  hyperboloid, 67
  shallow, 97
  spherical, 69
  tension, 67
Design modification, 484
  local approximation, 485
  sequential linear programming, 487
Design relation, 41
Design sensitivity analysis (DSA)
  adjoint variable method, 469
  continuum system, 471
  design derivative, 468
  design velocity field, 470, 475
  direct differentiation method, 469
  finite differences, 468
  material derivative, 472
  semianalytical method, 471
  stresses, 474
  structural volume, 474
Design space, 444
Design velocity field, 475
Diaphragm, 175
Dimple, 69
Disk, 400, 434–435
Ductile material, 25

Effective stress concentration factor, 36
Ellipse
  equivalent, 225
Ellipsoidal cavity, 348, 350
Elliptical cylindrical inclusion, 351, 352
Elliptical hole, 319–330
  crack, 215
  infinite row, 224
  internal pressure, 224
  optimal slot end, 330
  reinforcement, 225
  single, 215
Elliptical notch, 23
Equilibrium equations, 13, 446
Equivalent ellipse, 24, 225
Equivalent elliptical
  hole, 62
  notch, 62
Equivalent stress, 29, 193, 224, 246
Essential boundary conditions, 446

Factor of safety, 41
Failure theories, 24
  Coulomb-Mohr theory, 26
  Guest's theory, 28
  Internal Friction theory, 26
  Maximum Distortion Energy theory, 29
  maximum shear theory, 27, 28
  maximum stress criterion, 25
  Maxwell-Huber-Hencky theory, 29
  Mohr's theory, 26
  normal stress criterion, 25
  Octahedral Shear Stress theory, 29
  Rankine criterion, 25
  Treseca's theory, 28
  von Mises criterion, 28
Fatigue notch factor, 38, 386
Fatigue strength, 177
Fatigue stress, 178
Fedderson formula
  crack, 215
Fillet, 135, 143, 379, 479, 492
  bar in bending, 160
  bending, 137, 164–165
  circumferential, 142
  compound, 140, 145, 172–174
  conical, 142
  elliptical bending, 163
  flat tension bar, 152
  noncircular, 139
  parabolic, 141
  plate, 137
  pressure vessel, 158
  shoulder, 150
  stepped bar, 164–165
  stepped flat bar, 159
  stepped flat tension bar, 151, 156
  stepped tension bar, 156
  streamline, 141
  tapered, 142
  tension, 137
  tension bar, 152
  torsion, 166–174
  tube, 157
  variable radius, 139
Finite element method
  analysis, 445
  backward substitution, 460
  blending function method, 455
  Cholesky factorization, 460
  circular hole, 441
  connectivity matrix, 457
  element load vector, 450
  element stiffness matrix, 449
  forward substitution, 460
  Gauss elimination method, 460
  Gaussian quadrature, 455
  global kinematic matrix, 457
  h-refinement, 450
  hierarchical shape function, 455
  hp-refinement, 450
  incidence matrix, 457
  interpolation functions, 449
  isoparametric mapping, 454
  Jacobian Matrix, 454
  Lagrange element, 452
  locator matrix, 457
  mapping function, 454
  numerical integration, 455
  p-refinement, 450
  reduced stiffness matrix, 458
  serendipity element, 452
  shape functions, 449–450
  stress computation, 460
  subparametric mapping, 454
  superparametric mapping, 455
  system load vector, 458
  system stiffness matrix, 458
Finite element solution
  circular hole, 441
Flange
  circular with holes, 316
Form factor, 6

Fosterite model, 388
Fretting corrosion, 386

Gauss elimination, 460
Gaussian quadrature, 456
Gear, 385
Gear teeth, 377, 383
Gear tooth fillet, 412–415
Gordon solid, 479
Gordon surface, 479
Groove, 59
   bending, 134
   flat bottom, 70, 102,115, 125, 133
   hyperbolic, 69, 98, 121, 127
   tension, 69, 134
   U-shaped, 69, 99–101, 122–124, 128–131
   V-shaped, 132
Groove, bending
   circumferential, 72
   fatigue strength, 72
   flat-bottom, 73
   hyperbolic, 72
   U-shaped, 72
Gross cross-sectional area, 6

Helical spring
   cross section, 427
   rectangular, 427
   round wire, 397, 426
   square wire, 397, 426
   torsion, 428
   torsion spring, 398
   Wahl correction factor, 395, 426
Hessian matrix, 488
Heywood formula, 183, 227
Hole
   area ratio, 275, 277, 279
   circular, 175, 256–258
   circular pattern, 315–316
   circular pattern with internal pressure, 318
   circular with internal pressure, 317
   circular with lobe under tension, 331–332
   comparison of various shapes, 337
   cross hole, 236
   diamond pattern under tension, 313–314
   elliptical, 175, 211, 357
   elliptical in shear, 364–365
   elliptical in tension, 319–322, 330
   elliptical row in tension, 325–326
   elliptical under biaxial loading, 323–324
   elliptical with bead reinforcement, biaxial loading, 327–329
   elliptical with reinforcement, 365
   in bar, 245, 342, 363
   in beam, 239, 355–357
   in cylinder, 259–260
   in panel, 262–280, 282–283
   in plate, 240, 358–362
   in shear, 370–373
   in shear with reinforcement, 365, 367
   in sphere, 261
   in tube, 245, 342, 363
   in twisted plate, 374
   inclined, 231
   inclined in panel, 345–346
   inclusion, 175
   internal pressure, 199, 281
   intersecting, 236
   L hole, 236
   multiple, 175
   multiple in panel, 370
   multiple in plate, 360, 362
   multiple under biaxial loading, 301, 304–305
   multiple under tension, 299–300, 302–303
   nuclear reactor, 238
   ovaloid, 225
   pattern in shear, 371–373
   pattern under biaxial loading, 310
   patterns, 238
   rectangular, 175, 227
   rectangular in shear, 366
   rectangular pattern under tension, 312
   rectangular under biaxial loading, 333–336
   rectangular with rounded corners, 333–336
   reinforced, 188, 262–280
   shear, 243
   shell under torsion, 245
   square in shear, 367
   square pattern under biaxial loading, 309, 311
   square with reinforcement, 367
   square with reinforcement, biaxial loading, 338–339
   star-shaped, 238
   T hole, 236
   tension at angle to line of holes, 284
   tension parallel to line of holes, 282
   tension perpendicular to line of holes, 283
   three under biaxial loading, 306–308
   torsion, 243
   triangular, 228
   triangular under biaxial loading, 340–341
   tunnel, 236

Hole (*cont'd*)
   twisted plate, 244
   two under biaxial loading, 285–292
   two unequal holes in tension, 293–298
   two unequal holes under biaxial loading, 295
   two unequal in shear, 368–369
   volume ratio, 276, 278, 280
Hollow roller
   external concentrated loads, 437
   internal concentrated, 402, 436
Hook, 377

Impact
   test specimen, 112
Inclusion, 175
   cylindrical, 234
   ellipsoidal, 234
   elliptical cylindrical, 351–352
   modulus of elasticity, 176
   row, 352
   spherical, 234
Infeasible
   design, 444
Infinite plate
   with elliptical hole, 241
   with row of elliptical holes, 242
Intersecting holes
   cross hole, 237
   L hole, 237
   T hole, 237

Jacobian matrix, 454
Joint
   lug, 343–344
   multiple holes, 344
   pin, 229, 343–344

Keyseat, 377–378
   bending, 379, 408
   circular cross section, 408
   combined bending and torsion, 410
   end milled, 379
   fatigue failure, 381
   keyseat, 408
   semicircular, 409
   semicircular end, 408, 410
   sled-runner, 379
   torsion, 380, 409
Koiter formula
   crack, 215

Kuhn-Hardrath formula, 40
Kuhn-Tucker conditions, 489

Lagrange element, 453
Lamé solution, 402
Lamé's displacement equations, 446
Limit design factor, 42
Localized stress concentration, 15
Lug
   round-ended, 393, 423, 424
   square-ended, 392, 422, 424
Lug joint, 343–344, 377, 391

Maximum-shear theory, 25
Maximum-stress criterion, 25
Membrane, 175
Membrane analogy, 9
Minimum stress design, 492
   fillet, 492
Minimum weight design, 491
Mitchell formula, 227
Mohr's theory, 25

Net cross-sectional area, 6
Neuber's factor, 61
Neuber's formula, 61
Neuber's method, 20, 72
Neuber's solution, 70
Nominal stress, 4
Nonlocal stress concentration, 15–16
Notch, 59
   bending, 134
   deep, 120
   elliptical, 82, 118
   flat bottom, 65, 90–91, 94, 114
   hyperbolic, 81, 88, 03, 109, 116, 126
   multiple, 66
   rectangular, 117
   semicircular, 83, 92–96, 104–105, 111, 117, 119
   semielliptical, 113
   single semicircular, 63
   tension, 62, 134
   torsion, 134
   triangular, 117
   U-shaped, 63, 82, 94–89, 105–106, 108, 110
   V-shaped, 65, 87, 108
Notch sensitivity, 36, 177, 388
Notch, beams
   deep hyperbolic, 70
   flat bottoms, 71
   multiple, 71

Notch, beams (*cont'd*)
   semicircular, 70–71
   semielliptical, 71
   U-shaped, 70
   V-shaped, 70
Nozzle, 232
Numerical integration, 456
Nut, 387
Nuts and bolts, 377

Optimization, 484
   BFGS method, 489
   computational methods, 441
   conservative approximation, 489
   convex approximation, 489
   equality constraint, 490
   finite element method, 441
   Kuhn-Tucker conditions, 489
   local approximation, 485
   minimum stress design, 492
   minimum weight design, 491
   reciprocal approximation, 489
   sequential linear programming, 487
   sequential quadratic programming, 488
   shape, 441, 445
   sizing, 444
   topology, 445
   weight, 198
Ovaloid, 225

Panel, 59, 135, 175
   circular hole, 256
   circular hole in shear, 368–369
   circular pattern of holes, 315–316, 318
   diamond pattern of holes, 313–314
   diamond pattern of holes in shear, 373
   double row of holes, 208
   eccentric hole, 317
   elliptical hole, 319, 324, 330
   elliptical hole in shear, 364–365
   elliptical hole with bead reinforcement, 327–329
   elliptical in shear, 366
   elliptical row of holes, 325–326
   hole, 282–284
   hole with lobes, 331–332
   holes at an angle, 203
   holes in shear, 371–373
   inclined hole, 345–346
   multiple holes, 299–305
   offset circular hole, 257, 258
   pattern of holes in shear, 371–373
   pressurized hole, 281
   rectangular hole, 333–336
   rectangular pattern of holes, 312
   rectangular pattern of holes in shear, 372
   ring of holes, 209
   single elliptical hole, 215
   single row of holes, 207
   square hole in shear, 367
   square pattern of holes, 309, 311
   symmetrical pattern of holes, 208
   triangular pattern of holes, 306–308
   two holes, 285–298
   unequal holes, 206
   with hole and reinforcement, 262–280
   with hole in shear, 370
   pattern of holes, 310
   square hole, 338–339
   triangular hole, 340–341
   various shaped holes, 337
Photoelastic model, 388
Pin joint, 343–344
Pin
   hole joint, 393
Pinned joint, 229
Pit, 67
Plane strain, 10, 461
Plane stress, 10, 461
Plate, 175
   deep notch, 120
   elliptical hole, 361–362
   elliptical notch, 118
   hyperbolic notch, 116
   rectangular notch, 117
   semicircular notch, 117, 119
   triangular notch, 117
   twisted hole, 244
   twisted with hole, 374
   with a row of holes, 241
   with hole, 175, 240, 358–362
Plates, bending
   elliptical notch, 72
   hyperbolic notch, 72
   rectangular notch, 72
   semicircular notch, 72
Poisson's ratio, 11
Press-fitted member, 385
Pressure vessel, 158
   cylindrical, 403, 439
   fillet, 143
   nozzle, 232
   torispherical end, 403, 439
Principle of virtual work, 445, 447

Protuberance, 139
  trapezoidal, 154–155
Pulley, 385

Rectangular hole
  round-cornered, 227–228
Reducing stress concentration
  at shoulder, 147
Reinforced
  cylindrical opening, 232
Reinforced circular hole
  symmetric, 192
Reinforced hole, 188
  nonsymmetric, 191
  symmetric, 190
Reinforcement
  L section, 189
  optimal shape, 199
  shape factor, 197
Ring, 402
  external concentrated loads, 437
  internal concentrated, 436
Roller, 377
Rotating disk, 377, 400
  central hole, 434
  noncentral hole, 435

Saint Venant's semi-inverse method, 464
Sensitivity analysis, 467
Sequential linear programming, 487
Sequential quadratic programming, 488
Serendipity element, 453
Shaft
  U-shaped groove, 122–123
Shallow
  notch, 107
Shape factor
  reinforcement, 197
Shear
  flat bottom, 76
  holes, 370
  hyperbolic groove, 74
  hyperbolic notch, 74, 126
  panel with hole, 364–367
  panel with two holes, 368–369
  U-shaped groove, 74
  V-shaped groove, 76
Shear stress
  circular hole, 243
  elliptical hole, 243
  pattern of holes, 244

  rectangular hole, 244
  row of holes, 244
Shell
  elliptical hole, 261
  hole, 245
  hole in torsion, 375–376
  torsion, 245
  with hole, 259–260
Shoulder, 138
  narrow, 139
Shoulder fillet, 135
Shrink-fitted member, 377, 385
Sphere
  elliptical hole, 261
Spherical cavity, 349
Splined shaft, 377
  torsion, 383, 411
Stain-displacement relationship, 446
Strength theories, 24
  Coulomb-Mohr theory, 26
  Guest's theory, 28
  Internal friction theory, 26
  Maximum distortion energy theory, 29
  maximum shear theory, 27–28
  Maximum stress criterion, 25
  Maxwell-Huber-Hencky theory, 29
  Mohr's theory, 26
  Normal stress criterion, 25
  Octahedral shear stress theory, 29
  Rankine criterion, 25
  Treseca's theory, 28
  von Mises criterion, 28
Stress concentration
  analysis, 441
  design, 441
  multiple, 21
  normal stress, 3
  reduction, 174
  shear stress, 3
  torsion, 3
Stress concentration factor
  accuracy, 9
  brittle material, 37
  cavity, 232
  combined load, 32
  ductile material, 36
  effective, 36
  equivalent, 229
  fatigue notch factor, 38
  gray cast iron, 37
  gross cross-sectional area, 177

Stress concentration factor (*cont'd*)
   net stress, 177
   nominal stress, 177
   normal stress, 3
   principle of superposition, 32
   relationship to stress intensity factor, 50
   shear stress, 3
   torsion, 3
   wide panel, 227
Stress intensity factor, 50
   relationship to stress concentration factor, 50
Stress raisers, 14
Structural optimization, 444

T-head, 417
T-head factor, 389
Test specimen design, 76
Theory of elasticity
   circular hole, 180
   elliptical holes, 211
   infinite thin element, 180
   uniaxial tension, 180
   weak form, 447
Tooth fillet, 412–415
Torsion
   angle section, 433
   bar with fillet, 144
   box section, 433
   circumferential shoulder fillet, 144
   fillet, 166–171
   fillet, compound, 174
   flat-bottom groove, 76, 133
   groove, 128–131
   helical spring, 428
   hole in cylinder, 375–376
   hyperbolic groove, 74, 127
   hyperbolic notch, 74
   keyseat, 380, 409
   shaft, 172–173
   shell with hole, 245
   splined shaft, 383, 411
   U-shaped groove, 74
   V-shaped groove, 76, 132
Torsion and shear, direct, 464
   flat bottom groove, 76
   hyperbolic groove, 74
   hyperbolic notch, 74

   U-shaped groove, 74
   V-shaped groove, 76
Torsion bar, 464
Torsion spring, 398
Triangular hole
   rounded corners, 228
Tube, 157
   fillet, 169–171
   hole in torsion, 376
   transverse hole, 342
Tube in bending
   transverse hole, 363
Tube with a hole, 242
Tunnel
   cylindrical, 236
   hydrostatic pressure, 353–354
Turbine-blade fastening factor, 389
Twisted plate
   hole, 374

U-shaped member, 399, 431–432

Vessel
   cylindrical, 439
   pressure, 439
von Mises criterion, 25
von Mises stress, 193

Wahl correction factor, 426
Wahl factor, 395
Weld bead, 139
Wheel, 385
Whitworth thread, 388
Width correction factor, 215

Yield theories
   Coulomb-Mohr theory, 26
   Guest's theory, 28
   Internal friction theory, 26
   maximum distortion energy theory, 29
   maximum shear theory, 27–28
   maximum stress criterion, 25
   Maxwell-Huber-Hencky theory, 29
   Mohr's theory, 26
   Normal stress criterion, 25
   Octahedral shear stress theory, 29
   Rankine criterion, 25
   Treseca's theory, 28
   von Mises criterion, 28